Accession no.
01015291

KT-231-815

…HDRAWN

NUMERICAL MATHEMATICS AND SCIENTIFIC COMPUTATION

Parallel and Sequential Methods for Ordinary Differential Equations

KEVIN BURRAGE

Professor of Computational Mathematics
University of Queensland

LIBRARY
ACC. No. 01015291. DEPT.
CLASS No. 515.35 BUR
UNIVERSITY COLLEGE CHESTER
.0100833x

CLARENDON PRESS • OXFORD

1995

Oxford University Press, Walton Street, Oxford OX2 6DP
Oxford New York
Athens Auckland Bangkok Bombay
Calcutta Cape Town Dar es Salaam Delhi
Florence Hong Kong Istanbul Karachi
Kuala Lumpur Madras Madrid Melbourne
Mexico City Nairobi Paris Singapore
Taipei Tokyo Toronto

and associated companies in
Berlin Ibadan

Oxford is a trade mark of Oxford University Press

Published in the United States
by Oxford University Press Inc., New York

© Kevin Burrage, 1995

All rights reserved. No part of this publication may be reproduced, stored in a retrieval system, or transmitted, in any form or by any means, without the prior permission in writing of Oxford University Press. Within the UK, exceptions are allowed in respect of any fair dealing for the purpose of research or private study, or criticism or review, as permitted under the Copyright, Designs and Patents Act, 1988, or in the case of reprographic reproduction in accordance with the terms of licences issued by the Copyright Licensing Agency. Enquiries concerning reproduction outside those terms and in other countries should be sent to the Rights Department, Oxford University Press, at the address above.

This book is sold subject to the condition that it shall not, by way of trade or otherwise, be lent, re-sold, hired out, or otherwise circulated without the publisher's prior consent in any form of binding or cover other than that in which it is published and without a similar condition including this condition being imposed on the subsequent purchaser.

A catalogue record for this book is available from the British Library

Library of Congress Cataloging in Publication Data
(data available)

ISBN 0 19 853432 9

Typeset by the author using OzTeX

Printed in Great Britain by
Bookcraft (Bath) Ltd
Midsomer Norton, Avon

PREFACE

Introduction

As scientific technology becomes increasingly more sophisticated, the production of more data or the modelling of more complex systems requires computers with ever-increasing computational power. The performance of sequential machines is limited by immutable physical barriers such as the transmission speed of electrical signals and a limit on the physical spacing of the components of a computer. Unless radically new technology such as optical switching or Josephson junctions becomes feasible it would seem that small gains in computational performance will only eventuate at rapidly increasing costs.

An alternative approach is to develop systems based on some form of parallelism and which allows the distribution of both data and algorithms over large numbers of processors. The hope here is that not only can the numerical solution of problems be obtained more quickly than in a sequential computing environment but that problems that were previously insoluble because of either accuracy and/or size constraints can now be solved in a parallel environment. Parallel machines have the advantage of being scalable in which, theoretically, additional performance can be realized by adding more processors. This allows future upgrading without losing the massive investment in previous machines.

In recognition of the enormous amount of research now being devoted to scientific parallel computing, the main thrust of this monograph is to produce an up-to-date exposition of the current state of the art of numerical methods for solving differential equations in a parallel computing environment. Although the main focus will be on ordinary differential equation problems of initial value type (IVPs), some consideration will be given to boundary value problems (BVPs) and to both parabolic and hyperbolic partial differential equations. Furthermore, because linear algebra is an important component of the solution of differential equations, considerable attention is paid to linear algebra aspects.

Motivation

Ordinary differential equations arise in many fields of scientific endeavour ranging from environmental modelling, to quantum chemistry, to stochastic processes in economic problems and to problems in fluid flows, and invariably arise if there is motion or growth.

In many instances it may be desirable to

1. solve moderately sized problems in so-called "real time" when some simulation process is taking place,
2. solve a sequence of differential systems which arise from some form of parameter fitting within a minimization routine,
3. solve a stiff differential system with a vast number of defining equations which may have arisen from the application of the method of lines to a multidimensional time-dependent problem,
4. solve a system of equations over an enormously long time interval.

In these situations the application of parallel computing technology may well provide an accurate solution in a reasonable time frame.

The following example is presented to illustrate the magnitude of the types of problems that currently have to be solved when attempting to model ever-increasingly complex systems.

Thus consider the modelling of long-range transport of air pollutants in the atmosphere, which has been studied at the Danish Air Pollution Laboratory (see, for example, Zlatev (1988) and Zlatev and Berkowicz (1988)). If $c_s(\theta, t)$ is the concentration of the sth pollutant at space point θ and if $u(\theta, t)$, $v(\theta, t)$ and $w(\theta, t)$ are wind velocities along the x, y and z coordinate axes then this phenomenon can be modelled as

$$\begin{aligned}\frac{\partial c_s}{\partial t} &= -\left(\frac{\partial (uc_s)}{\partial x} + \frac{\partial (vc_s)}{\partial y} + \frac{\partial (wc_s)}{\partial z}\right) \\ &\quad + \frac{\partial (K_x \partial c_s/\partial x)}{\partial x} + \frac{\partial (K_y \partial c_s/\partial y)}{\partial y} + \frac{\partial (K_z \partial c_s/\partial z)}{\partial z} + E_s(\theta, t) \\ &\quad - (k_{1s} + k_{2s})c_s(\theta, t) + R_s(c_1, \ldots, c_q), \quad s = 1, \ldots, q.\end{aligned}$$

Here $E_s(\theta, t)$ represents the emission at space point θ and at time t for the sth pollutant, k_{1s} and k_{2s} are the dry and wet deposition coefficients, respectively, for the sth pollutant, $K(\theta, t)$ represents the diffusion coefficients along the three coordinate axes and R_s describes the chemical reactions concerning the sth compound (see Zlatev (1990)).

This model, with $q = 29$, has been studied at the Danish Air Pollution Laboratory. If the method of lines is used to solve this problem then four systems of IVPs are generated which are to be solved cyclically at each time integration step. If a $32 \times 32 \times 9$ equidistant grid is specified for this three dimensional problem then each system of IVPs contains 267,264 equations which are to be solved at each time step. In addition, this model has to be solved over a long time scale in order to study monthly and seasonal variations of the concentrations. Clearly, such a problem can only be solved efficiently in a parallel computing environment. Furthermore, if greater accuracy is required the spatial grid needs to be refined and this leads to even larger systems of equations.

Material covered

Because methods for solving differential equations in a parallel environment are very much machine dependent and since there is currently an enormously wide range of parallel architectures and environments, such parallel algorithms cannot hope to be understood in isolation. Consequently, Chapter 1 is devoted to a discussion of parallel computing in general: the different types of architectures, case studies of various machines and an introduction to a number of parallel computing paradigms.

In order to be able to compare both parallel and sequential differential equation methods, Chapters 2 and 3 of this monograph will be devoted to a summary of existing sequential methods for solving ordinary differential equations (mainly of initial value type). There are already a number of excellent books on this subject, including the works of Butcher (1987b), Hairer et al. (1987) and Hairer and Wanner (1991). The material in these two chapters cannot hope to cover this in any breadth because the main focus in this book is on parallel algorithms for differential systems. However, this material is given for two reasons. First it presents a different viewpoint and approach from most authors to the subject in terms of the development of numerical methods from a very general framework. Secondly some familiarity with this material is necessary if value judgements are going to be made about the efficacy of any proposed parallel algorithms – since their efficiencies are determined by comparing their performance with the most efficient sequential methods.

Because of the very nature of ordinary differential equations and stiff systems in particular, there is a strong connection between differential equation algorithms and linear algebra techniques. Therefore a complete chapter (Chapter 4) is devoted to material on parallel methods for the solution of linear systems. As with the material in the first three chapters the material in Chapter 4 is selective and is by no means comprehensive (there are excellent books devoted to this area alone, notably the monograph by Golub and van Loan (1989)). Nevertheless, Chapter 4 covers parallel direct methods for both full and banded systems as well as iterative techniques in some depth. In particular, some attention is paid to multigrid techniques since some very important work has been done in the area of the direct application of multigrid techniques to differential systems. In addition, some very new work on deflation techniques and adaptive preconditioning techniques is presented.

These first four chapters can be considered as introductory material to the main thrust of this monograph (although it is hoped that the reader will find much that is new and stimulating). The remainder of the material will actually be presented in two parts. Chapters 5 and 6 will be devoted, respectively, to direct methods and methods which exploit parallelism across the steps. Here the framework of the multivalue method will be used in order to give a general setting for this work. Some implementational work of these algorithms on the MasPar MP1 will also be presented.

Chapters 7, 8 and 9 will be devoted to a thorough exposition of waveform techniques for exploiting large-scale parallelism. Waveform relaxation algorithms were originally introduced for the modelling of integrated circuits and there is a monograph by White and Sangiovanni-Vincentelli (1987) on this topic. The emphasis in these three chapters will be different, however, focusing on the behaviour of waveform techniques in conjunction with a variety of numerical methods on very general classes of problems, including parabolic partial differential equations. In particular, Chapter 7 will give a discussion and analysis of dynamic iteration techniques for differential equations, while Chapter 8 will examine what happens when numerical methods are used in conjunction with waveform relaxation algorithms. Chapter 9 will present a thorough discussion of parallel implementation issues for differential equations. A code based on block Jacobi waveform relaxation and VODE (a BDF/Adams code) using message-passing paradigms will be described and some numerical results given. A number of conclusions and thoughts on future directions are given.

Readership

The material presented in this monograph is not overly mathematically demanding and the monograph is intended for final year undergraduate and postgraduate students and researchers in the field of the numerical solution of differential equations. Readers who have an interest in scientific parallel computing in general, and in particular, linear algebra, will find considerable new material.

The block Jacobi waveform relaxation code described in Chapter 9 is general enough to be run on a wide platform of distributed environments including the iPSC/860, Paragon and clusters of workstations. The code will be available by sending an email request to either kb@maths.uq.oz.au or bpohl@sam.math.ethz.ch.

Brisbane *Kevin Burrage*
January 1995

ACKNOWLEDGEMENTS

I would like to thank Oxford University Press for suggesting that I write a monograph on the topic of parallel methods for differential equations. I have found the years spent in researching and writing this monograph enormously enjoyable. The work was commenced while employed at the Centre for Mathematical Software Research, University of Liverpool, England, and I would like to thank Mike Delves for directing me into the area of parallel computing and giving me the opportunity to work in a stimulating environment.

There are many people who have helped me in the writing of this monograph. But my particular thanks go to Bert Pohl (SAM, ETH Zürich), Robert Chan and Philip Sharp (Department of Mathematics and Statistics, University of Auckland), Lawrence Lau and Alan Williams (CIAMP, University of Queensland) and Jocelyne Erhel (INRIA, Rennes). In particular, Bert Pohl spent 18 months working with me at the University of Queensland in developing the parallel code which is presented here and without his enormous programming efforts this monograph would be much less complete.

This monograph was completed while I was first Guest Professor in the Seminar für Angewandte Mathematik at ETH Zürich and later as a visitor in the numerics group at NTH Trondheim and finally as a visitor in the Department of Mathematics at Arizona State University. I would like to thank Rolf Jeltsch, Syvert Nørsett, Zdzislaw Jackiewicz and Rosie Renaut for providing the stimulating and rewarding environment necessary to finish the work.

In addition, I would also like to thank John Belward (CIAMP, University of Queensland), who is always so supportive and enthusiastic about the initiatives I have tried to take at that institution.

Finally, I would like to thank my mother, Peggy, and father, Stan, and my wife, Pamela, and my children, Matthew and Lauren, for providing the caring, supporting and stimulating environment that I needed to complete this work – and especially to thank Pamela for the support and encouragement she gave me while my health was not the best and I was struggling to make progress.

CONTENTS

LIST OF SYMBOLS

$:=$	value assigned by definition
$\approx$	numerical approximation
$\mathbb{R}, \mathbb{R}^+$	real (positive) numbers
$\mathbb{R}^m$	vector space of real m-vectors
$\mathbb{C}$	complex numbers
$\mathbb{C}^-, \mathbb{C}^+$	complex numbers with nonpositive (nonnegative) real part
$\mathbb{C}^m$	vector space of complex m-vectors
Z, Z^+	integers (positive) integers
$f, g, \ldots$	functions
$f^{(j)}$	jth derivative of f
J	Jacobian of f
$Q^\top$	matrix transpose, $Q^\top = [q_{ji}]$
Q^*, Q^H	conjugate transpose, $Q^* = [\bar{q}_{ji}]$
Q^{-1}	inverse of the nonsingular matrix Q
$Q \geq 0$	nonnegative matrix, $q_{ij} \geq 0$
$Q > 0$	positive matrix, $q_{ij} > 0$
(a, b, c)	tridiagonal matrix with vectors a and c on lower and upper diagonals, respectively, and vector b on the diagonal
I, I_n	identity matrix (dimension $n \times n$)
e_i	ith unit basis vector
0, $0_{r \times s}$	zero scalar, vector, or matrix (dimension $r \times s$)
e	vector $(1, \ldots, 1)^\top$
E	$(0\, e)$
$Q \otimes A$	tensor product

$\sigma(Q)$	spectrum of Q
$\lambda_{\max}(Q), \lambda_{\min}(Q)$	eigenvalue of Q having largest (smallest) modulus
$\rho(Q)$	spectral radius of Q, $\rho(Q) = \max\{\lvert\lambda_i\rvert,\ \lambda_i \in \sigma(Q)\}$
$\mu(Q)$	logarithmic norm of Q, $\lim_{\Delta\to 0^+} \frac{\lVert I+\Delta Q\rVert - 1}{\Delta}$
$\lVert \cdot \rVert_\infty$	L_∞ norm
$\lVert \cdot \rVert_1$	L_1 norm
$\lVert \cdot \rVert_2$	L_2 norm
$\lVert \cdot \rVert$	any of the above norms
$\langle \cdot, \cdot \rangle$	inner product
$\lVert u \rVert_T$	$\max_{x\in[0,T]} \lVert u(x) \rVert$
$\mathcal{P}_k$	set of polynomials of degree not exceeding k
Span S	Span of S
$K_k(Q,v)$	Krylov subspace of dimension k generated by Q and v
h	stepsize
$\mathrm{Int}(s)$	interior of S
$Re(z)$	real part of z
$\partial\Omega$	boundary of Ω
$\bar{\Omega}$	closure of Ω
$\mathcal{L}$	integral operator $\mathcal{L}y(x) = \int_0^x y(s)ds$
e^x	$\exp(x)$
$C[0,T]$	space of continuous functions on $[0,T]$
$L^p(\mathbb{R}^+, \mathbb{C}^m)$	space of complex-valued Lebesgue measurable functions which are pth power integrable
$F(t)(y)$	elementary differential
w	order of a method
$[t_1, \ldots, t_v]$	composite tree

1

ASPECTS OF PARALLEL COMPUTING

The main thrust of this monograph is to produce an up-to-date exposition of the current state of the art of numerical methods for solving IVPs in a parallel environment. As will be seen, any algorithmic implementation is architecture dependent and so before any attempts can be made in the development of such methods, the nature of the underlying computing environment should be considered. There is (at the time of writing in August 1994) an enormous variety of parallel computing architectures with completely different programming paradigms, and so the aim of this chapter is to acquaint the reader with the types of parallel machines that are now available and to present a general discussion on various aspects of parallel computing. Although two companies (Thinking Machines and Kendall Square) have only recently (late 1994) stopped trading, their products will still be discussed as their machines represent a significant step in the evolution of supercomputing.

This chapter will cover the following material:

- section 1.1: a classification of parallel architectures through the concepts of array computers, vector computers and multicomputers and the parallel computing paradigms, SIMD, MIMD and SPMD;

- section 1.2: an introduction to the CM5 (marketed by Thinking Machines Corporation), the KSR1 (marketed by Kendall Square Research), the MP1 (marketed by MasPar), the iPSC/860 and the Paragon (marketed by Intel), the T3D (marketed by Cray) and transputer-based machines (such as the Parsytec GC-series);

- section 1.3: a discussion on various techniques of parallelization such as load balancing and local communication and the different types of parallel paradigms – geometric, data, farm and hybrid;

- section 1.4: a discussion on computation versus communication issues as well as an introduction to parallel random access models and the importance of constructing generic parallel programming environments;

- section 1.5: a brief introduction to parallel languages including Fortran90 and High Performance Fortran;

- section 1.6: a discussion on software libraries for parallel machines and an introduction to the concept of BLAS (Basic Linear Algebra Subroutines) and the software packages LAPACK and SCALAPACK;
- section 1.7: an introduction to the concept of distributed and heterogeneous computing through software environments such as *p4*, *PVM* and NQS.
- section 1.8: a brief discussion on the nature and significance of parallel software tools;
- section 1.9: some comments on performance measures and the efficiency of parallel algorithms including the Amdahl and Gustafson laws, together with a brief analysis of the speed-up of a simple iterative method (the Jacobi method) on a distributed memory computer (the Paragon).

1.1 Parallel architectures

One of the few advantages that sequential machines have over parallel machines is that it is relatively easy to achieve portability of code, whereas adapting sequential code efficiently to a parallel environment can be very time-consuming. In addition, because of the wide variety of parallel architectures ranging from vector processor systems to massively parallel systems each with their own different memory management protocols, it is often extremely difficult to port a parallelized code from one parallel machine to another. On the other hand, parallel machines also have the advantage of being scalable.

A simple classification of this range of parallel systems would include the following characteristics:

Array computers – a large collection of processing elements (PEs) arranged in a mesh topology, with a possible wrap around. The same instruction is applied to different data on each PE and the computer is said to operate in SIMD (Single Instruction Multiple Data) mode. Many problems in modelling can be approximated by using a mesh for spatial discretization: such as fluid mechanics, stress analysis, spatial modelling and general problems where the computation and communications are regular. Due to the SIMD nature and relative simplicity of the design, very high performance can be obtained on these types of problems, but sometimes poor performance on problems that do not naturally fit the paradigm. Examples of such machines include the MasPar MP series and the CM2.

Multicomputers – a collection of homogeneous independent processors communicating using an interconnection network (see Figure 1.1). The memory associated with multicomputers can either be

- distributed (such as in the Paragon) in which the only main memory is that possessed by the processors,
- shared (such as in a Sequent Balance) in which the main memory is available globally to all processors,
- shared-distributed (such as in the Cray T3D MPP) which is a combination of the above.

These processors usually execute independently of each other and the computer is said to operate in MIMD (Multiple Instruction Multiple Data) mode. Peripherals supply mass storage and other functions, such as I/O, which can be a bottleneck. One solution to this is to attach disk arrays directly to the processors.

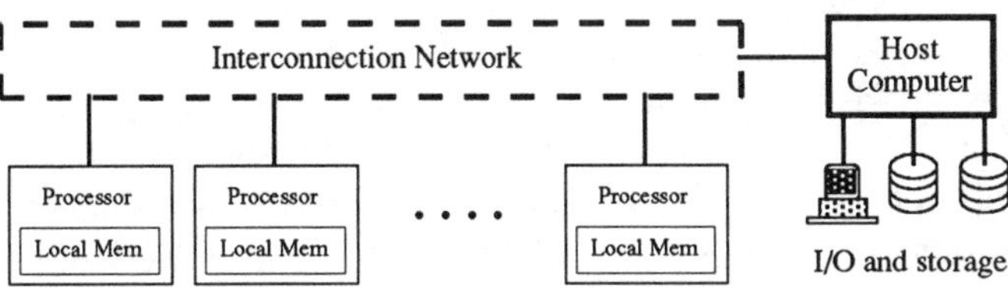

FIG. 1.1. general structure of a distributed memory multicomputer

(Multiple) vector computers – a vector pipe processes arrays, rather than a single data item, using a combination of fast registers and pipelining. A machine can consist of several vector processors sharing a common memory. Data is placed into a pipeline of processing stages in which, at each stage, the same instruction is executed on one particular data element. The data proceeds in a pipeline sequentially through the processing stages. Once a pipeline has been filled, each cycle produces a new result. However, there are overheads in filling and emptying the pipeline, so that this process is only really effective if the pipeline is suitably long. In addition, if the vector cannot fit into the register, it must be broken into suitable smaller pieces which can have some effect on performance. Because of their nature, vector computers perform very regular computations and are easy to program with compilers available which attempt to extract parallelism from existing sequential codes based on the examination of nested loops. An example here includes the Cray Y-MP series.

Distributed computing – the utilization of a number of distinct computers over a network. An example here could be a network of IBM RS/6000s in the P(arallel) V(irtual) M(achine) environment.

An attempt at classification of the various types of computer was first given by Flynn (1972). Although this has now been diversified it offers some insight into the nature of parallel processing. Flynn's taxonomy gives four basic categories:

SISD – single instruction-stream, single data-stream: the conventional sequential Von-Neumann architecture is an example of a SISD machine.

SIMD – single instruction-stream, multiple data-stream: here the same operation is applied to more than one data value simultaneously. This suits algorithms where the data is regular and the algorithm processes all the data in the same fashion. The machines which fall naturally into the SIMD category are vector and array processors.

MISD – multiple instruction-stream, single data-stream: there are no known systems that have been built fitting this characterization.

MIMD – multiple instruction-stream, multiple data-stream: this paradigm gives great flexibility since each processor can execute completely different processes with communication taking place between the processors using, for example, message passing or shared variables. Unfortunately, the MIMD approach is also the most difficult to program due to the complexity of synchronizing the processes and controlling the communications.

Recently, additional paradigms have been developed (such as SPMD and asynchronous SIMD) to cope with more complicated architectures.

SPMD – single program, multiple data-stream: here the same program is duplicated on all the processors and the data is divided into blocks which are independent of each other. Each block is computed at its own pace which makes it suitable for programs with many conditional branches or with irregular data. Many multicomputers are based on this programming model. This paradigm can be considered as a subset of the MIMD model.

ASIMD – asynchronous SIMD: a loosely synchronous or non-uniform data parallelism with some processor autonomy. This allows more flexibility when dealing with non-uniform algorithms (see Quinn (1987)). This autonomy can occur at four levels: execution, addressing, connection and I/O.

An alternative methodology for classification is based on the concept of granularity.

(i) Very coarse granularity: very powerful, and very expensive, supercomputers with a small number of high-performance (usually vector) processors. An example of such a machine is the Cray Y-MP series.

(ii) Coarse granularity: multicomputers based on a number of relatively powerful processors such as the Intel Paragon.

(iii) Fine granularity: many thousands of low-performance, small memory processors as in the CM2 or MasPar MP series. This architecture is suitable for regular data-intensive problems such as those arising in image processing and finite difference techniques.

1.2 Specific computers

In this section some introductory material on a number of parallel machines is given. These include the CM5 (Thinking Machines), the KSR1 (Kendall Square), the MP1 (Maspar), the iPSC/860 and Paragon (Intel), the T3D (Cray) and the Parsytec GC-series. The only reason for selecting these machines is that they represent a reasonably comprehensive and typical subset of parallel machines. Much of the information in this section has been gathered from various technical reports available through the vendors. There is also a very useful paper by Gates and Petersen (1993) which gives a technical description of a number of parallel computers.

CM5

The CM5 was developed by Thinking Machines Corporation (TMC) of Massachusetts, USA. The company previously produced the CM2 which had up to 16K simple processors in a SIMD configuration. A similar programming philosophy was applied to the development of the CM5, so that it primarily supports a SIMD data parallel programming model, with some rudimentary message-passing support.

The CM5 is a hybrid machine with each computation node initially based on a Sparc2 chip and four vector pipes. A peak performance of 128 Mflops per node is claimed. Additional processors control the I/O and system administration tasks. There are three separate networks managing data, control and diagnostics. The interconnection network is a proprietary fat-tree topology where the nodes are arranged in a hierarchical fashion with more bandwidth the higher in the tree.

The machine currently supports Fortran90 and C*. An interactive environment named PRISM allows the user to debug and profile the parallel programs. The CM5 has a comprehensive parallel scientific library that has been initially optimized for the CM2. TMC stopped trading at the end of 1994.

KSR1

The KSR1, developed by Kendall Square Research (KSR), was announced in June 1992. It is a new and novel architecture designed to hide the complications of interprocessor communications by automatically handling it in hardware using advanced caching technology.

The KSR1 processor is an in-house design with a peak speed of 20 Mips and 40 Mflops and is implemented using four chips. Up to 1088 processors can be arranged in a ring-of-rings structure in which up to 34 outer rings, each containing 32 processors, can be attached to an inner ring. The memory is globally addressable, that is, any processor can access any memory location in the system transparently to the user and this is essentially a shared-memory process. The all-cache memory configuration implies that multiple data copies and some asynchronous data dependency locks are unnecessary. Shared-memory machines have the additional advantage that porting codes is easier than with message-passing machines.

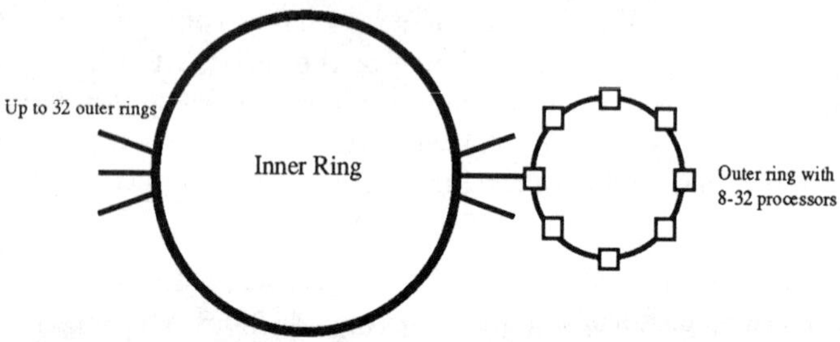

FIG. 1.2. network hierarchy of the KSR1 machine

The KSR1 was designed to be a general-purpose machine with Fortran, C and SQL (a relational database language) being supported. The KSR1 accepts standard Fortran77 code and has the advantage of executing in parallel existing programs if the loop iterations are independent of one another. The KSR1 offers an extensive range of parallel constructs including threads, parallel domains, tiling or any combination of these. KSR stopped trading at the end of 1994.

MasPar MP series

MasPar is a company which produces array computers, and the MasPar MP series is a SIMD machine which relies on data parallelism. The MP1 consists of a front-end Unix workstation which performs the serial part of the computation and a back-end Data Parallel Unit (DPU). The DPU consists of an Array Control Unit (ACU) and an array of processing elements. The ACU handles scalar variables program code. One processing element (PE) consists of a 4-bit processor with 64 Kbytes of memory and generates a sustainable performance of around 50 Kflops. However, machine configurations of between 1024 and 16K processors give a peak performance of up to 1.2 Gflops.

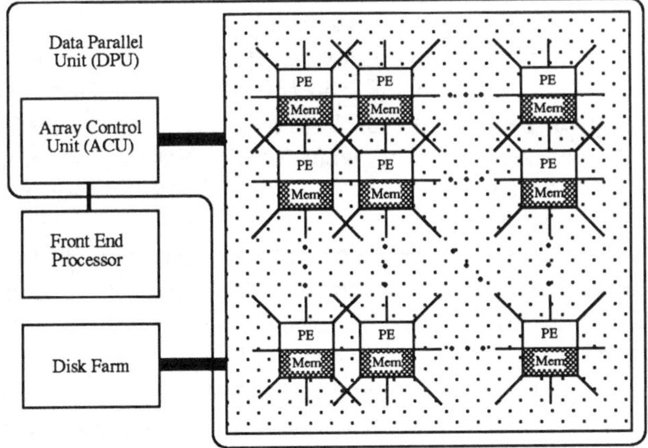

FIG. 1.3. architecture of the MP1

Communication is via the XNET, which is a lock-step interprocessor communication protocol through the eight nearest neighbours, or by the Global Router which allows for arbitrary processor-to-processor communication via a three-stage switch router. In the case of the Global Router, communication time is constant regardless of distance.

The MP1 supports two languages: MPL, an extended version of C, and MPFortran, based on the new Fortran90 standard (see Loveman (1993)). Programming in MPL is more efficient in a computational sense than programming in MPFortran but in this environment the programmer has to handle much of the low-level details of processor configuration and exchanging data between the front-end, ACU and DPU. On the other hand, the MPFortran compiler attempts to parallelize applications automatically by partitioning arrays and vectors on the DPU.

A graphical environment called MPPE allows interactive debugging of parallel programs, and a restricted version of the IMSL numerical library is available. In 1993, MasPar announced the availability of the MP2 system which is claimed to have between 1.5 and 4 times the performance of the equivalent MP1 machine.

iPSC/860 and Paragon

The Paragon is a scalable distributed-memory multicomputer from Intel, first delivered in September 1992. Its predecessor is the iPSC/860 and the Paragon's nodes are based on Intel's i860 RISC processor. The main difference between the Paragon and the iPSC/860 is a fast rectangular interconnection network and new operating system.

The Paragon consists of compute nodes, service nodes and I/O nodes which are uniformly integrated into a rectangular interconnection network. The compute nodes consist of a 50 MHz i860 XP microprocessor with a theoretical peak performance of 75 Mflops (see Esser and Knecht (1993)). The service nodes offer the capabilities of UNIX and make a traditional front-end computer unnecessary, while the I/O nodes interface to external networks and disk arrays.

The network for the Paragon provides the fast routing of messages with a bandwidth of 200 Mbytes/sec in each direction between nodes, while the iPSC/860 is a circuit-switched hypercube network. This represents the major difference between the two systems. Because of wormhole routing (see section 1.4) and a deterministic routing network, this bandwidth is claimed to be virtually independent of the distance between nodes for longer messages. The routing is done by routing chips each with five input and five output channels which connect to each of the four neighbouring nodes and the Network Interface Controller. The operating system allows the processor mesh to be partitioned into subsets which are restricted to particular users or jobs. The Paragon supports Intel's NX message-passing library, as well as other portable message-passing libraries such as P(arallel) V(irtual) M(achine) (see Geist et al. (1993)). Programming is done in Fortran77 or C. Optimized library routines are provided, including BLAS routines.

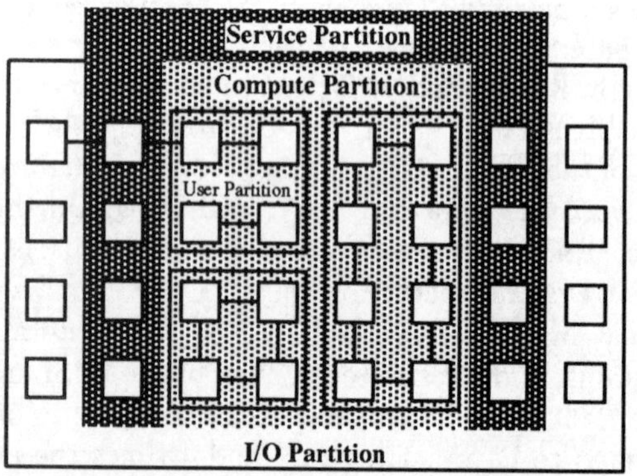

FIG. 1.4. layout of the Paragon

T3D

The Cray T3D project was announced in October 1992 and represented a branching out by Cray Research from the vector market into the parallel market. The architecture of the T3D is a three-dimensional torus and the compute nodes are based on the DEC-α chip. These processors can run at 200 MHz with a peak performance of 200 Mflops. Claimed latencies and interprocessor communication rates are very impressive being, respectively, 1.3 μsec and 300 Mbytes/sec. The machine currently requires a Cray vector machine such as a Y-MP as a front end.

The Cray T3D supports UNICOS and a variety of programming models including message passing, data parallel and hybrid models. Message passing is performed using *PVM*. The machine has a shared-memory capability and supports C, Fortran77 and a variant of Fortran90.

GC-series

Parsytec is a German company founded in 1985. Their GC series was initially based on the T9000 transputer. However, due to the unavailability of the T9000, Parsytec initially delivered the machine with a T805 transputer which a peak performance of about 1.5 Mflops. The central processing unit and the floating point unit of a transputer can operate concurrently, so that an address calculation and a floating point calculation can be executed simultaneously.

The T805 is a very under-powered processor and the interprocessor communication rate between these transputers is poor. Parsytec has announced a new series based on the T9000 transputer which has a peak performance of 25 Mflops and a clock speed of 50 MHz. However, at the time of writing the T9000 is more than a year overdue. Each of the four physical links on the T9000 supports an unlimited number of virtual links, and hardware on the T9000 can map the virtual links onto the physical links. A message-routing chip (the C104) containing 32 links has been developed which can send a message from any one of 32 transputers to any other. By configuring a number of C104s large numbers of direct communications are possible. However, as with the T9000, the C104 routers are currently not yet available.

The processors are connected in a three-dimensional cube topology and a gigacube of 64 transputers and eight C1O4 routers is the minimal building block. Thus the network topology can be viewed as a multilevel cluster with modular packaging and constant complexity. The operating system of the GC-series is called Parix and is a further development of the Helios-based Unix systems for older transputer technologies. The GC-series supports C, Fortran77 and Occam.

1.3 Techniques of parallelization

Perhaps one of the most important aspects of parallel processing is the mapping problem. That is the question of how to map a set of communication processes onto a network of processors as efficiently as possible. A parallel program can be expressed as a set of communicating sequential processes represented as a graph in which the edges represent the communication paths. In a similar way a network of processors can also be represented as a graph. The mapping problem is equivalent to embedding the software graph in the hardware graph subject to certain constraints such as those based on weight functions representing the execution time of each task.

In general, this minimization problem is NP-complete but there are efficient non-optimal approaches to the mapping problem. These include

(i) Polynomial time bound algorithms such as the critical path list scheduling algorithm (see Kohler (1975)) which lists the tasks in increasing critical time estimates with tasks being assigned to free processors under this ordering.

(ii) Simulated annealing algorithms (see Kirkpatrick et al. (1983)), in which an energy function related to the execution time is constructed. By mapping the hardware and communication characteristics onto a physical problem (see Fox and Otto (1986)), the concept of temperature and other physical entities can be attributed to the computing environments.

(iii) Evolutionary and genetic algorithms (see Mühlenbein et al. (1987)), in which evolutionary rules based on "survival of the fittest" are used to describe a particular minimization problem. This approach tends to be very problem dependent.

(iv) Dynamic mapping in which at a number of synchronizing steps, neighbouring processes exchange information about their load and probable position of all other processes. Processes causing high communication costs are distributed to neighbouring processes (see Boillat and Kropf (1990)).

A study of these approaches and others shows that there are certain basic principles that should be adhered to. These include

(a) load balancing – the even distribution (where possible) of the processing load over all processors;

(b) local communication – the placing of processes that communicate with one another onto processors that are connected by a direct link.

In an attempt to ameliorate the often onerous task of designing and then programming algorithms in a parallel environment, considerable attempts have been made to develop techniques which allow the automatic parallelization of sequential codes. This approach has met with variable success because it is both problem dependent and architecture dependent. These techniques include

(i) UNITY (see Chandy and Misra (1988)) – an environment for specifying parallel programs and which contains an implementation and specification language. UNITY attempts to find parallelism within the problem specification, whether it be synchronous or asynchronous, and then to partition automatically.

(ii) Automatic compiler techniques for vector computers which attempt to parallelize sequential codes by rewriting the outermost loops of loop-nests and inserting appropriate synchronizations (see Kuck (1977), for example).

The automatic parallelization of existing sequential code, or the construction of a programming environment which can extract parallelism automatically, is still very difficult. The major difficulties are the variability in parallel architectures and the fact that parallelism cannot be found if it is not identified in the algorithm. In addition, independent extensions to standard sequential languages such as LINDA (see Gelernter (1985), for example) require large overheads.

A recent approach to attempt to overcome these difficulties is based on functional languages and data flow architectures (Yuba et al. (1985)). For example, Large Grain Data Flow 2 (DiNucci and Babb (1990)) is intended to be an architecture-independent extension of existing languages, which is designed to help programmers split algorithms into disjoint sections. Functional languages use functional composition as the basic tool to transform or share data and data flow graphs specify an execution ordering based on function evaluation.

In spite of these developments most numerical parallel software is not parallelized automatically but is written directly based on certain standard approaches. Roughly speaking, four approaches can be identified: geometric parallelism, data flow parallelism, event parallelism and hybrid parallelism.

(i) Geometric parallelism

A problem often has an underlying regular (spatial) structure which allows subdivision into smaller regions (domain decomposition). Attempts are then made to preserve local communication so that only neighbouring processes communicate. Usually the communication load is proportional to the size of the boundary or overlap region, while the computational load is proportional to the size of the subregion. Examples of problems where a geometric approach to parallelism is fruitful occur in finite differencing schemes for solving elliptic partial differential equations (such as multigrid techniques on meshes of varying lengths) and in linear algebra applications such as cyclic reduction or tearing techniques. In the geometric approach there are often difficulties in terms of synchronization at the boundaries and with automatic load balancing.

(ii) Data flow parallelism

Here each processor executes a very small part of the program, and this results in a data flow between processors. In general this approach is flexible in allowing multiple connections and splitting/merging of the data as well as reducing the demand for memory since less data space is required for each processor. However, there are large communication loads on each processor compared with the computational load. If a code is subdivided too much, communication can saturate the system and the performance can be severely degraded.

(iii) Event or farm parallelism

Here processors carry out identical tasks on different sets of data with no communication between processors. Examples of such problems occur in Monte Carlo simulation where a number of random walks may be taking place in order to obtain a statistical solution (for example, the heat equation). A high efficiency can be achieved this way on appropriate problems and it is easy to port extant sequential software to this environment. This parallelism fits very well with the SPMD model.

(iv) Hybrid parallelism

There is a limit to the degree that data and code can be divided. Instead of relying on a single approach, a combination may prove to be more effective. A hybrid scheme allows more processors to be used before incurring significant overheads for large numbers of processors.

1.4 Communication and computation

In Flynn's interpretation of array processing, all processors in a network perform a series of computation cycles in which at the end of each cycle the processors synchronize and exchange data. This paradigm gives little flexibility and is inefficient. In transputer networks, however, because computation and communication can take place simultaneously, communication and computation can be overlapped. Since communication delays can cause processors to be idle, overlapping is beneficial if there is excess processor parallelism. In this case a number of processes, along with a scheduler, can be placed on each processor and communication occurs while the processes are being executed. The scheduler can postpone the execution of processes because of communication difficulties and activate others.

It should be noted that transputers and the programming language Occam developed by INMOS as the appropriate programming tool are consequences of Hoare's (1985) concurrency model of communicating sequential processes. In this model an application is decomposed into a collection of parallel processes communicating over channels. The processes cannot communicate by shared data, but rather only by sending messages along these channels. Even in the case with more than one process running on the same processor, communication is via channels. The advantage of this approach is that the same model of concurrency is used both within processors and between processes. Thus in the case of a transputer network running Occam, parallel processes exchange data simultaneously with data transfer occuring only when both processors are ready. Occam requires the programmer to make all parallelism explicit which gives the freedom to utilize all the different forms of concurrency, although at the same time it makes programming more difficult.

Thus not only do transputers possess communication and computation concurrency but some deadlock problems which naturally arise in a parallel environment are avoided. **Deadlock**, for example, can arise when a cycle of processors ends up in a state in which those in the cycle are all waiting on input from one another. In the case of the transputer, communication between transputers only takes place when both have signalled they are ready.

Deadlocking is one difficulty that arises in a parallel environment and not in a sequential environment; another is non-determinacy. Since information can arrive at the processors at unknowable times, especially in an asynchronous organization, verification of results is not always possible. Consequently, software validation is enormously more complicated in a parallel environment.

Where possible, communication networks should reflect the underlying nature of the problem as this simplifies the mapping of the algorithm to the network. Thus two and three dimensional rectangular networks can be directly embedded in physical space and are well-suited to problems arising from partial differential equations. However, the diameter (the distance between any two processors) of an n-dimensional mesh with p processors is $O(p^{\frac{1}{n}})$. In fact the most efficient network (that is, the one with the smallest diameter) is a hypercube of dimension $\log_2 p$ with $\log_2 p$ connections per processor and a diameter of $\log_2 p$.

With the development of networks that can be reconfigured dynamically by software commands, some of these network measures (such as diameter) can become blurred. This blurring is accentuated if message routing is through virtual channels, as in the T9000, or through network routing chips which support **wormhole routing** (see Freeman and Phillips

(1992)). A worm is just a message packet with a header which defines its destination and is processed by the routing chips in a pipelined manner. If the destination of a message consisting of N packets is p hops away then the theoretical time to deliver a message by wormhole routing is $p+N-1$ time units. Athas and Seitz (1988) claim that this approach makes the time to deliver messages of moderate size more or less independent of the path length.

Although there are so many variable factors involved in parallel computation (the number of physical links, static or dynamic reconfiguration, shared, distributed or shared-distributed memory, fine grain or coarse grain processing, input-output organization, the network topology, etc.), recent developments in through routing, wormhole routing and improved interprocessor communication rates suggest that standard environments are beginning to emerge. This is significant in that there has been considerable theoretical work in the development of standard models for parallel machines (for example, **PRAM** (see Cook (1985)). In the P(arallel) R(andom) A(ccess) M(achine) model it is assumed that a network of processors communicates synchronously and communicates via a global random access memory which is unbounded. Thus there is no interprocessor communication. There are modifications to this model based on how many processors can access memory at any given time. The advantage of this approach is that the parallel complexity is then based on the number of steps (or the depth of a circuit when the algorithm is embedded in an acyclic graph). Many papers have been written on the application of this model to various numerical applications such as the addition of n numbers, polynomial evaluation, matrix multiplication, etc. (see, for example, McColl (1989)). But, generally speaking, two general points can be observed when developing efficient parallel algorithms:

(i) the total cost (summed over all processors) may in general be much greater than the cost for a sequential implementation;

(ii) it is possible to construct theoretically efficient algorithms which are so fine grained that the number of processors grows very quickly. Matrix multiplication is a case in point. Clearly there is a point at which communication costs cannot be ignored.

Although the shared memory PRAM model is a very idealistic one it may well happen that with time and with some modifications it could become the basis of a general-purpose model for parallel computing (see Hey (1988), for example). In fact, Valiant (1988) has shown that it is possible to emulate shared memory models on distributed machines with only a constant factor of inefficiency. In some sense this remarkable result is a generalization of the well-known result of Turing, which gives a general-purpose model for sequential computing. This approach has been further refined and is called the bulk synchronous parallel model (Valiant (1990)).

The crucial factors in this model are:

Excess parallelism – in which a number of processes along with a scheduler are placed on each processor. Communication takes place while the processes are being executed with the proviso that the scheduler can activate and deactivate processes depending on communication difficulties.

Concurrent communication on all links – which can be achieved, for example, by networks such as the hypercube, with a communication delay of $O(\log p)$ and a maximum communication capacity assuming no message collisions. (Note that random routing can give a probabilistic worst case delivery time of $O(\log p)$, (see May (1989) for further discussions).) In the case of random routing a message is sent from processor p to processor q by sending the message by the shortest path to some randomly chosen processor and from there to q by again the shortest path.

With new developments such as fast routing chips, wormhole routing and improved interprocessor communication bandwidths, PRAM and similar models offer the hope of standard approaches to parallel programming. The advantages of this will be enormous, but most importantly it will impinge on the area of portability of software and consistent language development for parallel machines.

1.5 Parallel languages

Parallel computing and scientific parallel computing, in particular, will only develop if there is a uniformity and portability of both programming languages and operating systems across a wide range of parallel hardware. This seems likely if hardware and software considerations can be separated.

Until very recently, the situation in parallel computing curiously mirrored a similar situation thirty to forty years ago, when, in the infancy of serial computing, applications programmers were provided with virtually no software support. Algorithms and codes had to be developed from scratch and targeted to the unique features of the particular machine that was to be used. It was only with the establishment of software libraries that this wasteful approach (in both time and money) was discontinued. Not only this, the advent of numerical libraries provided a major impetus to the discipline of scientific computing. For it meant that researchers were free to concentrate on the problem to be solved rather than worrying about the nature of the numerical method needed to solve the problem.

The first numerical libraries developed in Fortran, such as NAG, IMSL and HARWELL, were multipurpose libraries covering a wide range of numerical algorithms in areas such as linear algebra, differential equations, quadrature and optimization. More recently, public domain software is now freely available over NETLIB and covers a large range of specific application areas. These packages include, for example,

(i) LINPACK (Dongarra et al. (1979)),

(ii) EISPACK (Smith et al. (1976)),

(iii) ODEPACK (Hindmarsh and Byrne (1983)).

A list of application areas can be obtained by sending the message

send index to **netlib@ornl.gov**

Because of the easy availability of such packages, it is now a relatively simple matter to include the appropriate piece of software in a driver program. However, in order for this progress to have occurred, certain standards such as robustness, accuracy, efficiency, utility, comprehensiveness and portability had to be imposed on the quality of software within a given library.

The lingua franca for the scientific community has been Fortran. There was a brief flirtation in the 1960s and 1970s with ALGOL 60 and recently C has become popular, but Fortran still remains the standard. The availability of a standard language which could run on a wide range of sequential machines has meant that software can be ported from machine to machine with very little change in the relative performance of the code. Because of the huge investment in both time and money, Fortran has been upgraded from 66 to 77 to 90 and High Performance Fortran. It is indisputable that the rapid development in the area of scientific problem solving in recent years has been due to hardware developments and the maintenance and portability of Fortran.

This has been recognized by those developing compilers for parallel machines and considerable effort has been spent in developing Fortran languages suitable for a wide variety of parallel environments. A brief discussion of Fortran90 and High Performance Fortran will now be given.

Fortran90 and High Performance Fortran

The new release of Fortran90 retains backwards compatibility and includes Fortran77 as a subset. Consequently, many existing Fortran programs can be reused in this environment. Fortran90 is intended to provide some degree of code portability across a range of parallel machines and there are now Fortran90 compilers for SIMD machines such as the MasPar, as well as workstations.

The array notation is the most significant new feature of Fortran90 and this allows manipulation of blocks of data at a time rather than individual data elements. Writing a program in Fortran90 makes it easier to map onto vector processors and other parallel machines since special compilers can automatically recognize and exploit some types of parallelism within the program by partitioning the data among the available processors.

Operations which involve combining arrays must have exactly the same shape, that is the same size and dimension, before being allowed in expressions. There are special subroutines to perform DOTPRODUCT(A, B) and MATMUL(A, B) for computing matrix-vector, vector-matrix and matrix-matrix products, depending on whether A or B are one or two dimensional arrays.

There are two special statements in Fortran90 which allow selective parallel array assignments – namely the WHERE and the FORALL statement. The WHERE statement is the parallel equivalent of the IF statement and the FORALL, the parallel version of the DO loop. The FORALL statement specifies array assignments on groups of array sections based on the masking of components. There are some limitations with these constructs. For example, only array assignments are allowed inside the WHERE construct and user-defined functions cannot be called from within the WHERE statement.

The Fortran90 constructs were originally designed for SIMD machines where the same operation is performed on a set of data. However, the new High Performance Fortran (HPF) standard (see Loveman (1993), for an excellent discussion) includes Fortran90 with additional extensions and has primitives to support message passing and other MIMD-style directives.

The HPF standard is intended to be architecture independent. The addition of compiler directives allows the user to tell the compiler how to distribute data to the processors in order to preserve data locality. Additional options allow for dynamic realignment and redistribution. This is done within the compiler by a three-level process in which certain data objects are aligned, distributed to abstract processes and then mapped to the physical processes (which can be defined by the user by a series of system inquiry intrinsics).

Loveman (1993) has described an HPF programming model in which the developer writes Fortran90 code in a SPMD style so that the code has a single thread of control and a global address space. HPF data mapping directives, which provide information about data locality and distribution, are then added and the code is then compiled through an architecture-specific compiler. The result in the case of RISC machines, vector computers, SIMD machines and MIMD machines is, respectively, pipelined superscalar code, vectorized code, single-threaded parallel code, and multi-threaded message-passing code. This is illustrated in Figure 1.5 (taken from Loveman (1993)).

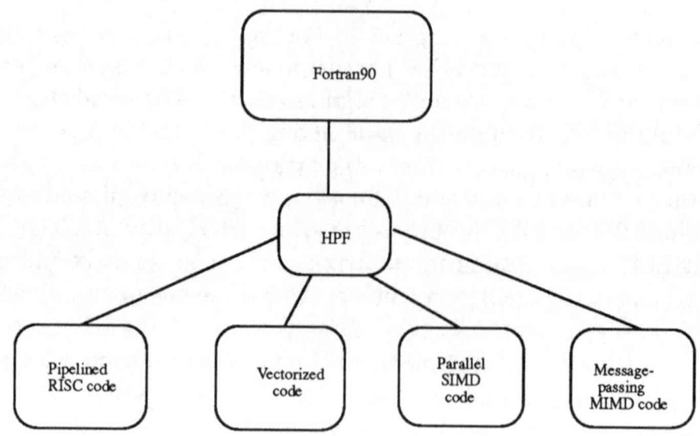

FIG. 1.5. HPF programming model

1.6 Parallel scientific libraries

LAPACK, SCALAPACK and BLAS

Recently, there has been some attempt towards developing a library of numerical software that performs reasonably well on a wide range of parallel architectures, but which is not necessarily optimal for a given architecture. The original project was called LAPACK and has been developed by Demmel et al. (1987). The idea behind LAPACK had been foreshadowed by a number of previous works (see Stetter (1979), for example).

LAPACK is a sophisticated extension (written in Fortran77) of the linear algebra packages LINPACK and EISPACK. These extensions involve, for example, the inclusion of higher-accuracy routines for singular value and eigenvalue problems as well as driver routines for linear systems. LAPACK focuses on systems of linear equations, linear least square problems, eigenvalue problems and singular value problems.

The most important facet of LAPACK is that it is designed as a source of building blocks for more sophisticated applications and library routines and is based on the concept of a BLAS routine (Basic Linear Algebra Subroutines). However, not all the routines have been written in this block form – singular value decomposition being a case in point. LAPACK performs well on vector machines, superscalar workstations and shared memory machines but currently gives poorer performance on massively parallel SIMD machines or distributed memory architectures (see Anderson et al. (1990)).

In order to exploit the performance available within LAPACK it is necessary that BLAS be efficiently implemented on the target machines along with the fine-tuning of such parameters as the blocking sizes. However, this functionality does give a degree of portability between parallel machines.

Three levels of BLAS can be identified:

1. vector–vector operations $y \to \alpha x + y$;
2. matrix–vector operations $y \to \alpha Ax + \beta y$;
3. matrix–matrix operations $C \to \alpha AB + \beta C$.

Here α and β are scalars, x and y are vectors, and A, B and C are matrices.

The only parallelism available through LAPACK is that inherent in the BLAS. Level 3 BLAS have a computation-to-communication ratio of $O(n^3)/O(n^2) = O(n)$, while for Level 1 BLAS and Level 2 BLAS this ratio is $O(1)$ and hence, in general, most of the speed-ups are obtained from parallelizing Level 3 BLAS. However, for certain vector machines Level 2 BLAS can achieve near optimal performance but this performance is limited by the amount of data that has to be moved between the different memory levels. In order to exploit the high performance available through the Level 3 BLAS many algorithms in LAPACK had to be restructured so that they operate on blocks of submatrices of the original matrix rather than on individual elements (see, for example, Anderson et al. (1990)).

Burrage et al. (1990) have implemented BLAS for T800 transputer networks and investigated appropriate blocksizes for a problem size of $n = 150$. With nine transputers an optimum blocksize seemed to be $n/3$. However, the speed-up in this case was only moderate (a factor of approximately 2). These effects have been studied in the book by Freeman and Phillips (1992), who develop a model for a local memory system in which there is a start-up time and a time to distribute data from a master processor to the slave processors. If these times are significant then there may be very little improved performance especially when implementing Level 1 and Level 2 BLAS. Even for Level 3 BLAS it is important in a distributed system for a complete matrix to be available on each processor if possible.

More recently, a scalable linear algebra library for distributed memory concurrent computers has become available (SCALAPACK (Choi et al. (1992)). This can be interpreted as a distributed LAPACK. The building blocks in SCALAPACK are BLAS and BLACS (Basic Linear Algebra Communication Systems) both of which are called by the parallel BLAS (PBLAS). There are some restrictions to the use of PBLAS (and hence SCALAPACK), with perhaps the most important being that data is assumed to be distributed in a block cyclic fashion. This has some restrictions on the types of matrix operations that can be performed. At the time of writing only seven, of the approximately 1000 routines available in LAPACK were available in SCALAPACK and these included LU and QR factorization routines as well as routines for reducing to Hessenberg, tridiagaonal and bidiagonal matrices.

Other libraries

There is a very wide variety in the quality of software currently available for parallel machines. Vector computers such as the Cray Y-MP and Convex C3 series have very rich and comprehensive libraries due to the fact that vector processor technology is the most mature of all the parallel computing technologies.

Thinking Machines Corporation has a fairly comprehensive software library with much of their software being ported from the CM2 to the CM5. MasPar's scientific library is not as comprehensive as that available for the CM5, and BLAS were only available for the MP1 at the end of 1992. However, in 1993 various subsets of the IMSL library were being ported to the MP1 and MP2. In the case of the Paragon, standard parallel linear algebra packages such as PBLAS and SCALAPACK are available at some installation sites (such as ETH Zürich). This highlights another difficulty in that even for the same machine there is often no uniformity of basic public software being available from site to site.

However, it is fair to say that scientific libraries for parallel systems are now expanding rapidly. To illustrate this last remark, the work undertaken by a large group of research workers at both Liverpool University and NAG to develop a numerical library for transputer networks can be mentioned. A Mark I version (in Occam) was produced in 1989 covering routines for dense linear systems, nonlinear equations and fast Fourier transforms. This has now been substantially upgraded to a Mark II and Mark III library available in Fortran and including additional routines in numerical quadrature, differential equations and eigenvalues.

The aims of this library are described fully by Delves and Brown (1988). It is intended that the transputer library have the same attributes and standards to software as traditional sequential libraries with a simple user interface, precompiled library facilities and comprehensive error mechanisms.

The efficiency of this numerical transputer library is very problem dependent and when large amounts of data have to be transferred from the host to the slave transputers, the performance can be severely degraded. In the case of linear algebra routines only the Level 3 BLAS will give enhanced performance. On the other hand if data is predistributed throughout the network, speed-ups are also obtainable with Levels 1 and 2 BLAS.

1.7 Heterogeneous and distributed computing

Because there are often severe restrictions on the amount of time a user can have on any supercomputer it is often not efficient to develop parallel code on a supercomputer. Fortunately, a number of computers, interconnected by a network, can also be used as a single distributed memory computer.

This can be achieved by software packages, such as *PVM* (Geist et al. (1993)) or *p4* (Butler and Lusk (1992)), that combine several computers (which may differ in speed, architecture or even data representation) as one single computational resource. Such an environment is called a **heterogeneous distributed computing** environment. The machines that can be combined with *PVM* or *p4* range from workstations to vector computers to SIMD machines to shared-memory and distributed-memory MIMD computers. Furthermore, the interconnecting networks between machines may be different, for example Ethernet or Internet. In addition, these environments can support computers with different data formats.

PVM, which stands for Parallel Virtual Machine, has been under development at Oak Ridge National Laboratory by Jack Dongarra et al. since 1989. Since the interconnecting networks are usually not very fast, *PVM* is mainly designed for programs that consist of subtasks that offer a large granularity of parallelism. *PVM* is based on the message passing model, allowing messages to be sent and received, and allowing barrier synchronization and broadcast. *PVM* offers automatic data conversion, if a heterogeneous network with machines that have different data representations is used.

PVM uses a daemon called *pvmd*, which is started by the command

pvm or *pvmd pvmdhosts_file.*

pvm opens a console in which a user can interactively add or delete hosts to the virtual machine and control the status of the processes.

After the *pvmd* daemon is started on each machine, the user can run *PVM* programs. These consist of standard Fortran77 or C code into which message passing routines contained in the *libpvm.a* library are inserted. Although the names and arguments of the Fortran and C routines differ, –C routines start with *pvm_* and the corresponding FORTRAN routines start with *pvmf*–their functionality is identical.

A similar package to *PVM* is *p4*, which stands for portable programs for parallel processors, developed at Argonne National Laboratory by Butler and Lusk (1992). It can be used with Fortran or C and also supports heterogeneous networks with automatic data conversion. Communication on distributed-memory machines is performed by message passing, whereas monitors or forks are used on shared-memory machines. *p4* accepts command line arguments which offer various levels of debugging and are very useful during program development.

Environments such as *p4* and *PVM* are very useful in a coarse-grained environment because they allow the programmer to divorce the communication issues from the underlying algorithm. This is particularly significant if the programmer can make use of extant sequential packages running in SPMD mode on each processor since the programmer only then has to focus on communication issues using a standard software environment such as *p4* or *PVM*. Unfortunately, these environments can suffer from a lack of uniformity and portability as has just been seen in the use of daemons in *PVM* and, in any case, many vendors often prefer to use their own message-passing libraries.

An advantage of heterogeneous environments is that subtasks can be placed on machines that are best suited to their solution. Another advantage is that the coding and debugging process can be done much more efficiently and cheaply than if only a supercomputer is available. Thus a typical debugging process using, for example, *PVM* would be

- to debug the program running serially on only one processor to eliminate computational and indexing errors;
- to debug the program running in parallel on a single workstation to eliminate errors in the message-passing calls and logic;
- to debug the program in parallel on a very low number of workstations to detect synchronization errors and deadlock;
- to migrate the program to a supercomputer.

In addition, the placing of jobs and subtasks in a heterogeneous environment can be achieved through the Network Queuing System (NQS) which is a collection of co-operating daemons configurable as a server/client environment and which automatically will transfer jobs or subtasks to the appropriate execution queues.

1.8 Parallel software tools

It is still the case that for many application programmers, distributed systems are difficult to program. The user must handle explicitly such issues as load balancing, balancing communication and computation, managing threads of control and managing the address space. These difficulties are much less severe in the case of shared-memory machines, for example.

In order to improve this situation, high-level languages such as High Performance Fortran (HPF) have been designed. HPF, for example, is based on a data parallel single-threaded model which, together with a standard message-passing interface, offers the promise of portability and ease of use. However, as noted in Clémencon et al. (1994), efficient portability is still not possible. This is because different vendors often support their own message-passing libraries and the characteristics of the different architectures, such as the communication/computation ratio, can dramatically affect program efficiency.

In addition to this, integrated tool environments that have been developed recently (see Turolte (1993) for an overview) are very rarely portable, often difficult to use, and often focus on micro effects rather than macro effects (see Decker et al. (1993), for a critique of extant distributed-memory tool environments).

In order to overcome some of these difficulties some attention is now being focused within the high-performance computing community on the emerging Message Passing Interface (MPIF (1993)). This interface consists of hardware-dependent operations partially supported by the operating system, operations for point-to-point communication of contiguous data such as blocking procedures, and global communication operations such as multicast. On the Intel Paragon, for example, at ETH Zürich, MPI has been implemented on top of the native NX communication library.

In order to exploit these developments a number of groups are developing integrated tool environments. In particular, the *Joint CSCS-ETH/NEC Collaboration in Parallel Processing* (Decker (1993)) is developing a sequence of tool prototypes in conjunction with application users. These tools support high-level and low-level programming based on HPF and MPI as well as interactive debugging and performance monitoring tools. A parallelization support tool acts as a parallelizing Fortran compiler accepting Fortran 77 annotated with compiler directives for data distribution and loop parallelization and generates parallel C code with communication primitives for execution. In order to evaluate this tool environment a suite of applications has been ported to parallel platforms using this environment. These applications include finite element applications, and multigrid and conjugate gradient codes, (see Clémencon et al. (1994) for a full description on the tool environment and the parallelization of applications).

Cheng (1993) has given an overview of existing parallel tools and integrated environments. These include Prism for debugging and visualization on the CM5, an interactive Fortran program browser and analyser (Forge) developed by Applied Parallel Research and the environment TOPSYS (Bemmerl and Bode (1991)) developed for a number of architectures and consisting of a parallel debugger, program visualizer and performance monitor. In addition, machines like the Paragon have tools such as SPV (Systems Performance Visualization tool) which allows node and communication and I/O activity to be displayed visually. However, SPV is somewhat restricted in that it does not differentiate between busy processor time and the wait/receive mode. A more sophisticated environment (ParaGraph) provides detailed visualization of interprocessor communication and performance analysis of programs.

In the discussion on the performance of waveform relaxation code presented in Chapter 9, ParaGraph will be used to demonstrate parallel efficiencies. ParaGraph can display, in many visual modes, processor utilization (idle, busy, I/O), interprocessor communication (through space-time diagrams and communication topologies), numerical statistics on loadings, and also provides a debugging environment.

1.9 Performance measures

Since programming in a parallel environment is a very much more complicated process than serial programming, performance models are often constructed in order to be able to judge the efficacy of a particular approach.

For a parallel environment, two quantities: speed-up (S) and efficiency (E) can be defined. Given a p-processor machine, then if $T(p)$ denotes the time taken to solve a given problem using p nodes, speed-up and efficiency are defined, respectively, as

$$S = \frac{T(1)}{T(p)}, \quad E = \frac{S}{p}. \tag{1.9.1}$$

E is usually strictly less than 1, because of load-balancing considerations and additional software and communication overheads, but in some rare cases superlinear speed-up is possible

Sophisticated timing models can be constructed taking into account start-up times, local and global communication times between processors and whether computation and communication can be overlapped (as is the case for transputers). However, even sophisticated models are at best very crude. For example, Fox et al. (1988) attempt a performance analysis of the efficiency of parallel algorithms based on the topological dimension of the physical system and the dimension of the processor network. The essence of this analysis is that, not surprisingly, large problems should be solved by a machine with a large number of processors and small problems by a machine with a small number of processors.

Another factor, known as Amdahl's law, states that if an innately sequential component takes a fraction θ of the time on a single processor then the maximum speed-up is $\frac{1}{\theta}$. However, as Fox et al. (1988) note, when designing an algorithm from scratch θ is usually small and has no significant effect. It is only when a parallelization of an extant sequential code is attempted that this factor may have some effect. Furthermore, this factor may become insignificant as the problem size increases.

In fact if the perspective of Amdahl's law is reversed, a more sensible measure of parallel performance can be obtained. This requires determining how long a given parallel program would take to run in a sequential environment (rather than the other way around) and is the basis of Gustafson's law (see Wilson (1993), for example). Thus if q is the (constant) time required for the sequential component of a given parallel program and $t(n, p)$

is the time required for the parallel component on a problem of size n with p processors then the speed-up counterpart of (1.9.1) is

$$S(n,p) = \frac{q + t(n,1)}{q + t(n,p)}. \tag{1.9.2}$$

By expanding $t(n,p)$ about $t(n,1)$ in terms of a Taylor series expansion, (1.9.2) can be written as

$$S(n,p) = 1 - (p-1)\frac{\partial t/\partial p}{q + t(n,p)},$$

and if $\partial t/\partial p < 0$ (so that an increase in the number of processors implies a decrease in time) and if $\partial t/\partial n > 0$ (so that an increase in the size of problem implies an increase in time) then any desired efficiency can be reached by increasing the problem size. For example, if $t(n,p) = \frac{n^r}{p}$ then

$$\lim_{n\to\infty} S(n,p) = p. \tag{1.9.3}$$

The interpretation of (1.9.3) is that speed-up can increase without bound for a large enough problem size. In other words, given any parallel environment it is hoped that much bigger problems can be solved than in a sequential environment in the same period of time.

In order to get some understanding of the behaviour of a parallel machine on a simple parallel algorithm, some results are given here on the implementation of the Jacobi iterative scheme for solving a dense linear system. The target architecture is the 96-node Paragon sited at ETH Zürich. The results are encapsulated in Figures 1.6 and 1.7.

The implementation was done using the sequential BLAS which are available in the **Kmath** library on the Paragon. These BLAS have been programmed in assembler code very efficiently to make use of the vector architecture of the Intel processor. As can been seen from Figure 1.6, a performance of approximately 40 Mflops can be obtained from one node and, although the claimed peak performance of one processor is 75 Mflops, a performance of 40 Mflops is almost optimal. It is interesting to note that if the Paragon BLAS are not used then typical performance from one node is of the order of 5 Mflops using Fortran77.

The performance results plotted in Figure 1.6 are typical of many parallel machines. For an apparently "embarrassingly parallel" algorithm close to linear speed-ups are obtained for a small number of processors. But as the number of processors is increased, communication costs start to dominate and the speed-up reaches a saturation level at which point increasing the number of processors is inappropriate. On the other hand, the effect of

Gustafson's law is clearly visible here in that as the problem size becomes bigger so does the performance improve.

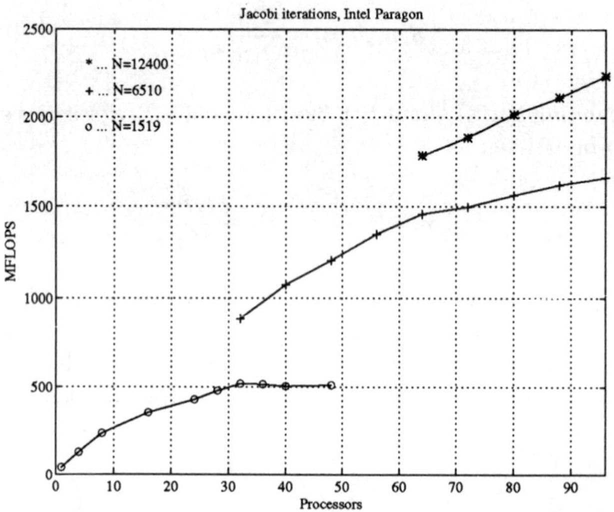

FIG. 1.6. Paragon performance in Mflops

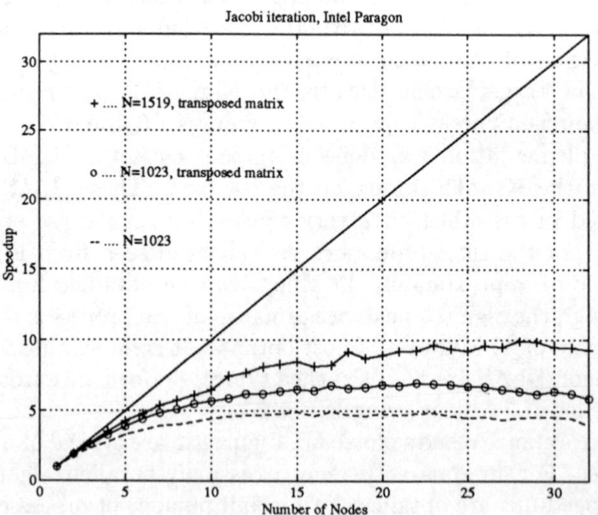

FIG. 1.7. speed-ups for the Jacobi algorithm

The final point to note here is that there is about a factor of two difference in performance depending on whether the matrix is distributed by rows or by columns (see Figure 1.7). This is to do with the caching and the fact that the Fortran compiler accesses elements in a matrix by columns rather than rows. Thus a column-based approach is preferable in terms of distributing the data.

These performance analyses are useful in giving guidelines for constructing parallel algorithms, but the concepts of speed-up and efficiency are of limited value. Essentially they compare the performance of the same algorithm in a one-processor and multiprocessor environment. This is often inappropriate. In order to judge the efficacy of a proposed parallel algorithm for solving a particular class of problems, such as a non-stiff IVP solver, it must be compared with the most efficient serial algorithm. The parallel and serial algorithm may be completely unrelated.

As a final remark in this chapter, it should be noted that it is possible to see examples of parallel algorithms which are published in respected scientific journals claiming linear speed-up when compared with the one-processor implementation. Whether deliberate or not, this is sophistry of the worst kind. Any proposed parallel algorithm should be compared with the most efficient serial codes and this is one of the reasons for the introduction to sequential methods in Chapters 2 and 3. It is impossible to give an intelligent discussion on parallel methods for a given subject area without relating it to the state of the art in the sequential arena.

2

AN INTRODUCTION TO SEQUENTIAL ODE METHODS

Since this monograph is concerned with an exposition of efficient methods for the numerical solution of systems of differential equations in a parallel computing environment, it is necessary to have a thorough insight into extant sequential methods. There is no space in this monograph to give a comprehensive discussion on all aspects of sequential algorithms for IVPs, rather, this monograph will attempt to give a flavour of this work and will emphasize techniques which have relevance to the development of parallel algorithms.

In spite of this caveat it will be seen that in many cases a general framework can be developed which allows the construction and analysis of either sequential or parallel methods from the same general class of methods, and wherever possible this approach will be adopted. Thus this chapter will offer a gradual introduction into the nature of ordinary differential equations of initial value type and present this framework which will allow an understanding of the structural properties of both efficient sequential and efficient parallel numerical methods.

The framework that will be presented is based on the concept of a multivalue method (sometimes called a general linear method). This framework is sufficiently general to be able to include most practical classes of algorithms that are studied, ranging from linear multistep methods to predictor-corrector methods to Runge-Kutta methods. Within this framework it is possible to present a general theory of order and stability which then particularizes to well-known classes of methods such as linear multistep methods and Runge-Kutta methods. This is the approach that will be taken in the next two chapters and although it is an unconventional one it is hoped that this will provide additional insight into the nature of differential equation algorithms and also suggest other approaches which may not have been so evident with the usual approach. Thus this chapter will deal with the following material:

- section 2.1: an introduction to the initial value problem and the boundedness of solution differences through the existence of a Lipschitz condition;
- section 2.2: a brief discussion on the nature of stiffness in terms of certain test problems and a one-sided Lipschitz condition;

- section 2.3: a discussion on the ways in which stiffness can arise in various application areas;
- section 2.4: an introduction to two elementary numerical methods, the Euler and implicit Euler methods and their behaviour on a linear test problem;
- section 2.5: an introduction to the concepts of multistep and multistage methods and their generalizations;
- section 2.6: an introduction to the class of linear multistep methods and a discussion on their implementation properties in terms of prediction-correction, variable coefficient methods and Nordsieck vectors;
- section 2.7: a presentation of the implementation of Runge-Kutta methods based on the structure of the Runge-Kutta matrix and including a discussion on efficient methods such as singly implicit and diagonally implicit methods;
- section 2.8: the interpretation and implementation of a general class of methods, known as multivalue methods, and its relationship to other well-known classes of methods;
- section 2.9: a brief discussion of other classes of methods which do not fit into the multivalue format, including Rosenbrock methods;
- section 2.10: problems which feature additional computational complexities such as discontinuities and algebraic constraints, and the ability of various codes to cope with these difficulties.

2.1 The initial value problem

An IVP can be written either in autonomous form

$$y'(x) = f(y(x)), \quad f : \mathbb{R}^m \to \mathbb{R}^m, \quad y(x_0) = y_0 \tag{2.1.1}$$

or in non-autonomous form

$$y'(x) = f(x, y(x)), \quad f : \mathbb{R} \times \mathbb{R}^m \to \mathbb{R}^m, \quad y(x_0) = y_0. \tag{2.1.2}$$

Non-autonomous problems can always be written as an autonomous system of equations of dimension one higher by the introduction of the equation

$$y'_{m+1}(x) = 1, \quad y_{m+1}(x_0) = x_0.$$

In addition, it should be noted that higher-order differential equations can always be written as a system of first-order equations by representing all necessary derivatives as additional components of the solution vector.

IVPs can arise in many different ways, especially when some form of modelling is taking place which involves growth or movement, and are usually classified as being stiff or nonstiff. A stiff problem can often arise when the underlying physical system has components with widely differing time constants. Of course such a classification is at best an imperfect one, because problems can be intermittently stiff and nonstiff.

Two examples of problems that can be classified as nonstiff and stiff, respectively, are the restricted three body problem and the laser oscillator model (see Byrne and Hindmarsh (1986), for example).

In the first problem, two bodies with normalized masses m_1, m_2 are in planar circular motion with another body of negligible mass moving in the same plane in an orbit around these two masses. The position of this body is described by

$$\begin{aligned} y_1' &= y_2 \\ y_2' &= y_1 + 2y_4 - m_1 \frac{y_1 + m_2}{r_1} - m_2 \frac{y_1 - m_1}{r_2} \\ y_3' &= y_4 \\ y_4' &= y_3 - 2y_2 - m_1 \frac{y_3}{r_1} - m_2 \frac{y_3}{r_2}, \end{aligned}$$

with

$$r_1 = ((y_1 + m_2)^2 + y_3^2)^{3/2}, \quad r_2 = ((y_1 - m_1)^2 + y_3^2)^{3/2}.$$

In the second problem, a pair of coupled equations represents a model of a ruby laser oscillator. If y_1 denotes the photon density and y_2 the population inversion the governing equations can be written as

$$\begin{aligned} y_1' &= -y_1(\theta y_2 + \delta) + \gamma \\ y_2' &= y_2(\Theta y_1 - \Delta) + \Gamma(1 + y_1), \end{aligned}$$

where

$$\theta = 1.5\times10^{-18}, \delta = 2.5\times10^{-6}, \gamma = 2.1\times10^{-6}, \Theta = 0.6, \Delta = 0.18, \Gamma = 0.016.$$

The three-body problem can be solved efficiently by an explicit method which in certain regions will take reasonably large stepsizes, while the laser problem is a difficult problem to solve because it is initially stiff and then oscillatory at a later stage. This latter remark is significant in that a problem does not have to be large (compare the problem described in the preface) to be computationally difficult to solve.

Before attempting to understand the nature of stiffness in more detail it is necessary to understand under what conditions unique solutions for (2.1.2) exist. In order to do this, the concept of a **Lipschitz condition** needs to be introduced.

Thus, if the range of integration of (2.1.2) is $I = [x_0, x_f]$ then f is said to satisfy a Lipschitz condition if f is continuous in $\mathbb{R}^m$ and

$$\| f(x,y) - f(x,z) \| \le L \| y - z \|, \quad \forall x \in I, \quad \forall y, z \in \mathbb{R}^m. \tag{2.1.3}$$

Here $\| \cdot \|$ will denote a standard inner product norm in $\mathbb{R}^m$ and $\langle \cdot, \cdot \rangle$ the corresponding inner product. Note also that f can satisfy a Lipschitz condition D on a subset of $\mathbb{R}^m$. Thus if (2.1.3) holds, then for any initial value there will exist a unique continuous and differentiable solution of (2.1.2). In addition, if f is differentiable with respect to y on D then the Mean Value Theorem implies that f will satisfy the Lipschitz condition with

$$L = \sup_{(x,y) \in D} \| \partial f(x,y)/\partial y \| .$$

Given that f satisfies a Lipschitz condition, then if y and z are any two solutions of (2.1.2) with different initial conditions, it is easily shown that

$$\| y(x_2) - z(x_2) \| \le e^{(x_2 - x_1)L} \| y(x_1) - z(x_1) \|, \quad x_2 > x_1. \tag{2.1.4}$$

A modification of this result, which allows the analysis of the stability of general nonlinear systems and which will prove useful when studying the behaviour of parallel waveform methods, is the nonlinear variation of constants formula due to Alekseev (1961) and Gröbner (1960).

Theorem 2.1.1. *If y and z are, respectively, the solutions of*

$$\begin{aligned} y' &= f(x,y), \quad y(x_0) = y_0 \\ z' &= f(x,z) + g(x,z), \quad z(x_0) = z_0, \end{aligned}$$

and $\partial f(x,y)/\partial y$ exists and is continuous then

$$\begin{aligned} z(x) = y(x) \quad &+ \int_0^1 \frac{\partial y}{\partial y_0}(x, x_0, y_0 + u(z_0 - y_0))(z_0 - y_0)du \\ &+ \int_{x_0}^x \frac{\partial y}{\partial y_0}(x, u, z(u))g(u, z(u))du. \end{aligned}$$

From this result, the so-called fundamental lemma (see Hairer et al. (1987)) can be proved.

Theorem 2.1.2. *If $z(x)$ is an approximate solution of (2.1.2) satisfying*

$$\begin{aligned} \| z(x_0) - y(x_0) \| &\leq \alpha \\ \| z'(x) - f(x, z(x)) \| &\leq \beta \\ \| f(x, z) - f(x, y) \| &\leq L \| z - y \|, \end{aligned}$$

then for $x \geq x_0$

$$\| y(x) - z(x) \| \leq \alpha e^{L(x-x_0)} + \frac{\beta}{L}(e^{(x-x_0)L} - 1).$$

2.2 Stiffness

It is extremely difficult to classify stiffness in a completely rigorous manner but considerable insight can be gained into the concept of stiffness from one simple problem. This is the so-called Prothero-Robinson problem (see Prothero and Robinson (1974))

$$y'(x) = q(y(x) - g(x)) + g'(x), \quad y(0) = y_0, \quad Re(q) \ll 0. \tag{2.2.1}$$

The solution to (2.2.1) can be written as

$$y(x) = g(x) + e^{qx}(y_0 - g(0)) \tag{2.2.2}$$

or

$$y(x + \theta) = g(x + \theta) + e^{q\theta}(y(x) - g(x)). \tag{2.2.3}$$

If g is a smooth function then (2.2.2) is said to consist of two components: a smooth or nontransient component $g(x)$ and a stiff, transient component $e^{qx}(y(0) - g(0))$. It is clear from (2.2.2) that, after a short time period, the transient will die away leaving only the smooth component. On the other hand, (2.2.3) shows that if the approximations to the solution are inexact, perturbations continually arise and complicate the numerical computation of the smooth component (see Dekker and Verwer (1984) for an excellent exposition of this topic). This behaviour is typical of a stiff problem and will be illustrated in the next section by two simple numerical methods.

Although problem (2.2.1) offers considerable insight into the nature of stiffness it does not give any information about when the general problem (2.1.2) is stiff. A more analytical approach is based on the concept of a Lipschitz condition and the result given in (2.1.4). Thus a problem is said to be stiff if $(x_f - x_0)L$ is large.

However, in many cases it may be difficult to know a priori the size of L. Furthermore, the bound given in (2.1.4) may not be a realistic one. On the other hand, stiffness can often be deduced from the nature of the physical system being modelled. Thus, for example, if the system has components with widely varying time constants then the model will often lead to a stiff system of linear ordinary differential equations.

One special class of problems that will be frequently investigated in this monograph is the class of problems of the form

$$y' = Qy + g(x), \tag{2.2.4}$$

where Q is a constant matrix. Such problems often arise when linear parabolic partial differential equations are semi-discretized in the spatial variables. Here $\| Q \|_2 \geq \rho(Q) \gg 0$ (where $\rho(Q)$ denotes the spectral radius of the matrix Q), so that if there exist some eigenvalues of Q such that $Re(\lambda_i) \ll 0$ and there also exist eigenvalues of moderate magnitude, then the problem will be stiff. The existence of eigenvalues with large positive real part is precluded as otherwise the problem is unstable.

In the case of nonlinear problems, stiffness is often described in terms of the range of the eigenvalues of the Jacobian of f, but, as noted originally by Vinograd (1952) and expounded in Dekker and Verwer (1984), this is not always a valid approach. When a numerical method is applied to a problem satisfying a Lipschitz condition it is natural to expect the numerical method to satisfy a discrete analogue of (2.1.4). However, because L is large for stiff problems, the discrete version of (2.1.4) offers little insight into the behaviour of a discrete method for such problems.

For this reason the concept of a one-sided Lipschitz condition has been introduced. A problem, given by (2.1.2), is said to satisfy a **one-sided Lipschitz condition** if there exists a **one-sided Lipschitz constant** ν such that

$$\langle f(x,y) - f(x,z), y - z\rangle \leq \nu \| y - z \|^2, \quad \forall x \in I, \quad \forall y, z \in \mathbb{R}^m. \tag{2.2.5}$$

Let $u(x) = \| y(x) - z(x) \|^2$, where y and z are any two solutions of (2.1.2) with different initial conditions. Then given a one-sided Lipschitz condition,

$$u'(x) = 2\langle y'(x) - z'(x), y(x) - z(x)\rangle \leq 2\nu u(x),$$

so that

$$(u(x)e^{-2\nu x})' \leq 0$$

and hence

$$\| y(x_2) - z(x_2) \| \leq e^{(x_2 - x_1)\nu} \| y(x_1) - z(x_1) \|, \quad x_2 > x_1. \tag{2.2.6}$$

It is often the case that it is possible to find values for ν which are moderate in size or even nonpositive for stiff problems. If $\nu = 0$ then problems satisfying (2.2.5) are called **monotonic** or **dissipative** and any two solutions y and z possess the contractive property

$$\| y(x_2) - z(x_2) \| \leq \| y(x_1) - z(x_1) \|, \quad x_2 > x_1. \tag{2.2.7}$$

The above formulation depends on the use of inner product norms, but more general results can be obtained through the use of a **logarithmic norm** which was first introduced by Dahlquist (1958). In the case of the linear test problem (2.2.4), the smallest possible one-sided Lipschitz constant $\mu[Q]$ for Q for a given inner product is the so-called logarithmic norm. The logarithmic norm of a matrix Q is defined as

$$\mu[Q] = \lim_{\Delta \to 0^+} \frac{\| I + \Delta Q \| - 1}{\Delta},$$

and Dahlquist (1958) has shown that (2.2.5) holds if f satisfies

$$\mu[f'(x,y)] \leq \nu, \quad \forall y \in \mathbb{R}^m, \quad \forall x \in I,$$

where $f'(x,y)$ denotes the Jacobian matrix of f.

As a consequence of this, it is easy to prove (Dekker and Verwer (1984))

Theorem 2.2.1. *Let $\| \cdot \|$ be a given norm on $\mathbb{R}^m$. If*

$$\mu[f'(x,y)] \leq \nu, \quad \forall y \in \mathbb{R}^m,$$

then for any two solutions y and z of (2.1.2)

$$\| y(x_2) - z(x_2) \| \leq e^{(x_2 - x_1)\nu} \| y(x_1) - z(x_1) \|, \quad x_2 > x_1.$$

For the $\| \cdot \|_1$, $\| \cdot \|_2$, $\| \cdot \|_\infty$ norms on $\mathbb{R}^m$,

$$\begin{aligned}
\mu_1[Q] &= \max_j (q_{jj} + \sum_{i \neq j} |q_{ij}|) \\
\mu_2[Q] &= \frac{1}{2}\rho(Q + Q^\top) \\
\mu_\infty[Q] &= \max_i (q_{ii} + \sum_{j \neq i} |q_{ij}|).
\end{aligned}$$

If, in addition, Q is a **normal** matrix (so $QQ^\top = Q^\top Q$) then

$$\mu_2[Q] = \max(Re(\sigma(Q))), \tag{2.2.8}$$

where $\sigma(Q)$ denotes the spectrum of Q.

2.3 The occurrence of stiffness

A very important application area in which stiffness can arise is in the modelling of gas mixtures in which the time scales of the reactants are much smaller than the movement times. Such areas include pollution models, nozzle design for rockets, combustion and nuclear reactions. Other areas include biological modelling, chemical kinetics and fluid flow.

The presence of stiffness in these problem areas often arises in complicated ways. Returning briefly to problem (2.2.1) it can be seen that there are two phases: a transient phase in which the transient component is still active and a smooth phase in which the numerical method must compute the nontransient while suppressing the transient. In the transient phase a small stepsize must be taken to follow the solution and so an explicit method should be used to keep implementation costs at a minimum; only in the smooth phase can the problem be said to be stiff and there a stable implicit method should be used.

This is an idealized situation. Problems can be intermittently stiff and nonstiff. For example, van der Pol's equation has a limit cycle that has this property. Other examples are the Belousov reaction system (see Prokopakis and Seider (1981), Field and Noyes (1974)) and a problem due to Dickinson and Gelinas (1976) which describes the reaction of atomic oxygen and ozone in the atmosphere. This latter problem is an example of a diurnal problem in which the rate constants vary diurnally, rising and decreasing rapidly at dawn and sunset, respectively. For such problems it is impractical to use a code which is either only stiff or nonstiff. Rather, the code should be able to switch backwards and forwards between stiff and nonstiff mode.

Even more complex problems are those based on combustion models (see Aiken (1985), for example). Here the explosive zone may be so small that a code can miss it all together. Furthermore, the eigenvalues associated with such problems can be both positive and negative, so that there is a degree of instability. In this case methods should be used which have no stability region in the right half plane, such as symmetric Runge-Kutta methods (see Chan (1989)).

Another difficult application area occurs when modelling large engineering systems such as nuclear reactors. Here a subset of the components can have highly oscillatory solutions and the numerical method must take an unnaturally small (for the rest of the system) stepsize to follow the oscillations. It may be possible to partition the system and treat the subsystems differently (see Hairer (1981) for example). But this partitioning may not be constant and requires monitoring. Furthermore, the coupling between components may be too complex to allow partitioning.

At this stage three points should be stressed.

1. It is impossible to develop a stability theory for numerical methods

sufficiently rich as to cover all the complexities and subtleties of real-life problems. Nevertheless, codes which have been developed (such as GEAR (Hindmarsh (1974)), ODEPACK (Hindmarsh and Byrne (1983)) and STRIDE (Butcher et al. (1979))) are based on these simplified stability models and seem to perform well on the types of complex problems mentioned above.

2. There are many special classes of problems which have their own peculiar attributes and, while it is desirable to develop general-purpose codes which are as robust as possible, there are still problem areas in which codes are designed for those problems only. Two areas, in particular, are chemical kinetics (see BELLCHEM (Edelson (1976)), KISS (Gottwald (1981)) and LARKIN (Deuflhard et al. (1981))) and structural engineering. This second category can give rise to highly oscillatory problems in which components of the solution change rapidly throughout the range of integration and cannot be damped. Here special methods are needed such as the multirevolutionary methods of Petzold (1981a) which attempt to follow the envelope of the rapidly varying components.

3. With the advent of highly efficient parallel computing environments, it is highly likely that the number of special-purpose codes for solving IVPs in a parallel environment will proliferate. This is due not only to the fact that there are a plethora of widely differing parallel environments for which codes and methods have to be structured differently to exploit the given parallel architecture, but also to the fact that depending on the structure of the problem itself (be it dense, sparse or banded) the algorithm may vary drastically. It is at this stage that the concept of waveform relaxation appears and a study of such structural approaches will occupy a significant amount of this monograph. In particular, time-dependent parabolic partial differential equations in which the spatial variables are discretized lead to large stiff systems which are block-banded, in which the bandwidth depends on the dimension of the underlying partial differential equation and the finite difference scheme. These problems are often **diagonally dominant** and so are amenable to special iterative techniques based, for example, on waveform relaxation (see White et al. (1985), for an excellent exposition).

As this monograph will focus, in part, on special methods for solving large well-structured problems the following example will be presented (see, for example, Aiken (1985)).

Example 2.3.1. Consider the numerical solution of the convective diffusion partial differential equation

$$u_t = u_{xx} - \alpha u_x, \quad u(x,0) = 0, \quad u(0,t) = 1, \quad u(1,t) = 0. \tag{2.3.1}$$

If u_{xx} is replaced by second-order differencing on a uniform grid with width $\Delta = \frac{1}{N+1}$ and u_x is replaced by the generalized upwind scheme

$$u_x|_{x=x_i} \approx \frac{1}{2\Delta}\left((1-\theta)u_{i+1} + 2\theta u_i - (1+\theta)u_{i-1}\right), \quad i = 1,\ldots,N,$$

where u_i is an approximation to $u(i\Delta, t)$, $i = 0,\ldots,N+1$, a system given by

$$\begin{aligned} u_i' &= \frac{1}{\Delta^2}(u_{i+1} - 2u_i + u_{i-1}) \\ &\quad -\frac{\alpha}{2\Delta}\left((1-\theta)u_{i+1} + 2\theta u_i - (1+\theta)u_{i-1}\right), \quad i = 1,\ldots,N \\ u_i(0) &= 0, \quad i = 1,\ldots,N, \quad u_0(t) = 1, \quad u_{N+1}(t) = 0 \end{aligned}$$

is obtained. This can be written in the form

$$u' = Qu, \tag{2.3.2}$$

where

$$Q = (N+1)^2 \begin{pmatrix} -2-2b\theta & 1-b\theta_1 & & & \\ 1+b\theta_2 & -2-2b\theta & 1-b\theta_1 & & \\ \ddots & \ddots & \ddots & & \\ & 1+b\theta_2 & -2-2b\theta & 1-b\theta_1 \\ & & 1+b\theta_2 & -2-2b\theta \end{pmatrix}, \tag{2.3.3}$$

with $\theta_1 = 1-\theta$, $\theta_2 = 1+\theta$ and $b = \frac{1}{2}\alpha\Delta$. Note that to obtain a compact representation of tridiagonal matrices with elements $c_2,\ldots,c_m$ on the upper diagonal, $b_1,\ldots,b_m$ on the diagonal and $a_2,\ldots,a_m$ on the lower diagonal, such matrices will be denoted by (a,b,c). Thus Q can be represented in compact form as $(N+1)^2(1+b\theta_2, -2-2b\theta, 1-b\theta_1)$.

Problems such as (2.3.1) which are solved by the method of lines invariably lead to stiff problems if N is moderately large. The reason for this can be seen by taking $\alpha = 0$ in (2.3.3). In this case the eigenvalues of Q are well known and given by

$$2(N+1)^2(-1+\cos(i\pi/(N+1))), \quad i = 1,\ldots,N \tag{2.3.4}$$

and lie in the interval $(-4(N+1)^2, 0)$. Thus the smaller the spatial grid, the stiffer the problem becomes.

Note that (2.2.8) and (2.3.4), with $\alpha = 0$, imply

$$\begin{aligned} \mu_2[Q] &= -2(N+1)^2(1-\cos(\pi/(N+1)) \\ \mu_\infty[Q] &= 0. \end{aligned}$$

2.4 Two elementary numerical methods

A general discrete method for solving an IVP generates a sequence of approximations to the solution at $x_1, x_2, \ldots, x_n, \ldots$. The quantity $h_n = x_n - x_{n-1}$ is called the stepsize for the interval $[x_{n-1}, x_n]$.

In its most general form a discrete method computes a vector of r, say, pieces of information at a point $x = x_{n+1}$ using k vectors of past information at the points $x_n, \ldots, x_{n-k+1}$ and possibly s additional **intermediate solution vectors** of information which are computed at the current step. Such a method will be called a k-step, r-value, s-stage method or more briefly a **multivalue method** and the vector consisting of r pieces of information will be called the **update vector**.

If, at any stage, any component of the update or intermediate solution vectors appears both as the update on the left-hand side of the equation describing the numerical method and on the right-hand side as a function evaluation, then such a method is said to be **implicit**; otherwise it is said to be **explicit**. Examples of an explicit method and an implicit method are the Euler and implicit Euler methods, respectively.

If applied to (2.1.2) with constant stepsize these two methods are represented by

$$y_{n+1} = y_n + hf(x_n, y_n) \tag{2.4.1}$$

$$y_{n+1} = y_n + hf(x_{n+1}, y_{n+1}), \tag{2.4.2}$$

respectively.

The quality of a numerical method is determined by a number of criteria such as order, stability and ease of implementation and these criteria will be discussed fully in Chapter 3. The methods given by (2.4.1) and (2.4.2) will be of order w, respectively, if

$$l_{n+1} = y(x_n + h) - y_{n+1} = y(x_n + h) - (y(x_n) + hf(\theta, Y)) = O(h^{w+1}),$$

where in the case of (2.4.1)

$$\theta = x_n, \quad Y = y(x_n)$$

and for (2.4.2)

$$\theta = x_{n+1}, \quad Y = y(x_n) + hf(x_{n+1}, Y),$$

with $y(x)$ denoting the true solution of (2.1.2) at the step point x. Here l_{n+1} represents the **local truncation error** (the difference between the computed solution and the exact solution) committed by either method over one step assuming exact initial values at the beginning of the step. For both methods a simple Taylor series expansion gives $w = 1$.

Although both methods are of order 1, there are considerable differences between them, both in terms of implementation and stability considerations. (2.4.2) is a nonlinear equation and requires a system of m nonlinear equations to be solved at each step using some variant of the Newton method. This can be seen by assuming, without loss of generality, an autonomous form for f and writing (2.4.2) in the form

$$F(y_{n+1}) := y_{n+1} - y_n - hf(y_{n+1}) = 0.$$

Applying some variant of the Newton method generates a sequence of approximations $\{y_{n+1}^{(k)}\}$ to y_{n+1}, given an initial prediction $y_{n+1}^{(0)}$, of the form

$$M(y_{n+1}^{(k+1)} - y_{n+1}^{(k)}) = -F(y_{n+1}^{(k)}) = -y_{n+1}^{(k)} + y_n + hf(y_{n+1}^{(k)}). \qquad (2.4.3)$$

Here $M = I_m - hJ$, I_m is the identity matrix of size m and J is some approximation to the Jacobian matrix of f associated with the problem. The solution of this system of equations involves $O(m^3)$ floating point operations for the LU factorization and $O(m^2)$ operations for the forward and back substitutions of the resulting linear systems. In this implicit case the additional cost of implementation is considerable, so is it justifiable?

The answer is a resounding yes. Applying (2.4.1) and (2.4.2) to (2.2.1) gives, respectively,

$$y_{n+1} = (1 + hq)(y_n - g_n) + g_n + hg_n' \qquad (2.4.4)$$

and

$$y_{n+1} = \frac{1}{1-hq}(y_n - g_n) + \frac{1}{1-hq}(g_n + hg_{n+1}' - hqg_{n+1}), \qquad (2.4.5)$$

where g_n denotes $g(x_n)$.

As in (2.2.1) the numerical approximation can be considered to consist of two components, and, just as in (2.2.2), in which the smooth component $y(x_n)-g(x_n)$ is rapidly damped by the negative exponential, a similar effect should occur in the numerical approximation. In the case of the implicit Euler method this term is damped by $\frac{1}{1-hq}$ for all positive h and so there is no restriction on h in terms of stability requirements since $|\frac{1}{1-hq}| \leq 1$, for all $Re(q) \leq 0$.

On the other hand, the Euler method is only damped if $|1 + hq| < 1$. With $Re(q) \ll 0$, stability imposes an unnatural restriction on the stepsize with h having to be much smaller than would be the case if only accuracy requirements were imposed. This is the fundamental difference between all explicit and implicit methods.

If $g = 0$, then (2.2.1) reduces to the so-called **standard linear test equation**

$$y' = qy. \tag{2.4.6}$$

Hence (2.4.1) and (2.4.2) reduce to

$$y_{n+1} = R(z)y_n, \quad z = hq \tag{2.4.7}$$

where $R(z)$ is $1 + z$ or $\frac{1}{1-z}$, respectively. $R(z)$ is called the **stability function** associated with the method. A stability region can be associated with any discrete method. In the case of a one-step method the **stability region** S is defined as

$$S = \{z \in \mathbb{C} : |R(z)| \leq 1\}. \tag{2.4.8}$$

More generally, for any r-value method, S is defined to be the set of complex z for which the update vectors do not grow in magnitude with respect to some norm as the method advances. A formal definition is given in Chapter 3.

For the Euler method, the stability region is the interior and boundary of the disk of radius 1 centred on $(-1, 0)$, while the stability region for the implicit Euler method is the exterior and boundary of the disk of radius 1 centred on (1,0) in the complex plane.

Let $\mathbb{C}^- = \{z \in \mathbb{C} : Re(z) \leq 0\}$. Then if

$$\mathbb{C}^- \subseteq S \tag{2.4.9}$$

a method is said to be ***A*-stable**. A-stability is a highly desirable property for any numerical method to possess when solving stiff problems. However, this property is not always possible to obtain because other qualities such as efficiency, robustness and reliability are also necessary and these can be in conflict with A-stability. This is the case for the class of methods based on backward differentiation formulae (BDF methods) of order greater than 2. Nevertheless, such methods have proved both popular and efficient solvers of stiff problems.

As a consequence of otherwise useful numerical methods not being A-stable, the property of **$A(\alpha)$-stability**, introduced by Widlund (1967), is a way of distinguishing between methods which are not A-stable. A method is said to be $A(\alpha)$-stable if when applied to (2.4.6) the numerical approximations do not grow in magnitude with respect to some norm for all $z \in S(\alpha) = \{z : Re(z) \leq 0, \arg(z) \leq \alpha\}$.

Thus the phenomenon whereby the nature of the problem can cause a numerical method to have its stepsize unnaturally restricted by stability rather than accuracy considerations is called stiffness. No explicit numerical methods can have an infinitely sized stability region (see Hairer et al. (1987)) and so they are unsuited for stiff problems.

2.5 A general framework for IVP methods

In the previous section, two very simple methods were introduced – one explicit and one implicit and their stability behaviour on a very simple test equation compared. It was seen that substantial differences can occur in terms of the stability of the method and how a problem can restrict the stepsize. Of course the stepsize can also be affected by the accuracy of the method itself, and since both of these methods are of order 1 they will in most cases be unsuitable for solving differential systems unless only very lax accuracy is required. Thus in order to construct families of higher-order methods with appropriate stability properties for the problem at hand, generalizations of the Euler method are needed.

There are clearly at least two approaches. The first possibility is to use information (either approximations to the solution and/or the derivatives) computed from previous steps to produce an approximation at the current step. The second possibility is to perform at each step additional function evaluations and to use this information (but not any past information) to compute a new approximation at the current step. These two approaches are characterized, respectively, by the names linear multistep methods and one-step methods and can be pictorialized by the following figure:

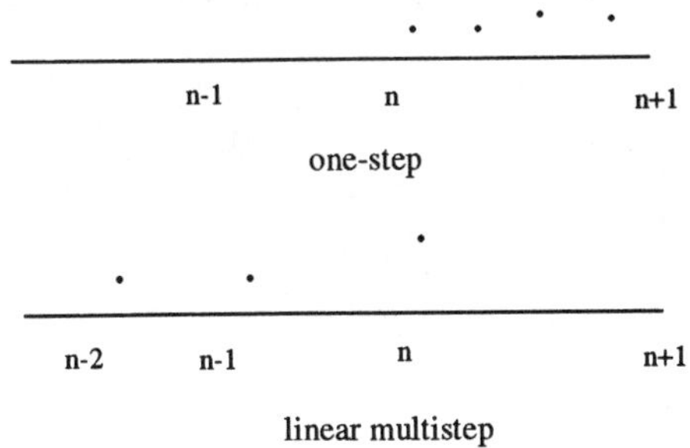

FIG. 2.1. one-step versus linear multistep

These two approaches will be discussed in some detail in the next two sections, while generalizations which will contain elements of both approaches will be analysed in section 2.8.

2.6 Linear multistep methods

The family of k-**step linear multistep methods** for solving (2.1.2) with fixed stepsize is given by

$$y_{n+1} = \sum_{j=1}^{k} \alpha_j y_{n+1-j} + h \sum_{j=0}^{k} \beta_j f(x_{n+1-j}, y_{n+1-j}). \tag{2.6.1}$$

If $\beta_0 \neq 0$ then the method is said to be implicit otherwise if $\beta_0 = 0$ it is explicit. Note that any k-step linear multistep method requires a starting procedure to generate $y_1, \ldots, y_{k-1}$ (approximations to the solution at $x_0 + h, \ldots, x_0 + (k-1)h$) in order for the computations to proceed.

An example of a nontrivial explicit linear multistep method is

$$y_{n+1} = y_n + \frac{h}{2}\left(3f(x_n, y_n) - f(x_{n-1}, y_{n-1})\right), \tag{2.6.2}$$

which uses derivative information from the previous two step points; while an example of a nontrivial implicit linear multistep method is

$$y_{n+1} = \frac{4}{3}y_n - \frac{1}{3}y_{n-1} + \frac{2h}{3}f(x_{n+1}, y_{n+1}), \tag{2.6.3}$$

which uses solution information from the previous two step points.

These two methods are representatives from two general classes of methods, known as Adams methods and Backward Differentiation Formulae (BDF methods), which can be generated directly by quadrature or derivative approximations. The general structure for the Adams classes of methods has

$$\alpha_1 = 1, \quad \alpha_j = 0, \quad j = 2, \ldots, k,$$

while BDF methods satisfy

$$\beta_0 \neq 0, \quad \beta_j = 0, \quad j = 1, \ldots, k.$$

In the case of an implicit linear multistep method, y_{n+1} is described by a nonlinear system of equations. This can be solved by a modified Newton approach which takes $O(m^3)$ per step if an LU factorization is performed at each step. In this case an algebraic system of the form

$$y_{n+1} = h\beta_0 f(x_{n+1}, y_{n+1}) + \Psi_{n+1}$$

must be solved for y_{n+1} at each step. These can either be solved by fixed-point iteration

$$y_{n+1}^{(p+1)} = h\beta_0 f(x_{n+1}, y_{n+1}^{(p)}) + \Psi_{n+1}, \quad p = 0, 1, \ldots \tag{2.6.4}$$

where $y_{n+1}^{(0)}$ is a predicted initial guess or by solving

$$F(y_{n+1}) := y_{n+1} - h\beta_0 f(x_{n+1}, y_{n+1}) - \Psi_{n+1} = 0$$

by some form of Newton iteration. The Newton iteration scheme takes the form

$$(I_m - h\beta_0 M)(y_{n+1}^{(p+1)} - y_{n+1}^{(p)}) = -F(y_{n+1}^{(p)}) \qquad (2.6.5)$$

where M is $f_y(x_{n+1}, y_{n+1}^{(p)})$ if the scheme is Newton's method, or M is an approximation to the Jacobian if a modified Newton's method is used.

Fixed-point iteration leads to the formulation

$$y_{n+1}^{(p+1)} = \sum_{j=1}^{k} \alpha_j y_{n+1-j} + h\sum_{j=1}^{k} \beta_j f(x_{n+1-j}, y_{n+1-j}) + h\beta_0 f(x_{n+1}, y_{n+1}^{(p)}),$$

which will converge to the unique solution of (2.1.2) provided

$$h < \frac{1}{|\beta_0|L},$$

where L is the Lipschitz constant for f.

For non-stiff problems this constraint on the stepsize is not as significant as the constraints imposed by accuracy. This suggests using an accurate guess for $y_{n+1}^{(0)}$ and one way of doing this is by the use of an explicit linear multistep method as a predictor and correction by the fixed-point iteration of the implicit corrector method given in (2.6.4). This process is called prediction-correction and the two underlying methods comprise a predictor-corrector pair. The general k-step **predictor-corrector** pair can thus be written as

$$\begin{aligned} \hat{y}_{n+1} &= \sum_{j=1}^{k} \hat{\alpha}_j y_{n+1-j} + h\sum_{j=1}^{k} \hat{\beta}_j f(x_{n+1-j}, y_{n+1-j}) \\ y_{n+1} &= \sum_{j=1}^{k} \alpha_j y_{n+1-j} + h\sum_{j=1}^{k} \beta_j f(x_{n+1-j}, y_{n+1-j}) + h\beta_0 f(x_{n+1}, \hat{y}_{n+1}). \end{aligned}$$

Predictor-corrector pairs are usually chosen so that the stability region of the pair is greater than that of the explicit predictor. There are various ways that predictor-corrector methods can be implemented. For example, the correction process can continue until some convergence criterion is satisfied and this is called **correcting to convergence**. However, this process is rarely used because of its unpredictability due to the fact that convergence may in some instances be very slow. A more acceptable approach is to state in advance the (fixed) number of corrections to be allowed in the iteration process.

The considerations so far have only covered a constant-stepsize implementation. In order to be efficient, any implementation must allow for variable stepsizes. One way of doing this is to try to estimate the local error at each step and control it by taking a stepsize such that the local error is less than some prescribed tolerance. The hope is that the global error will also be controlled. This is discussed more fully in Chapter 3.

There are additional difficulties in using multistep methods in variable stepsize mode, since the past values can no longer be used. Consequently, there are three approaches:

1. place restrictions on the stepsize changes;
2. interpolate past values;
3. use a variable step implementation in which the coefficients are functions of the stepsizes.

The first approach was one adopted by Krogh (1973) in which stepsize change is restricted to doubling or halving. Although this is a very elegant approach which means only using past information at $x_{n-1}, \ldots, x_{n-2k+1}$, Krogh (1973) noted that there is a substantial loss of accuracy with this approach compared with other techniques.

The second approach involves using polynomial interpolation of existing back data to approximate intermediate back data arising as a consequence of stepsize change. In the case of Adams methods, for example, only the f values need be interpolated and the interpolating polynomial can either be specified by the backward differences of f or by values of the interpolant and its derivative at the integration points.

This last approach was first proposed by Nordsieck (1962) and is equivalent to the representation of linear multistep methods in terms of a different basis. The basis used for a k-step method is the so-called **Nordsieck vector**

$$Z_n = (y(x_n), hy'(x_n), \ldots, \frac{h^{k-1}}{(k-1)!} y^{(k-1)}(x_n))^\top.$$

There is a linear transformation between Z_n and the backward difference representation B_n of the form

$$Z_n = DB_n,$$

where the matrix D is independent of the stepsize h. Consequently, a stepsize change $h_{n+1} = wh_n$ corresponds to the rescaling of Z_n by the diagonal matrix $\operatorname{diag}(1, w, \ldots, w^{k-1})$. Gear (1971a) used this Nordsieck approach in the implementation of Adams methods in the code DIFSUB (see Lambert (1991) for a thorough discussion of variable-stepsize implementations).

This third approach is the one used in codes such as EPISODE (Byrne and Hindmarsh (1975)) and ODEPACK, and was developed by Ceschino (1961), who derived variable-coefficient Adams methods of up to order 4, and was further developed by Krogh (1969). In this case a Newton's divided difference formulation must be used which was pioneered by Krogh (1974) in terms of Adams methods. Although this is probably a more robust approach, this still requires recalculation of the method parameters whenever the stepsize changes.

With interpolation, computational cost increases as the dimension size of the problem, m, increases, while for the variable-coefficient approach the computational cost is independent of m. For this reason, the variable coefficient approach is a more feasible one if m is large and more robust if there is frequent change of stepsizes.

2.7 Runge-Kutta methods

The fundamental difference between linear multistep and multistage one-step methods has already been alluded to. These classes can be considered to lie at opposite ends of a wide spectrum of numerical methods. Multistep methods use only the computed values for the solution and its derivative from previous node points while multistage, one-step methods introduce extra calculations within a step to calculate intermediate approximations to the solution and its derivative at various off-step points. A significant consequence of this is that one-step methods are trivial to implement in terms of a variable-stepsize implementation.

An important family of multistage, one-step methods is the family of **Runge-Kutta methods** so named because of the pioneering work of Runge (1895) and Kutta (1901). Runge (1895) generalized the Euler method by adding an additional Euler step at the midpoint of the interval, while Kutta (1901) derived a third-order method in which additional function evaluations are performed at the midpoint and at the end of the integration step and which can be written as

$$\begin{aligned} Y_1 &= y_n \\ Y_2 &= y_n + \frac{h}{2} f(x_n, Y_1) \\ Y_3 &= y_n + h(-f(x_n, Y_1) + f(x_n + h/2, Y_2)) \\ y_{n+1} &= y_n + \frac{h}{6}(f(x_n, Y_1) + 4f(x_n + h/2, Y_2) + f(x_n + h, Y_3)). \end{aligned}$$

Perhaps the most famous of all Runge-Kutta methods is the so-called classical method of order 4 discovered by Kutta (1901) which performs two additional function evaluations at the midpoint and another at the end of the integration step. This method can be written as

$$\begin{aligned}
Y_1 &= y_n \\
Y_2 &= y_n + \frac{h}{2} f(x_n, Y_1) \\
Y_3 &= y_n + \frac{h}{2} f(x_n, Y_2) \\
Y_4 &= y_n + h f(x_n + h, Y_3) \\
y_{n+1} &= y_n + \frac{h}{6}(f(x_n, Y_1) + 2f(x_n + h/2, Y_2) \\
&\quad +2f(x_n + h/2, Y_3) + f(x_n + h, Y_4)).
\end{aligned}$$

The general structure of an explicit Runge-Kutta method is given by

$$\begin{aligned}
Y_i &= y_n + h \sum_{j=1}^{i-1} a_{ij} f(x_n + hc_j, Y_j), \quad i = 1, \ldots, s \\
y_{n+1} &= y_n + h \sum_{j=1}^{s} b_j f(x_n + hc_j, Y_j).
\end{aligned} \tag{2.7.1}$$

Such a method is said to have s **stages**.

Until the early 1960s most of the work on Runge-Kutta methods focused on explicit methods, but Butcher (1963, 1964a) provided a general framework in which to study both explicit and Runge-Kutta methods. This entailed generalizing (2.7.1) to

$$\begin{aligned}
Y_i &= y_n + h \sum_{j=1}^{s} a_{ij} f(x_n + hc_j, Y_j), \quad i = 1, \ldots, s \\
y_{n+1} &= y_n + h \sum_{j=1}^{s} b_j f(x_n + hc_j, Y_j),
\end{aligned} \tag{2.7.2}$$

where $Y_1, \ldots, Y_s$ represent the intermediate approximations to the solution at $x_n + c_1 h, \ldots, x_n + c_s h$. The abscissae $c_1, \ldots, c_s$ usually lie in the interval $[0, 1]$ but this is not a necessity.

Such a method can be represented by the so-called **Butcher tableau**

$$\begin{array}{c|c} c & A \\ \hline & b^{\top} \end{array}, \tag{2.7.3}$$

where

$$c = (c_1, \ldots, c_s)^\top, \quad b^\top = (b_1, \ldots, b_s), \quad c = Ae, \quad e = (1, \ldots, 1)^\top.$$

In the case of an explicit method, A is strictly lower triangular and, in particular, $c_1 = 0$.

Using this notation, Euler's explicit method (2.4.1) can be written as a Runge-Kutta method in the form (2.7.3) with

$$\begin{array}{c|c} 0 & 0 \\ \hline & 1 \end{array}\;.$$

The two four-stage methods of order 4, the so-called classical and 3/8-methods of Kutta (1901) can be written in tableau form, respectively, as

$$\begin{array}{c|cccc} 0 & 0 & & & \\ \frac{1}{2} & \frac{1}{2} & & & \\ \frac{1}{2} & 0 & \frac{1}{2} & & \\ 1 & 0 & 0 & 1 & \\ \hline & \frac{1}{6} & \frac{1}{3} & \frac{1}{3} & \frac{1}{6} \end{array}$$

$$\begin{array}{c|cccc} 0 & & & & \\ \frac{1}{3} & \frac{1}{3} & & & \\ \frac{2}{3} & -\frac{1}{3} & 1 & & \\ 1 & 1 & -1 & 1 & \\ \hline & \frac{1}{8} & \frac{3}{8} & \frac{3}{8} & \frac{1}{8} \end{array}$$

In constructing these methods of order 4, Kutta used the fact that there is an intimate relationship between a quadrature order formula for approximating

$$\int_0^1 f(x)dx \quad \text{by} \quad \sum_{j=1}^{n} w_j f(x_j)$$

and the coefficients and abscissae for the update component. In particular, it can be seen that the methods given above are based on Simpson's rule and Newton-Cotes quadrature formulae, respectively.

In the case of implicit methods, it turns out that the structure of the **Runge-Kutta matrix** A is of the utmost importance in terms of obtaining efficient methods. In order to see this, a discussion on the implementation of Runge-Kutta methods is now given. In order to avoid some messy technical details the autonomous version (2.1.1) will be used in the following discussion.

Implementation

Let $Y = (Y_1^\top, \ldots, Y_s^\top)^\top \in \mathbb{R}^{sm}$ and let $F(Y) = (f(Y_1)^\top, \ldots, f(Y_s)^\top)^\top$; then any Runge-Kutta method can be written in tensor notation as

$$\begin{aligned} Y &= e \otimes y_n + h(A \otimes I_m)F(Y) \\ y_{n+1} &= y_n + h(b^\top \otimes I_m)F(Y). \end{aligned} \tag{2.7.4}$$

This system of sm nonlinear equations can be solved by using the Newton iteration scheme.

Thus, if at the end of one iteration, Y_i is replaced by $Y_i + \delta_i$ then

$$(I_s \otimes I_m - h\tilde{J})\delta = \Psi \tag{2.7.5}$$

where $\tilde{J}$ is a block matrix of order sm whose (i, j) element is $a_{ij} f'(Y_j), i, j = 1, \ldots, s$ and

$$\begin{aligned} \delta &= (\delta_1^\top, \ldots, \delta_s^\top)^\top, \quad \Psi = (\Psi_1^\top, \ldots, \Psi_s^\top)^\top \\ \Psi_i &= -Y_i + y_n + h\sum_{j=1}^{s} a_{ij} f(Y_j), \quad i = 1, \ldots, s. \end{aligned}$$

This can be simplified if the $f'(Y_j)$ are approximated by evaluating the Jacobian at a single point, y_n say. Thus by letting $J = f'(y_n)$ then (2.7.5) can be written as

$$(I_s \otimes I_m - hA \otimes J)\delta = \Psi. \tag{2.7.6}$$

If (2.7.6) is solved by an LU factorization followed by forward and back substitutions then the number of floating point operations to perform one iteration of the Newton process is, respectively,

$$O(cm^3/3), \quad O(dm^2), \quad \text{where} \quad c = s^3 \quad \text{and} \quad d = s^2.$$

This is approximately s^3 times the work for the equivalent implementation of a BDF method where there $c = d = 1$. Unless these factors can be improved, implicit Runge-Kutta methods are not a feasible option for stiff problems.

Fortunately, structures can be imposed on the Runge-Kutta matrix which reduce the computational cost of certain classes of implicit Runge-Kutta methods to approximately that of a linear multistep method and the nature of these structures depends on whether a sequential or parallel implementation is being considered.

Butcher (1976) proposed an ingenious technique for improving the computational efficiency of Runge-Kutta methods and it is based on transforming A to its Jordan canonical form. Thus suppose that there exists a nonsingular matrix T such that $TAT^{-1} = \Lambda$ and let $\bar{\delta} = (T \otimes I_m)\delta$, $\bar{Y} = (T \otimes I_m)Y$, $e \otimes \bar{y}_n = (Te) \otimes y_n$ and $\bar{F} = (T \otimes I_m)F((T^{-1} \otimes I_m)\bar{Y})$. Then (2.7.6) can be written as

$$(I_s \otimes I_m - h\Lambda \otimes J)\bar{\delta} = \bar{\Psi}, \tag{2.7.7}$$

where

$$\begin{aligned} \bar{\delta} &= (\bar{\delta}_1^\top, \ldots, \bar{\delta}_s^\top)^\top, \quad \bar{\Psi} = (\bar{\Psi}_1^\top, \ldots, \bar{\Psi}_s^\top)^\top \\ \bar{\Psi}_i &= -\bar{Y}_i + \bar{y}_n + h\sum_{j=1}^{s} \Lambda_{ij}\bar{f}(\bar{Y}_j), \quad i = 1, \ldots, s. \end{aligned}$$

In the case that A has only real eigenvalues and is similar to a diagonal matrix then

$$c = d = s.$$

On the other hand if A has a one-point spectrum so that A is similar to the matrix with λ on the diagonal and $-\lambda$ on the subdiagonal, then

$$c = 1, \quad d = s.$$

Such methods were called by Burrage (1978a) **singly implicit methods** (SIRKs). It can be seen from (2.7.6) that for singly implicit methods, the nonlinear equations are effectively decoupled and can be solved sequentially. Thus the computational costs for SIRKs in terms of one LU factorization and forward and back substitution in the Newton iteration process is given by

$$C_S = m^3/3 + sm^2 + O(m^2)$$

which compares favourably with the computational cost for BDF methods given by

$$C_B = m^3/3 + m^2 + O(m^2).$$

However, at each iteration of the transformed Newton procedure, Butcher's technique requires the Y and $F(Y)$ to be transformed and untransformed by the similarity transformation matrix. If l iterations are performed per step, the total cost for a singly implicit method over one integration step is approximately

$$C_S = m^3/3 + lsm^2 + 2ls^2m$$

while

$$C_B = m^3/3 + lm^2.$$

Thus only if m is large is C_S comparable with C_B.

Another way of overcoming the generally expensive implementational costs of Runge-Kutta methods is to impose a different structure on A, in which A is lower triangular. Such methods are called **diagonally implicit methods** and if all diagonal elements are equal they are called **singly diagonally implicit** (SDIRKs). Just as with SIRKs, $c = 1$ and $d = s$, but in this case no transformations are needed so that the computational cost associated with a SDIRK is

$$C_D = m^3/3 + lsm^2,$$

if l iterations are performed per Newton step.

An example of a SDIRK is the following family of two-stage methods of order 2 given by

$$\begin{array}{c|cc} \lambda & \lambda & 0 \\ 1-\lambda & 1-2\lambda & \lambda \\ \hline & \frac{1}{2} & \frac{1}{2} \end{array}$$

As well as transforming the method to its canonical form, it is also possible to transform problems with a full Jacobian, J, to an upper Hessenberg form. This was proposed by Enright (1978) in conjunction with multistep methods and by Varah (1979) for Runge-Kutta methods. This procedure is especially beneficial if J is kept constant over as many steps as possible.

2.8 Multivalue methods

The methods considered so far in this monograph have been either of Runge-Kutta or linear multistep type. Indeed, historically, the two most widely studied classes of methods for solving IVPs have been linear multistep and Runge-Kutta methods. These families have been gradually extended and a great deal of research has been done on a wide variety of methods including extrapolation (Bader and Deuflhard (1983)), Rosenbrock (Kaps and Wanner (1981), for example), semi-implicit (Hairer et al. (1982)) and multivalue methods (Burrage (1988)). However, it was not until the early 1960s that there was any concerted attempt to generalize these two methodologies and then new approaches were introduced independently by Butcher (1965a), Gear (1965) and Gragg and Stetter (1964).

A very natural approach and the one adopted by Butcher (1965a) was to add an additional stage or stages to the linear multistep format and this led to the method (Butcher (1965a)) given by

$$
\begin{aligned}
Y_1 &= y_{n-1} + \frac{h}{8}(3f(x_{n-1}, y_{n-1}) + 9f(x_n, y_n)) \\
Y_2 &= \frac{1}{5}(-23y_{n-1} + 28y_n) \\
&\quad + \frac{h}{5}(-26f(x_{n-1}, y_{n-1}) - 60f(x_n, y_n) + 32f(x_n + h/2, Y_1)) \\
y_{n+1} &= \frac{1}{31}(-y_{n-1} + 32y_n) + \frac{h}{5}(-f(x_{n-1}, y_{n-1}) \\
&\quad + 12f(x_n, y_n) + 64f(x_n + h/2, Y_1) + 15f(x_n + h, Y_2)).
\end{aligned}
$$

A different approach is to add additional information to a Runge-Kutta method and to allow this information to be carried from step to step. For example, Dekker (1981) generalized the concept of a diagonally implicit Runge-Kutta method to allow two pieces of information to be carried from step to step and constructed the following method of order 4:

$$
\begin{aligned}
Y_1 &= y_n - \frac{7}{6}z_n + \frac{2h}{3}f(x_n + h/2, Y_1) \\
Y_2 &= y_n + \frac{1}{6}z_n + \frac{2h}{3}(f(x_n + h/2, Y_1) + f(x_n + 5h/2, Y_2)) \\
y_{n+1} &= y_n + \frac{h}{2}(f(x_n + h/2, Y_1) + f(x_n + 5h/2, Y_2)) \\
z_{n+1} &= \frac{5}{11}z_n + \frac{h}{11}(-f(x_n + h/2, Y_1) + 7f(x_n + 5h/2, Y_2)).
\end{aligned}
$$

The natural generalization of both k-step linear multistep methods and s-stage Runge-Kutta methods which accommodates both the examples given above is the **multivalue method**. This approach allows the computation of internal approximations as well as the carrying of a vector of solutions from step to step. The general **s-stage r-value multivalue method** for solving (2.1.1) is given by

$$
\begin{aligned}
Y_i &= \sum_{j=1}^{r} a_{ij}^{(1)} y_j^{(n)} + h\sum_{j=1}^{s} b_{ij}^{(1)} f(Y_j), \quad i = 1, \ldots, s \\
y_i^{(n+1)} &= \sum_{j=1}^{r} a_{ij}^{(2)} y_j^{(n)} + h\sum_{j=1}^{s} b_{ij}^{(2)} f(Y_j), \quad i = 1, \ldots, r
\end{aligned}
\tag{2.8.1}
$$

and is characterized by the four matrices in the tableau

$$\begin{array}{c|c} A_1 & B_1 \\ \hline A_2 & B_2 \end{array}$$

In this formulation, Y_i $(i = 1, \ldots, s)$ are internal to each step and represent approximations to the solution at the off-step points $x_n + c_i h$, while the $y_i^{(n)}$ carry all the information from step to step. In order to start such a method r starting procedures $Q_1, \ldots, Q_r$ are needed to compute $y_1^{(0)}, \ldots, y_r^{(0)}$ from y_0. This can be done, for example, by using r Runge-Kutta methods. The implementation of the method is complete when the inverse of one of the starting procedures is applied to one of the components of $y^{(N)}$. (The **inverse of a method** is defined as that method that returns to the initial starting point if a method and its inverse are successively applied to an initial value.) The implementation of a multivalue method is represented pictorially in Figure 2.2. In this case the starting procedures have been labelled from $Q_0, \ldots, Q_{k-1}$ in order to show the relationship with linear multistep methods.

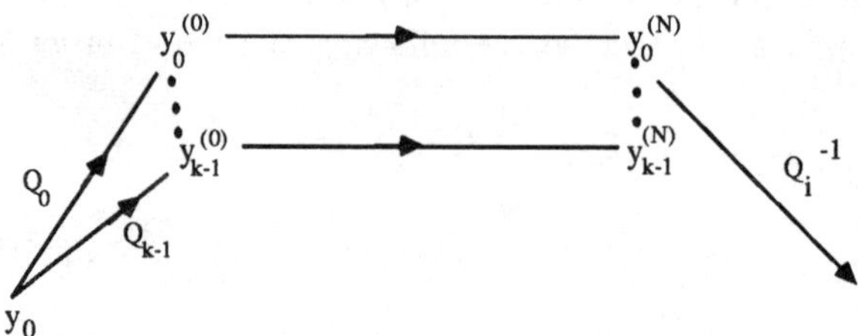

FIG. 2.2. implementation of a multivalue method

In order to be able to manipulate multivalue methods in a simple fashion when studying some of their order and stability properties, they will often be represented using a Kronecker tensor product notation:

$$\begin{aligned} Y &= (A_1 \otimes I_m) y^{(n)} + h(B_1 \otimes I_m) F(Y) \\ y^{(n+1)} &= (A_2 \otimes I_m) y^{(n)} + h(B_2 \otimes I_m) F(Y), \end{aligned}$$

where

$$Y = (Y_1^\top, \ldots, Y_s^\top)^\top, \quad y^{(n)} = (y_1^{(n)\top}, \ldots, y_s^{(n)\top})^\top,$$

and

$$F(Y) = (f(Y_1)^\top, \ldots, f(Y_s)^\top)^\top.$$

Two new and important classes of multivalue methods that extend the linear multistep and Runge-Kutta approach are multistep Runge-Kutta methods (MRKs) and Nordsieck methods. In the case of MRKs the $y_i^{(n)}$ are approximations to $y(x_{n+1-i})$ so this family can be written in the autonomous formulation as

$$\begin{aligned} Y_i &= \sum_{j=1}^{r} a_{ij} y_{n+1-j} + h \sum_{j=1}^{s} b_{ij} f(Y_j), \quad i = 1, \ldots, s \\ y_{n+1} &= \sum_{j=1}^{r} \alpha_j y_{n+1-j} + h \sum_{j=1}^{s} \beta_j f(Y_j). \end{aligned} \tag{2.8.2}$$

Remarks

1. (2.8.2) can be written in the multivalue format with

$$A_2 = \begin{pmatrix} \alpha_1 & \alpha_2 & \ldots & \alpha_r \\ 1 & 0 & \ldots & 0 \\ & \ddots & \ddots & \vdots \\ & & 1 & 0 \end{pmatrix}, \quad B_2 = \begin{pmatrix} \beta_1 & \ldots & \beta_s \\ 0 & \ldots & 0 \\ \vdots & & \vdots \\ 0 & \ldots & 0 \end{pmatrix}.$$

2. If $r = 1$, (2.8.2) gives the family of Runge-Kutta methods.
3. If $s = r + 1$ and A_1 and B_1 are partitioned as

$$A_1 = \begin{pmatrix} \alpha^\top \\ I_r \end{pmatrix}, \quad B_1 = \begin{pmatrix} \beta^\top \\ O_{r\times s} \end{pmatrix},$$

then (2.8.2) generates the family of r-step linear multistep methods.
4. The family of k-step linear multistep methods can also be written as a multivalue method with $s = 1$ and $r = 2k$ in which the $2k$ update vector is

$$(y_n, hf(y_n), y_{n-1}, hf(y_{n-1}), \ldots, y_{n-k+1}, hf(y_{n-k+1}))^\top.$$

5. The family of Nordsieck methods have the property that the $y_i^{(n)}$ are approximations to the scaled derivatives $\frac{h^{i-1}}{(i-1)!} y^{(i-1)}(x_n)$.

One reason for considering the more general family of multivalue methods is to attempt to overcome some of the existing difficulties with codes based on linear multistep or Runge-Kutta methods. The hope is that it will prove possible to construct methods which are efficient, avoid costly transformations, possess a high stage order and have suitable stability properties.

Two possible approaches to constructing efficient multivalue methods which do not involve the transformation techniques are based on generalizing the concepts of diagonal implicitness and backward differentiation formulae.

With this second approach, Cash (1980, 1983) has analysed BDF-type methods with $s = 2$ or 3 intermediate approximations per step. In this case there exist A-stable methods of orders 4 and 6, respectively, and considerable improvement over traditional BDF methods is obtained on stiff problems whose eigenvalues are close to the imaginary axis and the numerical performance is comparable otherwise.

On the other hand, by recognizing that the implementation of an implicit multivalue method is very similar to that of a Runge-Kutta method, it is natural to generalize the concept of single implicitness and diagonal implicitness to multivalue methods. Thus a multivalue method is said to be **explicit**, **diagonally singly implicit** or **singly implicit** if the matrix B_1 is, respectively, strictly lower triangular, lower triangular with constant value on the diagonal or if it has a one-point spectrum.

As will be seen in Chapter 3, because each multivalue method carries a vector of r pieces of information from step to step, multivalue methods can always be constructed with stage order r, whether they are explicit or implicit. In the explicit case a continuous extension of order r is thus immediately available without the high number of function evaluations per step associated with high-order explicit Runge-Kutta methods with interpolants; while in the diagonally implicit case the stiff order can always be r (as compared with one for DIRKs).

Burrage and Chipman (1989) have constructed both diagonally implicit MRKs and diagonally implicit Nordsieck methods of orders $r+2$ or $r+1$ with stage order r. In the case of diagonally implicit MRKs, Burrage and Chipman (1989) have found numerically, A-stable methods of orders up to 8 with stage order 6, while in the case of diagonally implicit Nordsieck methods, Burrage and Chipman (1989) have found methods of up to order 7 and stage order 5. However, in all cases the abscissae are well in advance of the update abscissae and this can be a drawback when solving discontinuous problems, for example. Thus given the existence of methods with suitable stability properties and small error coefficients, such methods are likely to be efficient and robust solvers of stiff problems. However, much more theoretical work is needed in this area.

On the other hand, there exist singly implicit multivalue methods with stage order $s+r-1$ (see Butcher (1981) and Schneider (1985), for example) and Burrage and Chipman (1985) have shown numerically that there exist A-stable, singly implicit, Nordsieck methods with stage order $s+r-1$ of up to order 9 (with $s = 9$, $r = 6$), which is a considerable improvement over SIRKs. Nevertheless, these approaches are unlikely to be competitive in a sequential environment because of the necessity of transforming and untransforming at each iteration step.

Other approaches that fit into this multivalue framework include the composite methods of Bickart and Rubin (1974), cyclic linear multistep methods with improved $A(\alpha)$-stability properties (Tendler et al. (1978)), the hybrid methods of England (1982) and more recently an implementation of multistep collocation methods of Radau type by Schneider (1993).

There has been considerable investigation into the suitability of various classes of multivalue methods for stiff and nonstiff methods. But until recently very few codes have been developed. One exception to this is an experimental code developed by van der Houwen and Sommeijer (1982) for solving problems arising from the semi-discretization of parabolic partial differential equations by the method of lines. This code is based on a special class of MRKs with extended stability intervals along the negative real axis.

In the case of explicit multivalue methods a number of approaches have been developed but in order to construct efficient methods considerable care must be taken in minimizing the local error coefficients, as is the case for Runge-Kutta methods. For example, Burrage and Sharp (1994) have analysed one-step, two-step and three-step Nordsieck methods in which s is at most 3. Method parameters have been selected on the basis of obtaining small truncation coefficients with reasonably large stability regions. On the other hand, Jackiewicz and Zennaro (1990) have constructed families of two-step explicit MRKs which appear to give better performance than the usual explicit Runge-Kutta methods with the same order.

2.9 Other approaches

As has been emphasised, linear multistep methods and Runge-Kutta methods can be considered as lying at opposite ends of the spectrum of multivalue methods and the general class of multivalue methods can be interpreted to lie in the space of methods in which the two orthogonal axes represent past values (P) and intermediate values (I).

In fact other classes of methods can be constructed which have a representation in a higher-dimensional space of methods, by introducing a third orthogonal axis based on higher derivatives of the solution (P). This will create a three-dimensional space of methods. Thus, for example, Taylor series methods lie on the P-axis in this three-dimensional space. An additional axis, based on the use of higher derivatives of f, can also be introduced (F), and a method lying in this four-dimensional space is called a multistep-multistage-multiderivative method. Examples of well-known methods that belong to this very general class of methods and are not representable as multivalue methods include Obreshkov methods (1942) and Turan methods (see Kastlunger and Wanner (1972) and Hairer and Wanner (1973)).

In this section two further classes of methods will be presented that do not fit into the multivalue format, but which have, nevertheless, been successfully used in solving both stiff and nonstiff problems, namely Rosenbrock and extrapolation methods.

Rosenbrock methods

Rosenbrock methods were first introduced by Rosenbrock (1963). The original idea of a Rosenbrock method is to insert an approximation to the Jacobian matrix into the numerical method directly and not apply the Newton iterative process. This technique leads to the solution of linear equations only.

The base method in this Rosenbrock process is usually a diagonally-implicit Runge-Kutta method. When applied to (2.1.1) it can be written in the general form (see Kaps and Rentrop (1979), for example)

$$\begin{aligned} k_i &= hf(y_n + h\sum_{j=1}^{i-1} \alpha_{ij}k_j) + hJ\sum_{j=1}^{i} \lambda_{ij}k_j, \quad i = 1, \ldots, s \\ y_{n+1} &= y_n + h\sum_{j=1}^{s} b_j k_j. \end{aligned}$$

If $\lambda = \lambda_{11} = \ldots = \lambda_{ss}$ then this method corresponds to a SDIRK method and only one LU factorization is needed per step. Kaps and Wanner (1981) have constructed A-stable methods up to order 6 with $J = f'(x_n, y_n)$. However, in the case that J is an approximation to the Jacobian the construction of high-order methods is more difficult because of the large number of order conditions. These methods are sometimes called ROW methods and their order properties have been studied by Steihaug and Wolfbrandt (1979).

Extrapolation methods

In attempting to compute an approximation to the solution of (2.1.2) at $x_n + H$ using an order w method, a sequence of approximations $T_{i1}, i = 1, \ldots, k$, can be computed by using a sequence of stepsizes $h_i = \frac{H}{p_i}, i = 1, \ldots, k$, and performing p_i steps with stepsize h_i. Assuming a global error expansion of the form

$$T_{i1} = y(x_n + H) + e_w(x_n + H)h_i^w + \cdots, \quad i = 1, \ldots, k,$$

successively higher-order approximations can be found by extrapolation based on either polynomial or rational extrapolation. The first extrapolation codes were developed by Bulirsh and Stoer (1966) and used rational extrapolation. However, Deuflhard (1983) has suggested that, in general, polynomial extrapolation is preferable.

In the case of symmetric methods, the global error expansion is in even powers of the stepsize and so each extrapolation improves the order by 2. In the case of symmetric methods a tableau of the form

$$\begin{pmatrix} T_{11} & & & & \\ T_{21} & T_{22} & & & \\ T_{31} & T_{32} & T_{33} & & \\ \vdots & \vdots & \vdots & \ddots & \\ T_{n1} & T_{n2} & T_{n3} & \dots & T_{nn} \end{pmatrix}$$

where

$$T_{j,k+1} = T_{j,k} + \frac{T_{j,k} - T_{j-1,k}}{(\frac{P_j}{P_{j-k}})^2 - 1}$$

is computed.

One of the first important advances in extrapolation techniques was the work of Gragg (1965) and his analysis of an explicit modified midpoint rule with a smoothing procedure, which has an even power expansion. Applying the modified midpoint rule over $2N$ steps and then performing a smoothing step gives

$$\begin{aligned} y_1 &= y_0 + hf(x_0, y_0) \\ y_{n+1} &= y_{n-1} + 2hf(x_n, y_n), \quad n = 1, \dots, 2N \\ \tilde{y}_{2N} &= \frac{1}{4}(y_{2N-1} + 2y_{2N} + y_{2N+1}). \end{aligned} \tag{2.9.1}$$

Gragg showed that this method has an even power expansion in the global error, but Stetter (1973) simplified the proof considerably by rewriting (2.9.1) as a multivalue method in one-step mode.

Equation (2.9.1) can be used to compute the T_{i1} with different stepsize sequences where the p_j are even. However, the application of the smoothing process is crucial to the efficacy of this method. It has also been used by Dahlquist and Lindberg (1973) in conjunction with the midpoint and trapezoidal rules when solving stiff equations. More recently, Chan (1989) and Butcher and Chan (1990) have presented a general theory which allows the smoothing of arbitrary Runge-Kutta methods. The crucial concept in this theory is that of the composition of two methods.

Thus if R denotes the s-stage method

$$\begin{array}{c|c} c & A \\ \hline & b^{\top} \end{array} \tag{2.9.2}$$

and $\hat{R}$ denotes the $\hat{s}$-stage method

$$\begin{array}{c|c} \hat{c} & \hat{A} \\ \hline & \hat{b}^\top \end{array} \tag{2.9.3}$$

then $R \circ \hat{R}$ denotes the application of method (2.9.2) followed by (2.9.3) and can be written as the $(s+\hat{s})$-stage method

$$\begin{array}{c|cc} c & A & \\ \hat{c}+\hat{e} & \hat{e}b^\top & \hat{A} \\ \hline & b^\top & \hat{b}^\top \end{array}$$

It is easily shown that if R is symmetric then so is $R^n = R \circ R \ldots \circ R$. One of the disadvantages of symmetric methods is that they provide no damping. For example, applying a symmetric Runge-Kutta method to the standard linear test problem gives

$$y_{n+1} = R(z)y_n, \quad R(z) = \frac{p(z)}{p(-z)},$$

where p is some polynomial. Hence as $z \to -\infty$

$$y_N \to (-1)^{sN} y_0.$$

Smoothing, as used by Dalhquist and Lindberg (1973), is used to provide a damping property. However, smoothing can destroy the even power expansion of the global error. Chan (1989) has shown that it is possible to choose a symmetrizing method which is applied at the end of an extrapolated step in such a way that the composite method, while no longer being symmetric, still possesses an asymptotic expansion of the global error in even powers of the stepsize. This remarkable result implies that symmetry and the property of even power expansion of the global error are not equivalent.

One particular class of these generalized symmetric methods has the form $\frac{1}{n}(R^{n-1} \circ \hat{R})$ where the symmetrizing method $\hat{R}$ is equivalent to the composition of two steps of the symmetric method R but with differing weights for the final update. This symmetrizer can be used, for example, when solving stiff problems, to improve the stability of the overall method (see Chan (1989)). The method $\hat{R}$ can be written as

$$\begin{array}{c|cc} c & A & 0 \\ c+e & eb^\top & A \\ \hline & b^\top - v^\top P & v^\top \end{array} \tag{2.9.4}$$

where P is the permutation matrix associated with the symmetric method (2.9.2). The vector $v^\top$ here is a free parameter.

Burrage and Chan (1993) have investigated the effect of smoothing on the midpoint, trapezoidal and two-stage Gauss methods. For problems of Prothero-Robinson type they showed that it is possible to construct smoothers for which there is no order reduction for the two-stage Gauss methods.

On the other hand, Auzinger and Frank (1988, 1989) have investigated the behaviour of the implicit midpoint and trapezoidal rules when applied to stiff problems. They have shown that for some classes of stiff problems the even power asymptotic error expansion is destroyed.

Codes based on extrapolation methods have the advantage that stepsize and order control can be implemented within a step in a very simple way, but the heuristics for implementing this control is often complicated and obtuse. Recently, Bader and Deuflhard (1983) have developed efficient extrapolation codes for both stiff and nonstiff problems.

2.10 Additional features of IVPs

In addition to stiff problems there are other classes of problems which can possess computational complexities. These include discontinuous problems, delay problems, differential-algebraic problems and problems whose Jacobians are sparse. An example of a discontinuous problem is that reported in Aiken (1985) by Thompson and Tuttle. They analysed a nuclear reactor coolant system in which difficulties can arise in the refill and reflood phase when the level of the water drops below the level of the core leaving the core immersed in superheated vapour. At this stage derivative discontinuities can occur.

How codes treat discontinuities depends on the nature of the discontinuity, for example, time events or state events, and whether the underlying method is a one-step method or multistep method. Automatic detection of state events is much more complicated than detection of time events. LSODAR, which is part of the ODEPACK series of routines, written by Hindmarsh and Petzold (in Hindmarsh and Byrne (1983)) has a root-finding capacity to find roots of a set of functions of the dependent and independent variables. Delay problems are discussed in detail in section 6.8 with respect to parallel methods.

Differential-algebraic equations (DAEs) arise naturally in many areas when constraints are imposed, such as control theory, network modelling and robotics. DAEs can be written in the general form

$$F(x, y(x), y'(x)) = 0 \tag{2.10.1}$$

or in a semi-explicit form

$$F(u', u, y, x) = 0, \quad G(u, y, x) = 0.$$

DAEs can be characterized by the index of the problem. The index can be thought of as the number of times needed to differentiate the algebraic constants to reduce the system to an explicit ODE. In general, the higher the index, the more complicated the problem. (See the book by Brenan et al. (1989) for an excellent discussion on DAEs.)

As Petzold (1981b) pointed out, DAEs are not IVPs and in many cases it may be neither appropriate nor possible to make this conversion. The code of DASSL by Petzold (1982) is a general code for solving DAEs. It is based on an idea of Gear (1971b) in which Euler's method is used to replace the derivative term in (2.10.1) to give a system of nonlinear equations at the integration points x_n

$$F(x_n, y_n, \frac{y_n - y_{n-1}}{h}) = 0. \tag{2.10.2}$$

This technique has been generalized to higher-order methods including BDF methods and Runge-Kutta methods. The BDF implementation is the corner stone of DASSL.

Another special class of problems that frequently arise are those in the linearly implicit form

$$A(x, y)y' = f(x, y). \tag{2.10.3}$$

Here A can be nonsingular or singular (in which case it is a DAE and great care must be taken to have consistent initial values). There are a number of packages available for this class of problems. They include LSODI (Hindmarsh (1980)) and more recently an experimental code RADAU5 by Hairer and Wanner (1991) in which the matrix A is assumed to be constant. This latter method is based on the three-stage Radau Runge-Kutta method of order 5, which will be described in Chapter 3.

A very important application area is, as mentioned previously, large sparse chemical kinetics problems. Some codes have automatic translators which translate the chemical reactions to a system of IVPs, for example BELLCHEM by Edelson (1976). Most recent packages incorporate sparse matrix packages from existing libraries. For example, LSODES (part of the ODEPACK library and written by Hindmarsh and Sherman) treats the Jacobian matrix in general sparse form. It can detect the sparsity structure automatically and uses a segment of the Yale Sparse Matrix Package. Other codes include SPARKS by Houbak and Thomsen (1979), STRIDE by Butcher et al. (1979) and SIMPLE by Nørsett and Thomsen (1986).

More recent packages include VODE (Brown et al. (1989)) and LSODPK (Brown and Hindmarsh (1986)). These are described more fully in Chapter 9 when they are used in conjunction with waveform relaxation algorithms. These packages build on the pioneering codes DIFSUB (Gear (1971a)), EPISODE (Hindmarsh and Byrne (1976)), LSODE (Hindmarsh (1980)) and ODEPACK (Hindmarsh and Byrne (1983)). The underlying methods are Adams linear multistep methods used in predictor-corrector mode in the nonstiff case and BDF methods in the stiff case.

Some of the codes mentioned above are type-insensitive and can switch backwards and forwards between stiff and nonstiff mode. However, there are a number of codes designed purely for nonstiff problems and include ODE (based on Adams methods) by Shampine and Gordon (1975), DVERK a routine by Hull et al. (1976) based on Verner's fifth and sixth-order pair of explicit Runge-Kutta methods, a code developed by Brankin et al. (1992) based on an order (7, 8) pair of explicit Runge-Kutta methods developed by Prince and Dormand (1981), Gragg's explicit modified midpoint rule with smoother (1965) and an extrapolation code by Bader and Deuflhard (1983).

More recent codes based on explicit Runge-Kutta methods perform local extrapolation, that is they advance on the higher-order method. In addition, continuous extensions for accurate output are provided, although this often requires additional function evaluations. Hairer et al. (1987) have performed a number of comparisons of codes based on Adams, Runge-Kutta and extrapolation methods and their conclusion is that high-order Runge-Kutta codes such as the Prince and Dormand (7, 8) pair are superior to extrapolation and Adams-based codes for smooth problems and on a wide range of tolerances. Only for very severe tolerances do extrapolation codes appear to be competitive with Runge-Kutta codes.

It is clear from the contents of this chapter that there is a wide range of sequential codes ranging from production codes with sophisticated user interfaces to experimental codes based on novel approaches. This range of codes can be general purpose or very specific in the types of problems that can be solved. However, as will be seen, in the parallel arena there are currently very few parallel codes. This is partly due to the newness of the area with suitable methods still in a state of flux, but it is also due to the fact that it is very difficult to provide a generic and sophisticated user interface over a widely varying range of architectures, operating systems and differential systems.

3

ORDER AND STABILITY – A GENERAL FRAMEWORK

Chapter 2 attempted to give an overview and present a general structure in which many classes of differential equation methods can be studied. This third chapter continues the review of sequential methods and describes in more detail the properties of the numerical methods introduced in Chapter 2. This will be done from the framework of a very general class of methods known as multivalue methods. These properties can then easily be particularized to various classes of methods such as linear multistep and Runge-Kutta methods by understanding how these classes of methods fit into the multivalue framework. This is a very natural approach and one that has not really been attempted before, simply because not all the theoretical tools had been sufficiently developed. Recent developments now allow a systematic treatment of multivalue methods. However, it is important not to lose sight of the historical developments which have nearly always involved a movement from the particular (linear multistep and Runge-Kutta methods) to the general (multivalue methods).

It should again be emphasized that the multivalue framework does not of course describe all methods for solving differential systems and excludes, for example, multiderivative methods and Rosenbrock-type methods. However, the parallel algorithms that will be introduced in Chapters 5 and 6 will mainly be multivalue-based and hence the somewhat restricted approach.

This chapter will cover the following material:

- section 3.1: the definitions of zero-stability, consistency and convergence as the crucial properties for any sensible differential equation method;
- section 3.2: the concept of order as applied to multivalue methods and the particularization to linear multistep and Runge-Kutta methods;
- section 3.3: an introduction to the concept of linear stability as applied to multivalue methods and the presentation of a number of stability and order results for various classes of methods including linear multistep and Runge-Kutta methods;
- section 3.4: the construction of high-order explicit Runge-Kutta pairs and their refinement based on FSAL designs and continuous extensions, with extensions to multivalue methods in general;

- section 3.5: an introduction to order simplifying assumptions for multistage methods and the construction of high-order and efficiently implementable methods of both Runge-Kutta and general multivalue type;
- section 3.6: the generalization of linear stability properties (A-stability) to nonlinear stability properties (algebraic stability) and their interrelationships;
- section 3.7: the importance of stage order and the effect of order reduction on stiff problems with extensions to singularly perturbed problems and differential-algebraic problems of index 1 and 2;
- section 3.8: variable stepsize implementations based on Richardson extrapolation, embedding and variable coefficient methods.

3.1 Zero stability, linear stability and convergence

In introducing Runge-Kutta and linear multistep methods in Chapter 2 it was seen that there is a fundamental difference between the two classes of methods, not only in terms of implementation but in terms of fundamental differences to very simple problems. In order to be able to understand these differences and others, the general framework of a multivalue method will be adopted which will require knowledge about the behaviour of A^n for large n where A is an arbitrary complex matrix.

Thus the following definitions and results are presented (the proofs can be found in Butcher (1987b), for example).

Definition 3.1.1. *A matrix A is convergent if $\| A^n \|_\infty \to 0$ as $n \to \infty$.*

Definition 3.1.2. *A matrix A is stable (or power bounded) if there exists a constant α such that $\| A^n \|_\infty \leq \alpha$ for all n.*

Definition 3.1.3. *A polynomial p satisfies the root condition if every zero of p lies in the closed unit disk with simple zeros on the boundary.*

Theorem 3.1.4. *The following statements, for matrices A and B, are equivalent:*

(i) A is convergent;
(ii) $\rho(A) < 1$;
(iii) there exists B similar to A such that $\| B \|_\infty < 1$.

Theorem 3.1.5. *The following statements, for matrices A and B, are equivalent:*

(i) A is stable;
(ii) the minimal polynomial of A satisfies the root condition;
(iii) there exists B similar to A such that $\| B \|_\infty \leq 1$.

Zero-stability

If any Runge-Kutta method is applied to the problem

$$y' = 0, \quad y(0) = y_0 \tag{3.1.1}$$

then

$$y_n = y_0, \quad \forall n,$$

and so the solutions are always stable.

On the other hand, if a linear multistep is applied to this problem in a fixed-step mode then a k-step linear difference equation of the form

$$y_{n+1} - \sum_{j=1}^{k} \alpha_j y_{n+j-1} = 0$$

results. In this case the stability of the numerical solutions depends on the zeros of the characteristic polynomial

$$p(z) = z^k - \alpha_1 z^{k-1} - \cdots - \alpha_k \tag{3.1.2}$$

associated with this difference equation.

In fact, this concept can be placed in a more general framework which includes the above cases as specific instances. Thus if a multivalue method given by (2.8.1) is applied to (3.1.1) then

$$y^{(n+1)} = A_2 y^{(n)} \tag{3.1.3}$$

and the numerical solutions will remain bounded if $\exists \alpha > 0$ such that

$$\|A_2^n\| \leq \alpha \text{ for any } n \in Z^+. \tag{3.1.4}$$

In other words, the numerical solutions of a multivalue method will remain bounded if A_2 is a stable matrix.

In the case of a linear multistep method this condition is equivalent to the so-called root condition for p (defined in Definition 3.1.3), since it can easily be shown that p is the minimal polynomial of A_2.

This condition that numerical solutions will remain bounded for the test problem (3.1.1) is known as **zero-stability**.

Consistency

Clearly, zero-stability is the minimum stability property that should be required of all differential equation methods. However, all methods should also be able to compute the exact solution of (3.1.1) at both the beginning and the end of the step. In the case of multivalue methods, Butcher (1966, 1987b) noted that this is equivalent to requiring the existence of a vector q_0 such that

$$A_1 q_0 = e, \quad A_2 q_0 = q_0, \tag{3.1.5}$$

where e is the unit vector in $\mathbb{R}^r$. This condition is known as the **preconsistency condition** and q_0 is called the **preconsistency vector** and is an eigenvector associated with the unit eigenvalue of A_2.

In the case of a Runge-Kutta method, all Runge-Kutta methods are automatically preconsistent, while a linear multistep method will, from (3.1.5), be preconsistent if, with $q_0 = e$,

$$\sum_{j=1}^{k} \alpha_j = 1. \tag{3.1.6}$$

The concept of preconsistency can be extended to the simple test problem $y'(x) = 1$ and to require appropriate behaviour of a numerical method for this problem. Thus Butcher (1966, 1987b) defines a method to be **consistent** if it is preconsistent and there exists a vector $q_1 \in \mathbb{R}^r$ such that

$$A_2 q_1 + B_2 e = q_1 + q_0, \tag{3.1.7}$$

and q_1 is known as the **consistency vector**.

The interpretation of (3.1.7) is that if

$$y^{(n)} = q_0 y(x_n) + q_1 h y'(x_n) + O(h^2)$$

then after a single step of the multivalue method (2.8.1)

$$y^{(n+1)} = q_0 y(x_{n+1}) + q_1 h y'(x_{n+1}) + O(h^2),$$

where the product of vectors is taken componentwise.

It is convenient to define the abscissae associated with (2.8.1) by

$$c = (c_1, \ldots, c_s)^\top = A_1 q_1 + B_1 e,$$

in which case

$$Y_i = y(x_n + c_i h) + O(h^2), \quad i = 1, \ldots, s.$$

In the case of an s-stage Runge-Kutta method, $q_0 = 1$ and $q_1 = 0$ and so the consistency condition is equivalent to

$$\sum_{j=1}^{s} b_j = 1. \tag{3.1.8}$$

In the case of a k-step linear multistep method $q_1 = (0, -1, \ldots, -k+1)^\top$ and so in this case consistency is equivalent to

$$\sum_{j=1}^{k} j\alpha_j = \sum_{j=0}^{k} \beta_j. \tag{3.1.9}$$

Of course other classes of methods such as predictor-corrector methods and multistep Runge-Kutta methods have appropriate consistency conditions based on (3.1.7) (see Butcher (1987b), for a comprehensive description of various classes of multivalue methods).

Convergence

Associated with the concepts of zero-stability and consistency is that of **convergence.** If a fixed stepsize implementation of a discrete numerical method is considered then in the limit that the stepsize h tends to zero, the numerical solutions should converge to the exact solution on some interval of integration I given that f is continuous and satisfies a Lipschitz condition.

In the case of a general class of methods such as multivalue methods there are some formal difficulties associated with convergence in terms of providing r initial approximations $y^{(0)} = (y_1^{(0)}, \ldots, y_r^{(0)})^\top$ from a set of starting procedures which will allow a multivalue method to proceed.

Thus Butcher (1987b) gives the following definition of convergence.

Definition 3.1.6. *A multivalue method is convergent for (2.1.1) on I if for any $x_0, x \in I$ and for $y^{(0)}(h)$ computed from x_0 such that for $i = 1, \ldots, r$*

$$\| y_i^{(0)}(h) - y(x_0) \| \to 0 \quad \text{as} \quad h \to 0$$

then

$$\| y_i^{(n)}((x - x_0)/n) - y(x) \| \to 0 \quad \text{as} \quad n \to \infty.$$

Butcher (1966, 1987b) has proved the following result by using the Lipschitz condition on f.

Theorem 3.1.7. *A multivalue method is convergent if and only if it is consistent and zero-stable.*

Thus the conditions of zero-stability and consistency are essential requirements for any numerical method. In fact the consistency condition is nothing more than an order condition and so to determine the appropriateness of methods for differential systems in terms of accuracy requirements a general formulation of order is necessary.

3.2 Order

Just as there are fundamental differences between linear multistep and Runge-Kutta methods in terms of their stability behaviour on very simple test problems, there are important differences when trying to judge their accuracy in terms of order properties. Whereas it is a simple matter (using quadrature techniques) to construct families of linear multistep methods with arbitrarily high orders, this is not necessarily the case for Runge-Kutta methods.

The criterion that is used in determining the accuracy of a method is the **order of consistency**. This is determined by examining the difference between the exact solution and the numerical approximation over one step, given that exact initial values are used for past information. This difference is known as the **local truncation error** and if this error behaves like $O(h^{w+1})$ then a method is said to be of **order** w. There are, however, some interpretational and implementational niceties with these definitions which must be further explored in the case of multivalue methods, as these methods require a set of starting procedures and carry more than one piece of information from step to step.

In the case of linear multistep methods the local error associated with (2.6.1) is given by

$$\begin{aligned} l_{n+1} &= y(x_0+(n+1)h) \\ &\quad -\sum_{j=1}^{k}\alpha_j y(x_0+(n+1-j)h)-\sum_{j=0}^{k}\beta_j y'(x_0+(n+1-j)h), \end{aligned}$$

which, by a Taylor's series expansion about x_n, gives

$$l_{n+1}=\sum_{p=0}^{\infty}C_p h^p y^{(p)}(x_n)/p!, \tag{3.2.1}$$

where

$$C_p = 1-\sum_{j=1}^{k}\alpha_j(1-j)^p - p\sum_{j=0}^{k}\beta_j(1-j)^{p-1}. \tag{3.2.2}$$

Thus the local truncation error is merely a linear combination of all the total derivatives of the exact solution. In the case of one-step methods there is no starting procedure so a Runge-Kutta method given by (2.7.2) will be of order w if

$$l_{n+1} = y(x_n+h) - y_{n+1} = y(x_n+h) - (y(x_n) + h\sum_{j=1}^{s} b_j f(Y_j)) = O(h^{w+1}),$$

where

$$Y_i = y(x_n) + h\sum_{j=1}^{s} a_{ij} f(Y_j), \quad i = 1,\ldots,s.$$

Here all possible partial derivatives of f appear in the local error expansion. Thus, for example, the classical Runge-Kutta method given in section 2.7 has to satisfy eight order conditions corresponding to all the elementary differentials of up to order 4 which can be enumerated as

$$f, \quad f'(f), \quad f''(f,f), \quad f'(f'(f)),$$
$$f'''(f,f,f), \quad f''(f,f'(f)), \quad f'(f''(f,f)), \quad f'(f'(f'(f))).$$

As the order of a Runge-Kutta method increases, the number of order conditions that have to be satisfied (corresponding to each elementary differential) increases dramatically. Table 3.1 shows how the number of order conditions increases as the order rises.

Table 3.1 *number of order conditions*

order	2	3	4	5	6	7	8	9	10
conditions	2	4	8	17	37	85	200	486	1205

Kutta (1901) managed to derive the eight order conditions for a general Runge-Kutta method of order 4 from first principles, but because there was no general framework in which to analyse new methods, very little was done in this area until the work of Gill (1951), Merson (1957) and Butcher (1963), while not until Huťa (1956) were sixth-order methods introduced. Then Butcher (1963, 1964a,b,c) provided a general framework in which to study Runge-Kutta methods and produced an order theory which relates each elementary differential associated with (2.1.1) to a rooted tree. This theory applies to both explicit and implicit Runge-Kutta methods and can be extended to more general classes of methods such as multivalue methods (Butcher (1972, 1974)). However, this extension is a nontrivial one.

Between 1972 and 1974 Butcher published a number of papers which provided a complete theory of order for multivalue methods. As Butcher (1974) later noted, the difficulty in constructing such a general theory was that merely trying to combine the order theories of linear multistep and Runge-Kutta methods would lead to inconsistencies because a linear multistep order theory requires a uniform accuracy over all approximation components while a Runge- Kutta order theory lacks a certain symmetry property. Thus Butcher (1972) expounded an algebraic theory of integration methods.

In this algebraic theory a general class of integration methods is related to a group (G) represented as a family of real-valued functions on a set of rooted trees, so that each integration method can be represented by a group element. From this, the very natural idea of composition of methods based on the product of group elements arises and order is defined as certain relationships between groups.

Butcher (1973) completed this general theory of order by showing how to write down the order conditions for any multivalue method. In order to do this, Butcher combined the ideas of effective order and the group theoretic approach and related order to the concept of starting procedures. The approach to order theory has been simplified by Hairer and Wanner (1974) with the introduction of Butcher series and this is the approach that will be taken here in studying the order properties of multistage methods in general. The material presented here is based on the work of Skeel (1976), Cooper (1978) and Burrage (1988).

Multivalue methods

Recalling that any implementation of a multivalue method can be visualized by Figure 2.2, the order of a method can be interpreted through the following diagram:

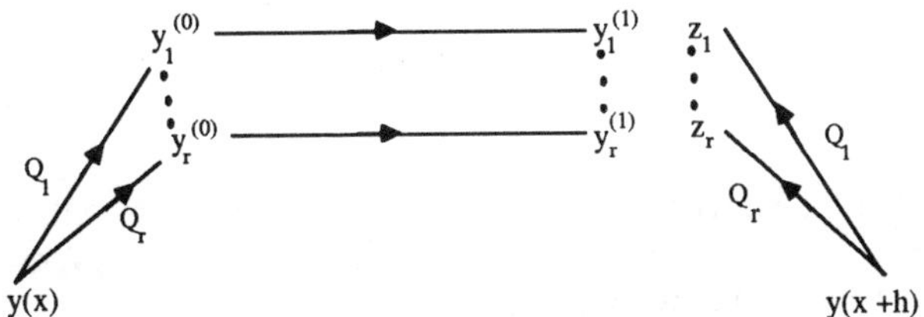

FIG. 3.1. order of a multivalue method

Thus a method is said to be of order w if starting from $y(x_n)$ the Taylor series expansion of (2.8.1) after one step coincides up to order w with the numerical results z_{n+1} obtained from applying the same starting procedures to $y(x_{n+1})$. Hence a multivalue method will be of order w if

$$z_{n+1} - y^{(n+1)} = O(h^{w+1}). \tag{3.2.3}$$

If $z(x,h)$ is the correct value function for (2.8.1) and $z_n = z(x_n,h)$ then the local error is given by

$$l_{n+1} = z_{n+1} - ((A_2 \otimes I)z(x_n,h) + h(B_2 \otimes I)F(Y)), \tag{3.2.4}$$

where

$$Y = (A_1 \otimes I)z(x_n,h) + h(B_1 \otimes I)F(Y). \tag{3.2.5}$$

In order to be able to relate the local truncation error to the method coefficients, a number of concepts need now to be introduced. Thus let T and T^* be the set of **rooted trees** and **monotonically labelled trees**, respectively. It is assumed that the empty tree ϕ is an element of T and that, when a tree is monotonically labelled, the integer 1 is attached to the root of the tree and thereafter the integer labels increase in size as the scanning moves up the branches of the tree away from the root. Without loss of generality it can be assumed that the differential equation to be solved is in autonomous form (2.1.1).

The elementary differentials $F(t)y$ associated with (2.1.1) are defined by

$$F(\phi)y = y, \quad F(\tau)y = f(y), \quad F(\dot{t})y = f^{(m)}(F(t_1)y,\ldots,F(t_m)y),$$

where $t = [t_1,\ldots,t_m]$ is the tree formed by joining the subtrees $t_1,\ldots,t_m$ by a single branch to a common root and where ϕ and τ represent the unique trees of order 0 and 1, respectively. In addition, $\rho(t)$ will denote the **order** of the tree t (the number of nodes of t), and τ^p will denote the **bushy tree** $[\tau,\ldots,\tau]$ of height 1 and order p.

As an example, Table 3.2 expresses all trees of up to order 4 in terms of τ as well as giving the corresponding elementary differential associated with the tree $t \in T$.

The crucial concept in the analysis of the order of multistage methods is that of a Butcher series. If $q : T \to \mathbb{R}$ is an arbitrary map, Hairer and Wanner (1974) call the series

$$B(q,y_0) = \sum_{t\in T*} q(t)[F(t)(y_0)]\frac{h^{\rho(t)}}{\rho(t)!} \tag{3.2.6}$$

a Butcher series. Thus, for example, the exact solution of (2.1.1) at x_0+kh is a Butcher series with $q(t) = k^{\rho(t)}$.

Suppose now that l_{n+1} and Y, given by (3.2.4) and (3.2.5), respectively, have Butcher series expansions given by, for $i = 1,\ldots,s$,

Table 3.2 *differentials*

order	t	differential
1	τ	f
2	$[\tau]$	$f'(f)$
3	τ^2	$f''(f,f)$
3	$[[\tau]]$	$f'(f'(f))$
4	τ^3	$f'''(f,f,f)$
4	$[\tau,[\tau]]$	$f''(f,f'(f))$
4	$[\tau^2]$	$f'(f''(f,f))$
4	$[[[\tau]]]$	$f'(f'(f'(f)))$

$$Y_i = B(k_i, y(x_n)), \quad l_{n+1} = B(e, y(x_n)), \tag{3.2.7}$$

where

$$l_{n+1} = \sum_{t \in T*} e(t)[F(t)(y(x_n))] \frac{h^{\rho(t)}}{\rho(t)!}. \tag{3.2.8}$$

Then in order to derive a formula for $e(t)$ the following theorem on the composition of Butcher series is quoted from Hairer et al. (1987).

Theorem 3.2.1. *Let $q : T \to \mathbb{R}$ and $p : T \to \mathbb{R}$ be two maps such that $p(\phi) = 1$; then the composition of two Butcher series is a Butcher series given by $B(q, B(p,y)) = B(pq, y)$.*

In order to be able to define $pq : T \to \mathbb{R}$ some more combinatorial definitions are required. For a given $t \in T$ let $\alpha(t)$ be the number of ways of monotonically labelling the nodes of t with the integers $1, 2, \ldots, \rho(t)$ starting with the root. For $t = [t_1, \ldots, t_m]$, Hairer and Wanner (1973) have shown

$$\begin{aligned} \rho(\tau) = 1, \quad \rho(t) &= 1 + \rho(t_1) + \cdots \rho(t_m) \\ \alpha(\tau) = 1, \quad \alpha(t) &= \binom{\rho(t) - 1}{\rho(t_1), \ldots, \rho(t_m)} \frac{\alpha(t_1) \ldots \alpha(t_m)}{v_1! v_2! \ldots} \end{aligned}$$

where, for some arbitrary enumeration of trees, v_i is the number of occurrences of the ith tree among $t_1, \ldots, t_m$.

Hairer and Wanner (1973) have also obtained a recursive formula for $\alpha(u,t)$ where $\alpha(u,t)$ is the number of ways of monotonically labelling the nodes of t such that the subtree u with the same root as t is labelled first. If $t = [t_1, \ldots, t_m]$ and $u = [u_1, \ldots, u_m]$ with u_i a subtree of t_i, where some of the u_i can be empty, then (u, t) represents a pair of trees in which u is a subtree of t with the same root. If $u = \phi$ then the pair (ϕ, t) is identified with t.

Let the notation

$$(u, t) = ((u_1, t_1), \ldots, (u_m, t_m))$$

express the fact that the pairs $(u_1, t_1), \ldots, (u_m, t_m)$ remain after the root of (u, t) and the adjacent arcs have been removed. Then Hairer and Wanner have shown that $\alpha(u, t)$ is given by

$$\binom{\rho(u) - 1}{\rho(u_1), \ldots, \rho(u_m)} \binom{\rho(t) - \rho(u)}{\rho(t_1) - \rho(u_1), \ldots, \rho(t_m) - \rho(u_m)} \frac{\prod_{j=1}^{m} \alpha(u_j, t_j)}{v_1! v_2! \ldots},$$

where $v_1, v_2, \ldots$ are the number of mutually equal pairs.

Using these results the composition $pq : T \to \mathbb{R}$ is defined as

$$pq(t) = \sum_{u \subseteq t} \binom{\rho(t)}{\rho(u)} \frac{\alpha(u, t)}{\alpha(u)} q(u) \prod_{v \subseteq t - u} p(v).$$

For each of the r components of a multivalue method there exists a correct value function $z(x, h) = (z_1(x, h), \ldots, z_r(x, h))^{\top}$, which possesses an expansion as a Butcher series

$$z_i(x, h) = B(q_i, y(x)), \quad i = 1, \ldots, r.$$

For a given tree t, let $q(t) = (q_1(t), \ldots q_r(t))^{\top}$ and for the k_i defined in (3.2.7) let $k(t) = (k_1(t), \ldots, k_s(t))^{\top}$. Then for any tree $t = [t_1, \ldots, t_m]$, (3.2.4), (3.2.5), (3.2.7) and (3.2.8) imply

Theorem 3.2.2. *A multivalue method is of order w if $e(t) = 0$, $\forall \rho(t) \leq w$, where for any tree $t = [t_1, \ldots, t_m]$*

$$e(t) = \sum_{u \subseteq t} \binom{\rho(t)}{\rho(u)} \frac{\alpha(u, t)}{\alpha(t)} q(u) - A_2 q(t) - \rho(t) B_2 \prod_{j=1}^{m} k(t_j),$$

where

$$k(t) = A_1 q(t) + \rho(t) B_1 \prod_{j=1}^{m} k(t_j), \quad k(\phi) = e.$$

In determining the order of a multivalue method there are two possible approaches:

(i) solve the order equations for the $q(t)$ to obtain as high an order as possible and then construct the starting procedure from these values, (see Burrage (1988), for example);

(ii) the values of the $q(t)$ are given in advance after the starting procedures have been fixed.

For example, in the case of MRKs (which include linear multistep and Runge-Kutta methods as special cases) the $q(t)$ are given by

$$\begin{aligned} q(\phi) &= (1,\ldots,1)^\top \\ q(t) &= (0,(-1)^p,\ldots,(1-r)^p)^\top, \quad \forall t, \rho(t)=p\geq 1, \end{aligned} \tag{3.2.9}$$

while for Nordsieck methods

$$\begin{aligned} q(t) &= e_{\rho(t)+1}, \quad \forall t \text{ with } \rho(t)=r-1 \\ q(t) &= 0, \quad \text{otherwise,} \end{aligned}$$

where $e_1,\ldots,e_r$ are the basis vectors for $\mathbb{R}^r$.

The study of the order of multivalue methods adds another degree of complexity if the $q(t)$ are unknown (see Burrage (1988), for example). However, if it is assumed that

$$q(t)=q(\tau^p), \quad \text{for all trees of order } p, \quad p=2,3,\ldots$$

then the analysis of order can be greatly simplified.

Thus denoting $A(w)$ by

$$A(w):\; q(t)=q(\tau^p), \quad \text{for all trees of order } p, \quad p=2,3,\ldots,w \tag{3.2.10}$$

then simplifying assumptions can be constructed based on the fact that

$$k(t)=k(\tau^p)=c^p, \quad \text{for all trees of order } p\leq w. \tag{3.2.11}$$

It is easily seen that this condition will imply

$$Y_i=y(x_n+c_ih)+O(h^{w+1}), \quad i=1,\ldots,s.$$

The condition $A(w)$ is a very natural one in that it implies that the r starting values are given by

$$\sum_{j=0}^{w} q_i(\tau^j)y^{(j)}(x_0)h^j+O(h^{w+1}).$$

If (3.2.10) holds along with the quadrature order conditions in (3.2.11) for the bushy trees ϕ, τ^p, $p=1,\ldots,w$, then a multivalue method is said to have **stage order** w. Thus introducing the relationships, which hold for $p = 0, \ldots, w$:

$$B(w): \quad A_2 q(\tau^p) = \sum_{j=0}^{p} \binom{p}{j} q(\tau^j) - pB_2 c^{p-1} \qquad (3.2.12)$$

$$C(w): \quad A_1 q(\tau^p) = c^p - pB_1 c^{p-1}, \qquad (3.2.13)$$

the following definition can be given.

Definition 3.2.3. *A multivalue method will have* **stage order** w *if* $A(w)$, $B(w)$ *and* $C(w)$ *hold.*

Given that $A(w)$ and $C(w)$ imply (3.2.11), it is a trivial matter using Theorem 3.2.2 to show that

Theorem 3.2.4. $A(w)$, $B(w)$ *and* $C(w-1)$ *imply order* w.

Theorem 3.2.4 can be used to construct higher-order methods with high stage order, but in some instances this may be inappropriate. Consequently, Burrage (1988) has proved a more general result by introducing a D-type assumption. Thus let

$$D(\mu, \xi): \; B_2 D_{s_1}^{m_1} B_1^{r_1} \ldots D_{s_k}^{m_k} B_1^{r_k} v_{\mu+p} = 0, \quad p = 1, \ldots, \xi \qquad (3.2.14)$$

for all nonnegative integers $r_1, \ldots, r_k, s_1, \ldots, s_k, m_1, \ldots, m_k$ such that

$$\sum_{j=1}^{k} r_j + \sum_{j=1}^{k} m_j s_j \leq \xi - p.$$

Here

$$\begin{aligned} x_p &= (x_1^{(p)}, \ldots, x_s^{(p)}) = A_1 q(\tau^P) + pB_1 c^{p-1} \\ v_p &= c^p - x_p, \quad D_p = \operatorname{diag}(x_1^{(p)}, \ldots, x_s^{(p)}). \end{aligned}$$

Theorem 3.2.5. $A(w-1)$, $B(w)$, $C(u)$, $D(u, w-u-1)$ *and* $(A_2 - I)(q(\tau^w) - q(t)) = 0, \quad \forall \rho(t) = w$, *imply order* w.

As an example to illustrate the above result, the following set of order conditions for order 4, assuming $A(4)$, is presented.

Example 3.2.1. Conditions for an order 4 method.

Let

$$v_2 = c^2 - A_1 q(\tau^2) - 2B_1 c$$

$$v_3 = c^3 - A_1 q(\tau^3) - 3B_1 c^2$$

and $C = \mathrm{diag}(c_1, \ldots, c_s)$; then the order conditions are given by

$$\begin{aligned} B(4):\ A_2 q(\tau^p) &= \sum_{j=0}^{p} \binom{p}{j} q(\tau^j) - pB_2 c^{p-1}, \quad p = 0, \ldots, 4 \\ B_2 v_2 &= B_2 C v_2 = B_2 B_1 v_2 = B_2 v_3 = 0. \quad \square \end{aligned}$$

By introducing the $r \times r$ matrix Q whose jth column is $q(\tau^{j-1})$, the $r \times s$ matrix $\tilde{Q}$ whose jth column is $q(\tau^{r+j-1})$ and the $s \times s$ matrices V, $\tilde{V}$, W and $\tilde{W}$ whose jth columns are, respectively, jc^{j-1}, c^j, $(r+j)c^{r+j-1}$ and c^{r+j}, it can be shown that $C(s+r-1)$ implies

$$\begin{aligned} A_1 &= (W - B_1 V) Q^{-1} \\ B_1(\tilde{V} - VQ^{-1}\tilde{Q}) &= \tilde{W} - WQ^{-1}\tilde{Q} \end{aligned} \qquad (3.2.15)$$

and the following results are true:

Theorem 3.2.6. *The maximum stage order of an implicit multivalue method is* $s + r - 1$.

Theorem 3.2.7. *The stage order of an explicit multivalue method cannot exceed* r.

These two results offer a simple approach to the construction of multivalue methods of high order by requiring that they have maximal stage order. This very general approach can now be particularized to linear multistep and Runge-Kutta methods.

Linear multistep methods

In the case of a k-step linear multistep method written in multivalue form with $s = k + 1$ and $r = k$ (see (2.8.2)), A-type simplifying assumptions automatically hold and satisfy (3.2.9). In addition, $B(w)$ and $C(w)$ are equivalent and imply (3.2.11). This in turn, from Theorem 3.2.2, implies

$$e(t) = e(\tau^p), \quad \forall t \text{ with } \rho(t) = p \le w.$$

Hence with (3.2.9) and $c = (1, 0, -1, \ldots, -k+1)^\top$ the following result holds:

Theorem 3.2.8. *A k-step linear multistep method is of order w if* $e(\tau^p) = 0$, $p = 0, \ldots, w$, *where*

$$e(\tau^p) = 1 - \alpha^\top q(\tau^p) - p\beta^\top c^{p-1}.$$

This analysis is of course equivalent to that given in (3.2.1) and (3.2.2). It can also be seen from this result that there is a fundamental difference between multistage methods and linear multistep methods. In the latter case order is equivalent to the annihilation of polynomials or quadrature conditions. In other words, for linear multistep methods stage order and order of consistency are equivalent.

Another way of constructing certain classes of linear multistep methods is to note that the solution of (2.1.2) satisfies

$$y(x_{n+1}) = y(x_n) + \int_{x_n}^{x_{n+1}} f(x, y(x))dx,$$

and the quadrature term can be approximated by an interpolating polynomial through the points

$$(x_i, f_i), \quad i = n-k+1, \ldots, n: \text{ Adams Bashforth (AB)}$$

or

$$(x_i, f_i), \quad i = n-k+1, \ldots, n+1: \text{ Adams Moulton (AM)},$$

where $f(x_i, y_i)$ is denoted by f_i.

Hence Adams methods satisfy

$$\alpha_1 = 1, \quad \alpha_j = 0, \quad j = 2, \ldots, k$$

and from Theorem 3.2.8 an Adams method will be of order w if

$$1 - p\sum_{j=0}^{k} \beta_j (1-j)^{p-1} = 0, \quad p = 1, \ldots, w.$$

Table 3.3 gives the coefficients (β_j) for the k-step AB and AM methods of order k and $k+1$, respectively, with $k = 1, \ldots, 4$:

In the case of BDF methods, which satisfy

$$\beta_j = 0, \quad j = 1, \ldots, k,$$

Theorem 3.2.8 implies that a BDF method will be of order w if

Table 3.3 *Adams method coefficients*

	AB				AM			
k	β_1	β_2	β_3	β_4	β_0	β_1	β_2	β_3
1	1				$\frac{1}{2}$	$\frac{1}{2}$		
2	$\frac{3}{2}$	$-\frac{1}{2}$			$\frac{5}{12}$	$\frac{8}{12}$	$-\frac{1}{12}$	
3	$\frac{23}{12}$	$-\frac{16}{12}$	$\frac{5}{12}$		$\frac{9}{24}$	$\frac{19}{24}$	$-\frac{5}{24}$	$\frac{1}{24}$
4	$\frac{55}{24}$	$-\frac{59}{25}$	$\frac{37}{24}$	$-\frac{9}{24}$				

$$1 - \sum_{j=1}^{k} \alpha_j (1-j)^p - p\beta_0 = 0, \quad p = 1, \ldots, w.$$

Another way of constructing BDF methods is to note that a polynomial p interpolates at the values $x_{n+1}, \ldots, x_{n+1-k}$ such that

$$p(x_{n+1-j}) = y_{n+1-j}, \quad j = 0, \ldots, k$$

with

$$p'(x_{n+1}) = f(x_{n+1}, y_{n+1}).$$

It is easy to show using this interpolational analysis (see Hairer et al. (1987), for example) that the BDF methods are given by

$$\sum_{j=1}^{k} \frac{1}{j} \nabla^j y_{n+1} = hf_{n+1},$$

where ∇ is the backward difference operator defined by

$$\nabla y_n = y_n - y_{n-1}.$$

Table 3.4 gives the coefficients α_j for the k-step BDF methods of order k for $k = 1, \ldots, 4$:

In the case of a variable stepsize implementation, the backward difference formulation must be replaced by a Newton divided difference formulation in which, with $f[x_n] = f_n$

$$f[x_n, \ldots, x_{n-j}] = \frac{f[x_n, \ldots, x_{n-j+1}] - f[x_{n-1}, \ldots, x_{n-j}]}{x_n - x_{n-j}}.$$

So the interpolating polynomial for $(x_{n-k+1}, f_{n-k+1}), \ldots, (x_m, f_m)$ is given by

Table 3.4 *BDF coefficients*

k	α_1	α_2	α_3	α_4	β_0
1	1				1
2	$\frac{4}{3}$	-$\frac{1}{3}$			$\frac{2}{3}$
3	$\frac{18}{11}$	-$\frac{18}{22}$	$\frac{2}{11}$		$\frac{6}{11}$
4	$\frac{48}{25}$	-$\frac{36}{25}$	$\frac{16}{25}$	-$\frac{3}{25}$	$\frac{12}{25}$

$$p(x) = \sum_{j=0}^{k-1+(m-n)} \prod_{i=0}^{j-1} (x - x_{m-i}) f[x_m, \ldots, x_{m-j}].$$

By performing a similar technique to that used before with $m = n$ or $n+1$ for AB and AM methods, it can be shown (see Krogh (1974)) that the variable step representations of AB and AM methods are, respectively,

$$\begin{aligned} y_{n+1} &= y_n + h_n \sum_{j=0}^{k-1} \gamma_j(n) F_j(n) \\ y_{n+1} &= y_n + h_n \sum_{j=0}^{k-1} \gamma_j(n) F_j(n) + h_n \gamma_k(n) \tilde{F}_k(n+1) \end{aligned}$$

where

$$\begin{aligned} \gamma_j(n) &= \frac{1}{h_n} \int_{x_n}^{x_{n+1}} \prod_{i=0}^{j-1} \frac{x - x_{n-i}}{x_{n+1} - x_{n-i}} dx, \quad j = 0, \ldots, k-1 \\ F_j(n) &= \prod_{i=0}^{j-1} (x_{n+1} - x_{n-i}) f[x_n, \ldots, x_{n-j}], \quad j = 0, \ldots, k-1 \\ \tilde{F}_k(n+1) &= \prod_{i=0}^{k-1} (x_{n+1} - x_{n-i}) f[x_{n+1}, \ldots, x_{n-k+1}]. \end{aligned}$$

On the other hand, the variable step representation of the class of BDF methods is given by

$$\sum_{j=1}^{k} h_n \prod_{i=1}^{j-1} (x_{n+1} - x_{n+1-i}) y[x_{n+1}, \ldots, x_{n-j+1}] = h_n f(x_{n+1}, y_{n+1}).$$

There are recursive formulas for computing $\gamma_j(n)$, $F_j(n)$ and $\tilde{F}_j(n)$, (see Hairer et al. (1987), for example), that avoid recalculating quantities that have not changed.

Runge-Kutta methods

In the case of a Runge-Kutta method, the correct value function $z(x,h)$ has a Butcher series given by

$$z(x,h) = B(q, y(x)),$$

where

$$\begin{aligned} q(\phi) &= 1 \\ q(t) &= 0, \quad \text{otherwise}, \end{aligned}$$

and the following result is a consequence of Theorem 3.2.2.

Theorem 3.2.9. *A Runge-Kutta method is of order w if and only if $e(t) = 0$, $\forall \rho(t) \le w$, where for any tree $t = [t_1, \ldots, t_m]$*

$$e(\phi) = 0, \quad e(t) = 1 - \rho(t) b^\top \prod_{j=1}^{m} k(t_j),$$

with

$$k(\phi) = e, \quad k(t) = \rho(t) A \prod_{j=1}^{m} k(t_j).$$

Table 3.5 gives formulas for the $e(t)$ for all trees of order 4 or less.

Table 3.5 *error terms*

order	t	e(t)
1	τ	$1 - b^\top e$
2	$[\tau]$	$1 - 2b^\top c$
3	τ^2	$1 - 3b^\top c^2$
3	$[[\tau]]$	$1 - 6b^\top Ac$
4	τ^3	$1 - 4b^\top c^3$
4	$[\tau, [\tau]]$	$1 - 8b^\top CAc$
4	$[\tau^2]$	$1 - 12b^\top Ac^2$
4	$[[[\tau]]]$	$1 - 24b^\top A^2 c$

The simplifying assumptions introduced in (3.2.11), (3.2.12) and (3.2.13) and the concept of stage order can be particularized to Runge-Kutta methods trivially. In this case $B(w)$ and $C(w)$ are equivalent to

$$B(w): \quad b^{\mathsf{T}} c^{p-1} = \frac{1}{p}, \quad p = 1, \ldots w, \tag{3.2.16}$$

and

$$C(w): \quad Ac^{p-1} = \frac{c^p}{p}, \quad p = 1, \ldots w. \tag{3.2.17}$$

The following results are thus trivial consequences of the preceding material on multivalue methods.

Theorem 3.2.10. *A Runge-Kutta method satisfying $C(w)$ and $B(w)$ has stage order w.*

Theorem 3.2.11. *The maximum stage order of an s-stage Runge-Kutta method is s.*

Theorem 3.2.12. *Both $C(w-1)$ and $B(w)$ or $B(w), C(u)$ and $D(u, w-u-1)$ imply order w.*

In a similar way, the order conditions for a method of order 4 given in Example 3.2.1 can be particularized to the Runge-Kutta case.

Example 3.2.2. Order conditions for a Runge-Kutta method of order 4.

$$\begin{aligned} B(4): b^{\mathsf{T}} c^{p-1} &= 1/p, \quad p = 1, \ldots, 4 \\ b^{\mathsf{T}}(c^2 - 2Ac) &= b^{\mathsf{T}} C(c^2 - 2Ac) = 0 \\ b^{\mathsf{T}} A(c^2 - 2Ac) &= b^{\mathsf{T}}(c^3 - 3Ac^2) = 0. \end{aligned}$$

3.3 Linear stability

It is very natural to extend the concept of zero-stability via the simple test equation given in (2.4.6). The reason for doing this is that it provides information about the behaviour of numerical methods when applied to linear systems of the form (2.2.4). If (2.2.4) has a matrix Q whose eigenvalues have all negative real part then this test equation can be used to model the effects of stiffness on a numerical method. Unfortunately, it may be hard to extrapolate these results in a meaningful way if the problem is a nonlinear one. Although the rationale for studying the behaviour of numerical methods on linear problems is that a nonlinear problem can be linearized in terms of its Jacobian, it is well known through Vinograd's problem (see Dekker and Verwer (1984), for example) that linearization may not provide any useful information about the behaviour of the actual solution. Nevertheless, understanding the behaviour of a numerical method on problems such as (2.2.4) is still considered to be a crucial facet in designing methods suitable for solving stiff problems, for example.

Thus if a multivalue method given by (2.8.1) is applied to the standard linear test problem (2.4.6) with a constant fixed stepsize then

$$\begin{aligned} y^{(n+1)} &= M(z)y^{(n)}, \quad z = qh \\ M(z) &= A_2 + zB_2(I - B_1 z)^{-1}A_1. \end{aligned} \tag{3.3.1}$$

The **stability region** associated with this method is defined as

$$S = \{z \in \mathbb{C} : M(z) \text{ is power bounded}\}.$$

Thus a multivalue method is zero-stable if

$$0 \in S$$

and *A*-**stable** if

$$\mathbb{C}^- \subseteq S. \tag{3.3.2}$$

Note that it is always understood that $\mathbb{C}$, the set of complex numbers, includes the accumulation point at infinity.

A suitable damping property known as *L*-stability can also be introduced. Thus a multivalue method is said to be *L*-**stable** if it is *A*-stable and $M(z)$ has vanishing eigenvalues at infinity. Denoting $M(\infty)$ by

$$M(\infty) = A_2 - B_2 B_1^{-1} A_1$$

the following result is a trivial consequence of this definition and (3.3.1).

Theorem 3.3.1. *A multivalue method is L-stable if it is A-stable and* $\rho(M(\infty)) = 0$.

Of course, if the implementation is not a fixed-stepsize one then this definition has to be modified slightly to take into account the stability of products of matrices. In this case

$$y^{(n+1)} = M_n y^{(n)}$$

where M_n has the same meaning as $M(z)$, but now the coefficients are functions of the stepsize ratios. Hence stability definitions have to be extended to the requirement that there exist $\| \cdot \|$ and finite α such that

$$\| M_{n+j} \dots M_n \| \leq \alpha, \quad \forall n \text{ and } j \geq 0.$$

Clearly it can be difficult to get accurate bounds on acceptable stepsize ratios; see Grigorieff (1983), for example.

A number of reasons were given in Chapter 2 for the construction of a general class of methods such as multivalue methods, and some of these reasons revolve around implementational issues. However, another reason is that such methods can be used to break the first and second Dahlquist barriers. In particular, the Daniel-Moore conjecture states that the maximum order of an s-stage multivalue method which is A-stable is $2s$. This was proved in the order stars paper (Wanner et al. (1978)) and is the reason why the maximum order of an A-stable linear multistep method is 2 (it can be written as a multivalue method with $s = 1$).

From both a numerical and analytical viewpoint it is difficult to compute the stability region S of any multivalue method based on the requirement that for any $z \in S$, the eigenvalues of $M(z)$ must have modulus less than or equal to one with eigenvalues of modulus one being non-defective. However, a more suitable result based on the Maximum Modulus principle can be developed to establish sufficient conditions for A-stability. This was done by Chartier (1993) for a particular class of multivalue methods known as parallel one-block methods, but the result is easily extended to multivalue methods in general.

Thus by noting that $\rho(M(z))$ will be analytic in the left half complex plane if the spectrum of B_1 has positive real part, then by the Maximum Modulus principle, $\rho(M(z))$ will take its maximum on the imaginary axis. Suppose now that a multivalue method is zero-stable and let

$$\pi(w, z) = \det(M(z) - wI) \tag{3.3.3}$$

denote the characteristic polynomial of $M(z)$. Then if $\pi(w, z)$ has all its zeros inside the unit disk, then such a polynomial is said to be a **Schur polynomial.**

Consider now the transformation

$$w = \frac{1 + \lambda}{1 - \lambda}$$

which maps the boundary of the unit disk to the imaginary axis and the interior of the unit disk to the left half plane $Re(\lambda) < 0$. Then $\pi(w, z)$ will be a Schur polynomial if and only if $\Pi(\lambda, y)$ has all zeros lying strictly within the left half plane, where

$$\begin{aligned} \Pi(\lambda, y) &= (1 - \lambda)^{2r} \pi((1 + \lambda)/(1 - \lambda), iy) \pi((1 + \lambda)/(1 - \lambda), -iy) \\ &= a_0 \lambda^{2r} + a_1 \lambda^{2r-1} + \cdots + a_{2r}. \end{aligned}$$

This latter condition can be investigated via the Routh-Hurwitz condition. Define the $2r \times 2r$ matrix G_{2r} given by

$$G_{2r} = \begin{pmatrix} a_1 & a_3 & a_5 & \cdots & a_{4r-1} \\ a_0 & a_2 & a_4 & \cdots & a_{4r-2} \\ 0 & a_1 & a_3 & \cdots & a_{4r-3} \\ 0 & a_0 & a_2 & \cdots & a_{4r-4} \\ \vdots & \vdots & \vdots & & \vdots \\ 0 & 0 & 0 & \cdots & a_{2r} \end{pmatrix},$$

where $a_j = 0$, $j > 2r$ and a_0 is assumed to be positive. Then $\Pi(\lambda, y)$ will satisfy the Routh-Hurwitz condition if all the leading principal minors of G_{2r} are positive.

Note that this analysis only leads to a sufficient result for A-stability, since it is possible for methods to be A-stable with the matrix B_1 having a zero eigenvalue. The trapezoidal method which can be written as the Runge-Kutta method

$$\begin{array}{c|cc} 0 & 0 & 0 \\ 1 & \frac{1}{2} & \frac{1}{2} \\ \hline & \frac{1}{2} & \frac{1}{2} \end{array}$$

is a case in point.

An alternative approach to that outlined above is to apply the **Schur criterion** directly to (3.3.3). Thus if $\pi(w, z)$ is written as

$$\pi(w, z) = a_r w^r + \cdots + a_0$$

then π will be a Schur polynomial iff

$$|a_0| < |a_r|$$

and

$$\tilde{\pi} = \frac{1}{w}(\pi^*(0, z)\pi(w, z) - \pi(0, z)\pi^*(w, z))$$

is also Schur, where $\pi^*(w, z)$ is given by

$$\pi^*(w, z) = \bar{a}_0 w^r + \cdots + \bar{a}_r.$$

The stability interval can also be studied by applying the Routh-Hurwitz condition to the polynomial

$$\Pi(\lambda, z) = (1 - \lambda)^r \pi((1 + \lambda)/(1 - \lambda), z) = a_0\lambda^r + a_1\lambda^{r-1} + \cdots + a_r,$$

so that

$$G_r = \begin{pmatrix} a_1 & a_3 & a_5 & \cdots & a_{2r-1} \\ a_0 & a_2 & a_4 & \cdots & a_{2r-2} \\ 0 & a_1 & a_3 & \cdots & a_{2r-3} \\ 0 & a_0 & a_2 & \cdots & a_{2r-4} \\ \vdots & \vdots & \vdots & & \vdots \\ 0 & 0 & 0 & \cdots & a_r \end{pmatrix}$$

is required to have all its leading principal minors positive. This general approach can now be particularized to linear multistep and Runge-Kutta methods as will now be shown.

Linear multistep methods

If the multivalue method (2.8.1) is a k-step linear multistep method and if Y_n is used to represent the vector $(y_n, \ldots, y_{n-k+1})^\top$ then (3.3.1) becomes

$$Y_{n+1} = C(z)Y_n, \quad z = qh. \tag{3.3.4}$$

Here $C(z)$ is a companion matrix of order k whose elements in the first row are

$$\frac{\alpha_j + \beta_j z}{1 - \beta_0 z}, \quad j = 1, \ldots, k.$$

It is sometimes convenient to define

$$\rho(w) = 1 - \sum_{j=1}^{k} \alpha_j w^j, \quad \sigma(w) = \sum_{j=0}^{k} \beta_j w^j \tag{3.3.5}$$

and hence define the **stability polynomial**

$$\pi(w, z) = \rho(w) - z\sigma(w). \tag{3.3.6}$$

It is clear from (2.8.2) and (3.3.4) that $\pi(w, z)$ is the minimal polynomial associated with $C(z)$. Hence from Theorem 3.1.5, the stability region associated with any linear multistep method is the set of z for which $\pi(w, z)$ satisfies the root condition.

Example 3.3.1. In order to illustrate the above approach, a real stability interval of the explicit Adams Bashforth method of order 2 will be determined. In this case

$$y_{n+1} = y_n + \frac{h}{2}\left(3f(y_n) - f(y_{n-1})\right)$$

and so

$$\pi(w, z) = w^2 - (1 + 3z/2)w + z/2$$

so that

$$\Pi(\lambda, z) = (2 + 2z)\lambda^2 + (2 - z)\lambda - z.$$

Applying the Routh-Hurwitz criterion to $\Pi(\lambda, z)$ with $k = 2$ gives

$$a_j > 0, \quad j = 0, 1, 2$$

and so the real interval of absolute stability is $[-1, 0]$.

It has already been seen that if a linear multistep method is implicit then the maximum attainable order is $2k$, otherwise if it is explicit then the maximum order is $2k - 1$. Unfortunately, these high-order methods are computationally useless because they are not zero-stable. Dahlquist (1956) has proved the following order barrier for zero-stable methods.

Theorem 3.3.2. *The maximum order of an implicit zero-stable method is $k + 2$ if k is even and $k + 1$ if k is odd; while the maximum order of an explicit zero-stable method is k.*

A k-step zero-stable method of order $k + 2$ is known as an optimal method and it can be shown that all optimal methods have stability regions which only contain the origin or do not contain the negative real axis in the neighbourhood of the origin. Thus for all intents and purposes the order bound in Theorem 3.3.2 is essentially $k + 1$ for implicit methods.

In the case of the Adams method ρ has a simple zero at 1 and $k - 1$ zeros at 0, and so this class is automatically zero-stable. The BDF methods on the other hand are not guaranteed to be zero-stable for all k but are zero-stable for $k \leq 6$.

Adams Bashforth methods are suitable for nonstiff methods because of the above-mentioned property. But in the case of implicit linear multistep methods, their appropriateness for stiff problems is restricted by Dahlquist's second barrier theorem (1956).

Theorem 3.3.3. *The maximum order of an A-stable linear multistep method is 2.*

Because of this result, the property of $A(\alpha)$-stability (discussed in section 2.4) is a way of distinguishing between methods of order higher than 2 that are not A-stable. In the case of BDF methods of order k the $A(\alpha)$-stability angle gets progressively smaller as k increases so that for $k = 5$ and 6, $\alpha = 52°$ and $18°$, respectively. As mentioned in Chapter 2, BDF methods are the basis of many stiff multistep codes such as LSODE and VODE. In the case of nonstiff problems, an explicit method can be applied directly (as in the case of Adams Bashforth methods) or used in conjunction with an implicit method in a predictor-corrector setting. In this latter case if the predictor-corrector pair is not iterated to convergence then the stability region of the method is determined by both the predictor and the corrector.

If predictor-corrector pairs are to have any value, then the stability region of the pair should be greater than that of the explicit predictor. This is explored in Table 3.6 for Adams predictor-corrector pairs of orders 3 and 4. Here I denotes the real stability interval. Note that although

Table 3.6 *stability intervals*

Order	Mode	I
3	P	[-0.55, 0]
3	C	[-6, 0]
3	PECE	[-1.8, 0]
4	P	[-0.3, 0]
4	C	[-3, 0]
4	PECE	[-1.3,0]
4	PEC	[-0.16, 0]
4	PECECE	[-0.9, 0]

the order 4 PEC mode saves on a function evaluation compared with the PECE mode, it has a considerably worse real stability interval, even when I is scaled by the number of function evaluations (E).

Runge-Kutta methods

If the multivalue method which leads to (3.3.1) is the Runge-Kutta method (2.7.2) then

$$y_{n+1} = R(z)y_n, \quad R(z) = 1 + b^\top z(I - Az)^{-1}e. \tag{3.3.7}$$

Here $R(z)$ is called the **stability function** associated with (2.7.2). Using the Binomial theorem, R can be expanded to give

$$R(z) = 1 + \sum_{j=0}^{\infty} b^\top A^j e z^{j+1}$$

and, since, in the case of an explicit method $A^s = 0$, the stability function of any s-stage explicit Runge-Kutta method is a polynomial of at most degree s.

Furthermore, if a Runge-Kutta method (explicit or implicit) is of order p then $R(z)$ must be an approximation to e^z of order p and so

$$R(z) = 1 + z + \cdots + \frac{z^p}{p!} + O(z^{p+1}). \tag{3.3.8}$$

The following result is a trivial consequence of (3.3.8) and the previous remark.

Theorem 3.3.4. *An explicit s-stage Runge-Kutta method cannot be $A(\alpha)$-stable. Furthermore, the maximum order of an explicit method is s.*

In the case of an implicit method, no particular structure is imposed on A and applying Cramer's rule to (3.3.7) gives a general form for R, namely

$$R(z) = \frac{\det(I - zA + eb^{\top})}{\det(I - zA)}.$$

Hence, for any implicit method, the associated stability function is a rational function. A general analysis of the order properties of implicit methods will be given in section 3.5, but if a method of order $2s$ exists then $R(z)$ must be an approximation to the exponential function of order $2s$ and in this case

$$R(z) = \frac{p(z)}{p(-z)}, \quad p(z) = \sum_{j=0}^{s} \frac{\binom{s}{j}}{\binom{2s}{j}} \frac{z^j}{j!}, \tag{3.3.9}$$

which is the (s, s)-Padé approximation to the exponential function of order $2s$.

The general form of a stability function is

$$R(z) = \frac{p_k(z)}{q_m(z)}, \quad k, m \leq s, \tag{3.3.10}$$

where p_k and q_m are polynomials of degree k and m, respectively. If the order of this approximation to e^z is $k+m$ then (3.3.10) is the (k,m)-Padé approximation.

Ehle (1969) conjectured that a Runge-Kutta method whose stability function is a (k, s)-Padé approximation is A-stable if and only if

$$s - 2 \leq k \leq s.$$

This was proved by Wanner et al. (1978) in an elegant theory called order stars in which the boundedness of $R(z)/e^z$ rather than $R(z)$ is studied. Ehle (1969) also proved that the $(s-1, s)$ and $(s-2, s)$ Padé approximations give rise to Runge-Kutta methods which are L-stable. (Note that from Theorem 3.3.1 an A-stable Runge-Kutta method will be L-stable if $b^{\top} A^{-1} e = 1$.)

In determining whether a Runge-Kutta method is A-stable, the standard procedure is to use the Maximum Modulus principle. Thus if

(i) $R(z)$ is analytic for $Re(z) \leq 0$

(ii) $|R(iy)| \leq 1, \quad \forall y \in \mathbb{R}$

then the corresponding method will be A-stable.

By writing

$$R(iy) = \frac{N(iy)}{D(iy)}$$

and defining the so-called E-polynomial

$$E(y) = |D(iy)|^2 - |N(iy)|^2$$

then (ii) is equivalent to $E(y) \geq 0, \forall y \in \mathbb{R}$.

There is in fact a simple relationship between the order of a method and the E-polynomial associated with the method and is based on the fact that R is an order p approximation to the exponential function. Thus it can be shown that for any method of order p

$$E(y) = \sum_{j=1+[p/2]}^{s} \alpha_j y^{2j}.$$

A consequence of this result is that it is easy to check the positivity of the E-polynomial for high-order methods. In particular, methods of order $2s-2$ or more have an E-polynomial of the form $\alpha_s y^{2s}$.

Cooper (1986) and Scherer and Türke (1987) have independently given algebraic characterizations of A-stability. They have shown that a method with a non-degenerate stability function is A-stable if and only if there exists a symmetric positive definite matrix R such that

$$Re = b, \quad RA + A^\top - bb^\top \geq 0. \tag{3.3.11}$$

As will be seen in section 3.6, if $R = B = \text{diag}(b_1, \ldots, b_s)$ then a method satisfying (3.3.11) is said to be **algebraically stable**.

3.4 Explicit methods

The construction of high-order explicit methods is a completely different process to the construction of high-order implicit methods for multistage methods. In the general multivalue case the maximum stage order of any explicit method is r (Theorem 3.2.7), while for explicit Runge-Kutta methods the restriction is even more severe since the first internal stage is an Euler step and so the maximum stage order is 1. Consequently, simplifying assumptions other than those based on high stage order are necessary.

Butcher (1964c) was one of the first to attempt a methodology for constructing high-order explicit Runge-Kutta methods. He has constructed s-stage methods satisfying

$$C(1), \quad e_k^\top Ac = c^2/2, \quad k = 3, \ldots, s-1$$

$$b_2 = 0, \quad b^\top A = (b^\top - b^\top C)$$

and proved the following results in (1964c), (1965b) and (1985).

Theorem 3.4.1. *There are no explicit s-stage Runge-Kutta methods of order p with $s = p$ for $p \geq 5$.*

Theorem 3.4.2. *There are no explicit s-stage Runge-Kutta methods of order p with $s = p + \delta$ ($\delta = 1$ or 2) for $p \geq 6 + \delta$.*

Table 3.7 shows the relationship between the number of stages (s) and the order (p) of an explicit method for explicitly constructed methods.

Table 3.7 *order of explicit methods*

s	1	2	3	4	5	6	7	11	17
p	1	2	3	4	4	5	6	8	10

In the case of explicit Runge-Kutta methods, high-order methods are usually presented as a pair of embedded formulae which take into account a variable stepsize implementation and this will be the approach adopted here. The technique of embedding an s-stage method of order w in an $(s+1)$-stage method of order $w+1$ was first considered by Sarafyan (1966), Fehlberg (1968) and England (1969). Thus if

$$\begin{array}{c|c} c & A \\ \hline y_{n+1} & b^\top \end{array}$$

represents the embedded method, then

$$\begin{array}{c|c} c & A \\ c_{s+1} & \alpha^\top \\ \hline \hat{y}_{n+1} & \hat{b}^\top \end{array}$$

where

$$\alpha^\top = (a_{s+1,1}, \ldots, a_{s+1,s+1}), \quad \hat{b}^\top = (\hat{b}_1, \ldots, \hat{b}_{s+1})$$

represents the error-estimating method. In the case of an explicit implementation $a_{s+1,s+1} = 0$.

Hence

$$\begin{aligned} T_n &= \hat{y}_{n+1} - y_{n+1} \\ &= h\left(\sum_{j=1}^{s}(\hat{b}_j - b_j)f(Y_j) + \hat{b}_{s+1}f(Y_{s+1})\right) \end{aligned}$$

is an estimate of the local error in the lower-order method.

This embedded pair is often written as

$$\begin{array}{c|c} c & A \\ c_{s+1} & \alpha^{\top} \\ \hline & b^{\top} \\ & \hat{b}^{\top} \end{array} \tag{3.4.1}$$

The solution can now be advanced on the lower-order or higher-order method (the latter approach is called local extrapolation).

Most extant explicit Runge-Kutta codes are based on the pioneering work of Fehlberg who constructed pairs of embedded methods. Fehlberg (1968) derived pairs of order (5, 6), (6, 7), (7, 8) and (8, 9) with 8, 10, 13 and 17 stages, respectively. Since all these methods were intended to advance the numerical approximation using the lower-order method, Fehlberg minimized the truncation coefficients of the lower-order formula. Consequently, these methods often underestimate the local error.

Verner (1978) derived (5, 6), (6, 7), (7, 8) and (8, 9) pairs which, unlike the pairs of Fehlberg, have a reliable error estimation and are designed to be implemented using local extrapolation. The (5,6) pair is the basis of the code DVERK which is recommended for most nonstiff problems when high accuracy is not required. Later Prince and Dormand (1981) derived (5, 6) and (7, 8) pairs with a reliable error estimate and small error coefficients in the higher-order method. The (7, 8) pair has 13 stages and is appropriate for stringent error tolerances.

If method (3.4.1) is represented as an $(s+1)$-stage method then it is generally desirable for embedded pairs to have the property that

$$e_{s+1}^{\top}A = b^{\top}, \tag{3.4.2}$$

since the final derivative evaluation within the step can be used in the next step. An example of such a method is a (4, 5) pair given in Dormand and Prince (1980):

$$
\begin{array}{c|ccccccc}
0 & & & & & & & \\
\frac{1}{5} & \frac{1}{5} & & & & & & \\
\frac{3}{10} & \frac{3}{40} & \frac{9}{40} & & & & & \\
\frac{4}{5} & \frac{44}{45} & -\frac{56}{15} & \frac{32}{9} & & & & \\
\frac{8}{9} & \frac{19372}{6561} & -\frac{25360}{2187} & \frac{64448}{6561} & -\frac{212}{729} & & & \\
1 & \frac{9017}{3168} & -\frac{355}{33} & \frac{46732}{5247} & \frac{49}{176} & -\frac{5103}{18656} & & \\
1 & \frac{35}{384} & 0 & \frac{500}{1113} & \frac{125}{192} & -\frac{2187}{6784} & \frac{11}{84} & \\
\hline
 & \frac{35}{384} & 0 & \frac{500}{1113} & \frac{125}{192} & -\frac{2187}{6784} & \frac{11}{84} & \\
 & \frac{5179}{57600} & 0 & \frac{7571}{16695} & \frac{393}{646} & -\frac{92097}{339200} & \frac{187}{2100} & \frac{1}{40}
\end{array}
$$

Researchers have suggested a number of criteria for constructing efficient high-order explicit Runge-Kutta pairs and although a reasonably large stability region is necessary, a reliable error estimate and small error coefficients seem to be crucial in the efficacy of any pair. For any method of order w the local truncation error can be written as

$$l_{n+1} = \sum_{\rho(t)=w+1}^{\infty} \frac{h^{\rho(t)}}{\rho(t)!} e(t)F(t)y(x_n),$$

and Shampine (1985) suggests criteria which involve minimizing the norm of the error coefficients associated with all trees of order $w+1$ (this is a measure of the size of the principal error coefficient) subject to certain conditions on the norm of the error coefficients associated with all terms of order $w+2$. However, Sharp (1991) has shown that these criteria are less meaningful for high orders. It has also been suggested that as few as possible of the $e(t)$ should be zero in order not to misrepresent the true error behaviour by a spurious zeroing-out of some of the elementary differentials associated with f.

More recently, considerable effort has been spent using symbolic manipulation packages such as MAPLE in constructing efficient pairs and then comparing their efficacy with other efficient pairs such as the (5,6) and (7,8) pairs of Dormand and Prince on the DETEST package (Hall et al. (1973)) program. In particular, Verner and Sharp (1991) have constructed efficient Runge-Kutta pairs of orders 5 and 6 based on a FSAL design in which the approximation of order 6 may be propagated using eight stages and the extra stage required to obtain the error estimating approximation of order 5 may be re-used following any successful step. Verner and Sharp (1991) have also generalized this approach to higher orders. These new methods compare favourably with the Dormand and Prince pairs on DETEST. An example of a FSAL (5,6) pair from Verner and Sharp (1991) is

$$
\begin{array}{c|ccccccccc}
0 & & & & & & & & & \\
\frac{1}{9} & \frac{1}{9} & & & & & & & & \\
\frac{1}{6} & \frac{1}{24} & \frac{1}{8} & & & & & & & \\
\frac{1}{4} & \frac{1}{16} & 0 & \frac{3}{16} & & & & & & \\
\frac{5}{8} & \frac{5}{8} & 0 & -\frac{75}{32} & \frac{75}{32} & & & & & \\
\frac{2}{3} & \frac{374}{1539} & 0 & -\frac{44}{57} & \frac{4880}{4617} & \frac{640}{4617} & & & & \\
\frac{7}{8} & -\frac{30023}{14592} & 0 & \frac{10353}{1216} & -\frac{35035}{5472} & -\frac{70}{171} & \frac{315}{2564} & & & \\
1 & \frac{70169}{18620} & 0 & -\frac{1914}{133} & \frac{38592}{3325} & \frac{1392}{665} & -\frac{243}{100} & \frac{432}{1225} & & \\
\hline
 & \frac{53}{700} & 0 & 0 & \frac{1264}{3375} & \frac{128}{675} & \frac{81}{500} & \frac{128}{875} & \frac{7}{135} & \\
 & \frac{137}{2100} & 0 & 0 & \frac{32}{75} & -\frac{64}{75} & \frac{27}{20} & -\frac{64}{425} & 0 & \frac{2}{15}
\end{array}
$$

Verner (1992) has continued this analysis by attempting to produce classification schemes for constructing explicit Runge-Kutta methods based on the Runge-Kutta matrix having a certain structure.

Most modern codes now have continuous extensions (see Horn (1983) and Enright et al. (1986), for example), which produce continuous approximations to the solution at $x_n + \theta h, \quad 0 < \theta \leq 1$. If the Runge-Kutta matrix A is to be independent of θ, this technique requires additional function evaluations, unless the order is low or the pair has extra stages. The advantage of such a method is that dense output is provided. The order of continuous extension methods is easily analysed using a slight modification of the theory that led to Theorem 3.2.9. In this case $e(t)$ is modified so that

$$e(t) = \theta^{\rho(t)} - \rho(t) b^{\top} \prod_{j=1}^{m} k(t_j)$$

for $t = [t_1, \ldots, t_m]$ with

$$k(\phi) = e, \quad k(t) = \rho(t) A \prod_{j=1}^{m} k(t_j).$$

Owren and Zennaro (1991, 1992) have constructed continuous Runge-Kutta methods of order 5 with eight stages. Santo (1991) has shown that at least eleven stages are needed for continuous methods of order 6, while Verner (1992) has suggested that there may be as much as a 50% additional cost over conventional pairs to obtain continuous approximations.

Some of these drawbacks can be overcome if more general multistage methods are considered. The original approaches of Butcher (1965a) and Gear (1965) were to add one or two additional stages to the linear multistep format, but more recently general multivalue methods have been considered with stage order r. If r is large enough this approach allows for dense output without additional function evaluations. However, there are additional difficulties in knowing what criteria should be used in attempting to minimize the local truncation errors, and this relates to how the local error of a multivalue method should be estimated. This has been considered by Skeel (1976) and Hairer et al. (1987) and will be further discussed in section 3.8.

3.5 Implicit methods

The construction of high-order methods is facilitated by the use of simplifying assumptions. In addition to (3.2.12) and (3.2.13) the following simplifying assumption for Runge-Kutta methods can be defined

$$D(w) : pb^\top C^{p-1} A = b^\top - b^\top C^p, \quad p = 1, \ldots, w. \tag{3.5.1}$$

Using the B-type, C-type and D-type conditions, Butcher (1964a) has shown

Theorem 3.5.1. *$B(w), C(\xi)$ and $D(\mu)$ imply order w where $w \leq \min\{\xi + \mu + 1, 2\xi + 2\}$.*

Corollary 3.5.2. *$B(s+t)$ and $C(s)$ imply order $s+t$.*

Implicit methods satisfying $C(s)$ and $B(s+t)$ are called collocation methods of order $s+t$ (see Wright (1970)). These methods are characterized by

$$\begin{aligned} a_{ij} &= \int_0^{c_i} \frac{p(x)}{(x-c_j)p'(c_j)} dx, \quad i,j = 1, \ldots, s \\ b_j &= \int_0^1 \frac{p(x)}{(x-c_j)p'(c_j)} dx, \quad j = 1, \ldots, s, \\ p(x) &= \prod_{j=1}^{s} (x - c_j) = A_0 P_s(x) + \cdots + A_{s-t} P_t(x) \end{aligned} \tag{3.5.2}$$

and where the $P_j(x)$ are the Legendre polynomials of degree j orthogonal on $[0,1]$ satisfying

$$P_j(0) = 1, \quad P_j(1) = (-1)^j, \quad P_j(x) = \sum_{k=0}^{j} \binom{j}{k} \binom{j+k}{k} (-x)^k.$$

Collocation methods of the form (3.5.2) are easily constructed from underlying quadrature formulas such as Gaussian, Radau and Lobatto quadrature (see Butcher (1964b), for example). These methods include:

(i) Gauss methods of order $2s$, with $A_0 = 1$;

(ii) Radau IIA methods of order $2s - 1$ with $A_0 = 1$, $A_1 = -1$ and $c_s = 1$;

(iii) Lobatto IIIA methods of order $2s - 2$ with $A_0 = 1, A_1 = 0$, $A_2 = -1$ and $c_1 = 0$ and $c_s = 1$.

The requirements of maximal stage order can be weakened to still obtain high-order methods. These methods include:

(iv) Radau IA methods of order $2s-1$ satisfying $C(s-1)$, $D(s)$ and $B(2s-1)$ with $c_1 = 0$;

(v) Lobatto IIIC methods of order $2s - 2$ (Chipman (1971)) satisfying $B(2s-2), C(s-1)$ and $D(s-1)$ and $e_s^\top A = b^\top$.

Note that Runge-Kutta methods satisfying the condition

$$e_s^\top A = b^\top \iff b_j = a_{sj}, \quad j = 1, \ldots, s \tag{3.5.3}$$

will be called **stiffly accurate** and such methods will be discussed further in section 3.7.

The above five classes of methods can be shown to be A-stable for all values of s, with all but the Gaussian methods and the Lobatto IIIA methods being L-stable. Examples of the above five classes of methods are presented with $s = 2$ and 3:

$$\begin{array}{c|cc}
\frac{3-\sqrt{3}}{6} & \frac{1}{4} & \frac{3-2\sqrt{3}}{12} \\
\frac{3+\sqrt{3}}{6} & \frac{3+2\sqrt{3}}{12} & \frac{1}{4} \\
\hline
 & \frac{1}{2} & \frac{1}{2}
\end{array}$$

Gauss, s=2

$$\begin{array}{c|ccc}
\frac{5-\sqrt{15}}{10} & \frac{5}{36} & \frac{10-3\sqrt{15}}{45} & \frac{25-6\sqrt{15}}{180} \\
\frac{1}{2} & \frac{10+3\sqrt{15}}{72} & \frac{2}{9} & \frac{10-3\sqrt{15}}{72} \\
\frac{5+\sqrt{15}}{10} & \frac{25+6\sqrt{15}}{180} & \frac{10+3\sqrt{15}}{45} & \frac{5}{36} \\
\hline
 & \frac{5}{18} & \frac{4}{9} & \frac{5}{18}
\end{array}$$

Gauss, s=3

$$\begin{array}{c|cc} \frac{1}{3} & \frac{5}{12} & -\frac{1}{12} \\ 1 & \frac{3}{4} & \frac{1}{4} \\ \hline & \frac{3}{4} & \frac{1}{4} \end{array}$$

Radau IIA, s=2

$$\begin{array}{c|ccc} \frac{4-\sqrt{6}}{10} & \frac{88-7\sqrt{6}}{360} & \frac{296-169\sqrt{6}}{1800} & \frac{-2+3\sqrt{6}}{225} \\ \frac{4+\sqrt{6}}{10} & \frac{296+169\sqrt{6}}{1800} & \frac{88+7\sqrt{6}}{360} & \frac{-2-3\sqrt{6}}{225} \\ 1 & \frac{16-\sqrt{6}}{36} & \frac{16+\sqrt{6}}{36} & \frac{1}{9} \\ \hline & \frac{16-\sqrt{6}}{36} & \frac{16+\sqrt{6}}{36} & \frac{1}{9} \end{array}$$

Radau IIA, s=3

$$\begin{array}{c|cc} 0 & 0 & 0 \\ 1 & \frac{1}{2} & \frac{1}{2} \\ \hline & \frac{1}{2} & \frac{1}{2} \end{array}$$

Lobatto IIIA, s=2

$$\begin{array}{c|ccc} 0 & 0 & 0 & 0 \\ \frac{1}{2} & \frac{5}{24} & \frac{1}{3} & -\frac{1}{24} \\ 1 & \frac{1}{6} & \frac{2}{3} & \frac{1}{6} \\ \hline & \frac{1}{6} & \frac{2}{3} & \frac{1}{6} \end{array}$$

Lobatto IIIA, s=3

$$\begin{array}{c|cc} 0 & \frac{1}{4} & -\frac{1}{4} \\ \frac{2}{3} & \frac{1}{4} & \frac{5}{12} \\ \hline & \frac{1}{4} & \frac{3}{4} \end{array}$$

Radau IA, s=2

$$\begin{array}{c|ccc} 0 & \frac{1}{9} & \frac{-1-\sqrt{6}}{18} & \frac{-1+\sqrt{6}}{18} \\ \frac{6-\sqrt{6}}{10} & \frac{1}{9} & \frac{88+7\sqrt{6}}{360} & \frac{88-43\sqrt{6}}{360} \\ \frac{6+\sqrt{6}}{10} & \frac{1}{9} & \frac{88+43\sqrt{6}}{360} & \frac{88-7\sqrt{6}}{360} \\ \hline & \frac{1}{9} & \frac{16+\sqrt{6}}{36} & \frac{16-\sqrt{6}}{36} \end{array}$$

Radau IA, s=3

$$\begin{array}{c|cc} 0 & \frac{1}{2} & -\frac{1}{2} \\ 1 & \frac{1}{2} & \frac{1}{2} \\ \hline & \frac{1}{2} & \frac{1}{2} \end{array}$$

Lobatto IIIC, s=2

$$\begin{array}{c|ccc} 0 & \frac{1}{6} & -\frac{1}{3} & \frac{1}{6} \\ \frac{1}{2} & \frac{1}{6} & \frac{5}{12} & -\frac{1}{12} \\ 1 & \frac{1}{6} & \frac{2}{3} & \frac{1}{6} \\ \hline & \frac{1}{6} & \frac{2}{3} & \frac{1}{6} \end{array}$$

Lobatto IIIC, s=3

An important class of implicit Runge-Kutta methods is the class of symmetric methods, characterized by a global error expansion in even powers of h. The order of a symmetric method is even and, in particular, any symmetric collocation method has symmetric abscissae and weights. A class of symmetric methods can be characterized algebraically (see Stetter (1973)) by assuming that there exists a permutation matrix P such that

$$\begin{aligned} b^\top &= b^\top P \\ A + P^\top A P &= e b^\top . \end{aligned}$$

The Gauss and Lobatto IIIA methods are examples of symmetric methods.

In the case of more general multistage methods, Burrage (1988) and Lie and Nørsett (1989) have shown that s-stage, r-step MRKs can attain an order of $2s + r - 1$ with a stage order $s + r - 1$. However, in the case of Nordsieck methods, Burrage (1988) has shown that if $r > 2$ the maximum order is $s + r - 1$ while if $r \leq 2$ the maximum order is $2s$.

The result for MRKs is entirely consistent with that for Runge-Kutta methods (with $r = 1$). It holds because there is an analogue of the D-type simplifying assumptions for Runge-Kutta methods given in (3.5.1) which holds for MRKs and enables the construction of high-order MRK collocation methods with stage order $s + r - 1$. This is expressed in the following results which appears in Burrage (1988).

Theorem 3.5.3. *For MRKs, $B(s+r-1+t)$ and $C(s+r-1)$ imply order $s+r-1+t$.*

Theorem 3.5.4. *The maximum order of a MRK is* $2s + r - 1$.

The collocation methods of the form (2.8.2) satisfying Theorem 3.5.3 satisfy for $i = 1, \ldots, s$

$$\sum_{j=1}^{r} a_{ij}(1-j)^p = c_i^p - p\sum_{j=1}^{s} b_{ij}c_j^{p-1}, \quad p = 0, \ldots, s+r-1 \tag{3.5.4}$$

and

$$\sum_{j=1}^{r} \alpha_j(1-j)^p = 1 - p\sum_{j=1}^{s} \beta_j c_j^{p-1}, \quad p = 0, \ldots, s+r-1+t. \tag{3.5.5}$$

Equation (3.5.5) can be considered as an extension of Hermite interpolation in which the basis functions are

$$l_i(x) = C_i \prod_{j=1}^{r}(x - x_j) \prod_{j \neq i}^{s}(x - c_j)^2$$

where

$$x_j = j - r, \quad j = 1, \ldots, r.$$

These polynomials of degree $2s + r - 2$ must satisfy, in addition,

$$l_i(c_i) = 1, \quad l_i'(c_i) = 0, \quad i = 1, \ldots, s$$

. By taking logarithms of the l_i, this last condition can be written as

$$\sum_{j=1}^{r} \frac{1}{c_i - x_j} + \sum_{j \neq i}^{s} \frac{2}{c_i - c_j} = 0, \quad i = 1, \ldots, s-1.$$

When deriving methods of maximal order there is no unique solution for the abscissae. In fact it can be shown that there are $\binom{s+r-1}{r-1}$ distinct solutions and it is natural to choose those solutions for which the abscissae lie in $[0, 1]$. These methods of maximal order cannot be A-stable and as for Runge-Kutta methods it is natural to reduce the order by one to $2s + r - 2$ in an attempt to construct methods which are stiffly accurate with $c_s = 1$. Such methods are said to be of Radau type (Schneider (1993)). Since the last row of A and B are, respectively, $\alpha^\top$ and $\beta^\top$ these methods can be characterized in the form

$$c^\top \mid A \mid B.$$

It is easily shown that with $s = r = 2$ and $c_s = 1$ then the choice of $c_1 = \frac{\sqrt{17}-1}{8}$ leads to an A-stable Radau MRK of order 4 characterized by

$$A = \begin{pmatrix} 1.04671554852736471 & -0.04671554852736471 \\ 1.02010509586877204 & -0.02010509586877204 \end{pmatrix}$$

$$B = \begin{pmatrix} 0.4004407511365936 & -0.0567680964617507 \\ 0.7707238584700291 & 0.2091710456611988 \end{pmatrix}. \tag{3.5.6}$$

Schneider (1993) and Hairer and Wanner (1991) have investigated the order and stability properties of Radau MRKs with $s = 3$. For example, the three stage Radau MRK with $r = 2$ of order 6 with

$$c_1 = 0.177891722985607, \quad c_2 = 0.673235257220651$$

is A-stable, while for $r \leq 16$ there exist $A(\alpha)$-stable Radau MRKs with $\alpha > 84°$.

Efficiently implementable methods

In spite of the fact that there exist s-stage implicit Runge-Kutta methods of order $2s$ with excellent stability properties these methods have been rarely used in practice. The main reason for this is the cost of implementation compared with codes based on BDF methods or certain classes of efficient Runge-Kutta methods.

Nevertheless, singly implicit methods (SIRKs) do have other substantial advantages over BDF methods. These are expressed in the following results whose proofs are in Nørsett and Wolfbrandt (1977) and Burrage (1978b), respectively.

Theorem 3.5.5. *The maximum order of an s-stage method with real eigenvalues is $s + 1$.*

Theorem 3.5.6. *There exist SIRKs with stage order s in which $c_j/\lambda, j = 1, \ldots, s$, are the zeros of the sth-degree Laguerre polynomial given by $L_s(z) = \sum_{j=0}^{s} (-1)^j \binom{s}{j} z^j / j!$.*

Corollary 3.5.7. *If, in addition to the assumptions of Theorem 3.5.6, $L'_{s+1}(\frac{1}{\lambda}) = 0$, the SIRKs are of order $s + 1$.*

In addition to the above properties, it has been shown by Wolfbrandt (1977) that there exist L-stable SIRKs (with $L_s(\frac{1}{\lambda}) = 0$) of orders 1 to 6 and 8. SIRKs are used as the basis of STRIDE (Butcher et al. (1979)) which allows the use of methods from orders 1 to 15, all being $A(\alpha)$-stable with $\alpha > 83^o$. In addition, $\frac{1}{\lambda}$ is always chosen to be a zero of L_s, since this guarantees that the method is stiffly accurate.

For a given s-stage method there are s possible values for λ and the exact choice should be based on stability and principal error considerations. Error estimates can be provided either by the process of embedding or by the use of the collocating polynomial. Numerical testing, for example by Gaffney (1981) and Addison (1984), has shown that STRIDE is more robust than BDF-based codes because of its superior stability properties, but is generally slower.

Butcher (1979) has constructed the transformation matrix T (see (2.7.7)) and its inverse explicitly in the case of singly implicit collocation methods. He has shown with $\xi_i = c_i/\lambda$, $i = 1, \ldots, s$, that

$$T_{ij} = L_{j-1}(\xi_i), \quad i, j = 1, \ldots, s$$

and

$$T_{ij}^{-1} = \left(\frac{\xi_j L_{i-1}(\xi_j)}{s^2 L_{s-1}(\xi_j)^2}\right), \quad i, j = 1, \ldots, s.$$

Two important SDIRKs are the three-stage, A-stable SDIRK of order 4 given by

$$\begin{array}{c|ccc} \lambda & \lambda & 0 & 0 \\ \frac{1}{2} & \frac{1}{2}-\lambda & \lambda & 0 \\ 1-\lambda & 2\lambda & 1-4\lambda & \lambda \\ \hline & u & 1-2u & u \end{array}$$

where

$$\lambda = \frac{1}{2} + \frac{\sqrt{3}}{3}\cos\frac{\pi}{18}, \quad u = 1 - \sqrt{3}/2 \tag{3.5.7}$$

and the three-stage, stiffly accurate, L-stable method of order 3 given by

$$\begin{array}{c|ccc} \lambda & \lambda & 0 & 0 \\ c & c-\lambda & \lambda & 0 \\ 1 & 1-a-\lambda & a & \lambda \\ \hline & 1-a-\lambda & a & \lambda \end{array}$$

where $\lambda^3 - 3\lambda^2 + 3\lambda/2 - 1/6 = 0$, $\lambda \approx 0.435867$ and

$$c = \frac{\lambda^2 - \frac{3}{2}\lambda + \frac{1}{3}}{\lambda^2 - 2\lambda + \frac{1}{2}}, \quad a = \frac{\lambda^2 - 2\lambda + \frac{1}{2}}{c - \lambda}. \tag{3.5.8}$$

SDIRKs have been studied by a number of authors including Nørsett (1974), Alexander (1977), Cash (1979) and Cameron (1983), who applied them to differential-algebraic problems. Cooper and Sayfy (1979) have constructed a six stage, order 5 SDIRK. A number of codes, based on SDIRKs, have been developed including SIRKUS (Nørsett (1974)), DIRK (Alexander (1977)), a code by Cash and Liem (1980) and SIMPLE (Nørsett and Thomsen (1986)). Error estimates can be provided (see, for example Cash (1979)) by the process of embedding one SDIRK inside another.

These codes perform well on stiff problems if low accuracy is required but they suffer from a certain inflexibility in that they are generally of fixed order. More importantly, they suffer from an even more serious disability – their stage order is only 1. However, continuous extensions can be provided for DIRKs in much the same way as they are provided for explicit methods.

In an attempt to overcome the low stage order properties of DIRKs recent attention has focused on the construction of diagonally implicit multivalue methods which can always have a stage order of r. Some of this work was alluded to in Chapter 2. In addition, Butcher (1993a, 1993b, 1994) and Butcher and Jackiewicz (1993) have introduced a new family of diagonally implicit multivalue methods known as DIMSIMs which can be used in stiff or nonstiff mode.

The starting values for DIMSIMs of order w satisfy $A(w)$ (see (3.2.10)). Methods constructed in Butcher and Jackiewicz (1993) satisfy $B(w)$ and $C(w-1)$ with either $s=r=w-1$ or $s=r=w$. The vector of abscissae c allows a DIMSIM to represent a linear multistep approach if $c=(-(s-2),-(s-3),\ldots,0,1)^{\top}$ or a Runge-Kutta approach if the $c_i \in [0,1]$. Various low-order methods up to order 4 are given in Butcher and Jackiewicz (1993) and their linear stability properties are studied.

An example of a class of DIMSIMs of order 3 with $s=r=2$ is

$$\begin{pmatrix} A_1 & B_1 \\ A_2 & B_2 \end{pmatrix} = \begin{pmatrix} 1 & 0 & \lambda & 0 \\ 0 & 1 & 0 & \lambda \\ v & 1-v & av & bv \\ v & 1-v & av-b & bv+1+b \end{pmatrix}$$

where

$$a = 1/2+\lambda, \quad b = 1-a, \quad v = \frac{5-12\lambda}{6(1-2\lambda)}, \quad \lambda \neq 1/2.$$

Note that because this method decouples the internal components that define the $f(Y_i)$, this method is amenable to a parallel implementation in either stiff or nonstiff mode.

Although diagonally and singly implicit methods are the cheapest (computationally speaking) of the class of Runge-Kutta methods, they are not the only possibilities. Butcher's technique outlined in (2.7.5) applies to any implicit method, but if the order of a Runge-Kutta method is greater than

$s+1$ then some of the eigenvalues are necessarily complex so that Λ will consist of some 2×2 blocks of the form

$$\begin{pmatrix} \alpha & -\beta \\ \beta & \alpha \end{pmatrix}.$$

This is the case, for example, of the three-stage, stiffly accurate Radau IIA method of order 5 (which has been implemented by Hairer and Wanner (1991)) for which Λ takes the form

$$\begin{pmatrix} \gamma & & 0 \\ 0 & \alpha & -\beta \\ 0 & \beta & \alpha \end{pmatrix}.$$

In this case two linear systems of dimension m and $2m$ must be solved. Alternatively, the real $2m \times 2m$ linear system could be solved as an m-dimensional complex system.

The standard way of performing complex multiplications consists of four real multiplications. Consequently, if s is even, the total cost for implementing an s-stage Radau IIA method over one integration step (given l Newton iterations per step) is given by

$$C_R = 2(s-1)m^3/3 + lsm^2 + 2ls^2m, \tag{3.5.9}$$

while if s is odd, the cost is given by

$$C_R = (2s-1)m^3/3 + lsm^2 + 2ls^2m. \tag{3.5.10}$$

If s and m are small then the high-order methods such as the Radau IIA methods are competitive, but as the dimension of the problem becomes large it can be seen that Radau IIA methods become approximately $2s-2$ or $2s-1$ times as expensive as singly implicit methods (although their order is approximately double).

Although multivalue methods, in general, have high order and good stability properties there are few successful implementations of such methods. One possible candidate for efficient implementation is the family of s-stage k-step multistep Runge-Kutta methods. As has already been remarked, Burrage (1988) has constructed methods of order $2s+k-1$ and stage order $s+k-1$. However, these methods of maximal order are not stable at infinity for $k > 1$. Schneider (1993), by weakening the order by one, has constructed stiffly accurate methods of order $2s+k-2$, with $c_s = 1$, with excellent stability properties. These methods can be considered as generalizations of both the Radau IIA methods ($k = 1$) and BDF methods ($s = 1$).

As a particular class of such methods, Schneider chooses $s = 3$ and constructs methods of order $k + 4$ (up to order 12), which are $A(\alpha)$-stable for $\alpha > 87°$. The coefficient matrix for these methods has two complex eigenvalues and one real eigenvalue so that the implementation costs are similar to those in the RADAU5 code, in that a transformation within the Newton process allows a linear system of dimension $3m$ to be split into a system of dimension m and dimension $2m$. However, because stepsize changes are more expensive (due to interpolation) numerical tests presented in Schneider (1993) indicate that one step of the multistep Radau code is approximately twice as expensive as one step of the RADAU5 code. Nevertheless, the advantage of the multistep approach is that very high orders are possible, so that in many cases the multistep Radau code is more efficient than RADAU5 or LSODE at stringent tolerances when solving stiff problems.

This work represents one of the first successful implementations of a multistage multistep method. More improvements could be expected in terms of improved order and stepsize selection. In particular, since this code should reduce to a RADAU5 or BDF implementation (with $s = 1$ or $k = 1$, respectively) then the code should perform equally as well as either of these two implementations for low-precision computation.

As well as transforming the method to its canonical form, it is also possible to transform problems with a full Jacobian, J, to an upper Hessenberg form. This was proposed by Enright (1978) in conjunction with multistep methods and by Varah (1979) in conjunction with Runge-Kutta methods. This procedure is especially beneficial if J is kept constant over as many steps as possible.

This last remark leads to the point that if an inexact Jacobian is used (as it often is because of efficiency considerations), then the convergence behaviour of the Newton process is no longer quadratic. In fact Verwer (1981) in an analysis of diagonally implicit methods and Dekker and Verwer (1984) have noted that for some classes of linear problems, such as the variable-coefficients Kreiss problem (Kreiss (1978)), the stability properties of the underlying method can be contaminated by the behaviour of the modified Newton method.

Thus Dekker and Verwer (1984) report on an analysis of STRIDE on the Kreiss problem

$$y'(x) = \begin{pmatrix} \cos x & -\sin x \\ \sin x & \cos x \end{pmatrix} \begin{pmatrix} -\frac{1}{\epsilon} & 0 \\ 0 & -1 \end{pmatrix} \begin{pmatrix} \cos x & \sin x \\ -\sin x & \cos x \end{pmatrix} y(x)$$

in which, as $\frac{1}{\epsilon} \to \infty$, the variable order code is observed to degenerate to the first-order implicit Euler method. The problem here appears to be due to the fact that the modified Newton method often does not converge for large values of the stiffness parameter. On the other hand if an exact Jacobian is used at every step convergence takes place in one iteration.

This problem is a very difficult one for a multistage method because the Jacobian matrix is varying extremely quickly and multistage methods are nearly always implemented with the Jacobian matrices evaluated at $Y_1, \ldots, Y_s$ being approximated by a single Jacobian evaluated at say y_n. In this case, this strategy is not appropriate. It is interesting to note that codes based on BDF methods solve this problem very efficiently since they are only single-stage methods.

The main point to observe from this discourse is that an implicit multistage method used in conjunction with the modified Newton process can be represented as a Rosenbrock-type method in which an approximation to the Jacobian matrix is inserted into the numerical method directly. This technique leads to the solution of just linear equations, rather than nonlinear systems.

The extension of nonlinear stability theory to ROW methods has been considered by Hundsdorfer (1981), Vaneslow (1979), Hairer et al. (1982) and Hundsdorfer (1984). In fact Rosenbrock methods do not have strong nonlinear stability properties, as has been pointed out in some of the above papers. This suggests that nonlinear stability may not be such a significant criterion when a numerical method is implemented efficiently, since a modified Newton process is a necessity for efficient implementation (see Dekker and Verwer (1984), for an excellent discussion on this).

3.6 Nonlinear stability

As has already been seen, many high-order methods can be A-stable, but in fact they may satisfy even stronger stability properties. Thus Burrage (1978a) considered the application of a Runge-Kutta method to the non-autonomous linear problem

$$y'(x) = q(x)y(x), \quad Re(q(x)) \leq 0 \tag{3.6.1}$$

and defined a method to be **AN-stable** if $|y_{n+1}| \leq |y_n|$ when applied to (3.6.1) with an arbitrary $q(x)$ satisfying $Re(q(x)) \leq 0$. An application of a Runge-Kutta method given by (2.7.2) to (3.6.1) gives

$$y_{n+1} = M(Z)y_n, \quad Z = \text{diag}(z_1, \ldots, z_s), \quad z_j = hq(x_n+c_i h), \quad j = 1, \ldots, s,$$

where

$$M(Z) = 1 + b^\top Z(I - AZ)^{-1}e. \tag{3.6.2}$$

Here $M(Z)$ is a multilinear rational function in $z_1, \ldots, z_s$.

In a similar way an application of a multivalue method given by (2.8.1) to (3.6.1) gives

$$y^{(n+1)} = M(Z)y^{(n)}, \quad M(Z) = A_2 + B_2Z(I - B_1Z)^{-1}A_1,$$

where

$$Z = \text{diag}(z_1, \ldots, z_s), \quad z_i = hq(x_n + hc_i), \quad i = 1, \ldots, s.$$

Butcher (1987a) has defined a multivalue method to be AN-stable if products of matrices $M(Z)$ are power bounded (since different values of Z may occur at each integration step), or equivalently, if there exists a norm such that

$$\| M(Z) \| \leq 1, \quad \forall Z = \text{diag}(z_1, \ldots, z_s) \text{ with } Re(z_i) \leq 0, \quad i = 1, \ldots, s.$$

This represents a completely natural generalization of AN-stability as applied to Runge-Kutta methods to multivalue methods.

Burrage and Butcher (1979) showed that A-stability and AN-stability are not equivalent and that if the c_j are distinct then a Runge-Kutta method is AN-stable if and only if

$$B \geq 0, \quad M = BA + AB^\top - bb^\top \geq 0, \quad B = \text{diag}(b_1, \ldots, b_s). \tag{3.6.3}$$

In actual fact the condition $B \geq 0$ should be replaced by the condition $B > 0$ if irreducible methods are considered. (An s-stage Runge-Kutta method is said to be **reducible**, if a p-stage method with $p < s$ can reproduce exactly the same numerical results as the s-stage method for any differential equation problem.)

Although a study of linear stability gives much insight into the suitability of stable methods for stiff problems it is not the complete answer. A consequence of stiffness is the existence of a large Lipschitz constant, but for many stiff problems one-sided Lipschitz constants ν can be found which are small in magnitude and may even be negative. Thus a suitable nonlinear test problem is one satisfying (2.2.5).

Dahlquist (1975) was the first to study systematically the behaviour of a class of discrete methods, called one-leg methods, to nonlinear dissipative problems. These methods represent a nonlinear variant of linear multistep methods. In order to do this Dahlquist introduced the concept of a G norm in which

$$\langle u, v \rangle_G = u^H G v.$$

A method is then said to be **G-stable** if, when applied to a dissipative problem, there exists $G > 0$ such that

$$\| Y_{n+1} \|_G \leq \| Y_n \|_G,$$

where Y_{n+1} represents the vector of information that is carried from step to step. Dahlquist (1978) has proved the remarkable result that A-stability and G-stability are equivalent.

Butcher (1975) generalized the concept of G-stability to Runge-Kutta methods. A method is said to be **B-stable**, if for any dissipative problem, two solution sequences computed by the same Runge-Kutta method satisfy

$$\| y_{n+1} - z_{n+1} \| \leq \| y_n - z_n \| .$$

The corresponding property for non-autonomous problems is called **BN-stability**. Denoting the condition (3.6.3) by the term **algebraic stability** the following result appears in Burrage and Butcher (1979) and Crouzeix (1979).

Theorem 3.6.1. *If the c_j are distinct, AN-stability $\Leftrightarrow$ BN-stability $\Leftrightarrow$ algebraic stability.*

As noted in Dekker and Verwer (1984), the proof given in Burrage and Butcher (1979) is not complete because they make specific assumptions about the solvability of the nonlinear equations that determine the intermediate approximations when solving dissipative problems. Hundsdorfer and Spijker (1981b) have shown that algebraic stability and S-irreducibility (in the sense of Hundsdorfer and Spijker) are not sufficient for the existence of a unique solution for all dissipative problems.

However, in the case of **nonconfluent** methods (methods with distinct abscissae), it has now been demonstrated that AN-stability, BN-stability and algebraic stability are equivalent. Furthermore, Hundsdorfer and Spijker (1981a) have shown that for methods which are S-irreducible BN-stability and B-stability are equivalent. Burrage (1978c) has shown that many high-order methods such as the Gauss, Radau IA, Radau IIA and Lobatto IIIC methods are algebraically stable. In addition, he has shown that for any nonconfluent algebraically stable method of order p

$$\text{rank}(M) = s - [p/2].$$

In particular, the Gauss methods have a stability matrix M which is the zero matrix.

Burrage (1982) and Hairer and Wanner (1981) have obtained conditions which relate algebraic stability to the simplifying assumptions. In particular, Burrage (1982) has shown

Theorem 3.6.2. *An algebraically stable method satisfying $B(k)$ and either $C(k)$ or $D(k)$ must be of order $2k-1$.*

Hairer and Wanner (1981) have shown

Theorem 3.6.3. *An algebraically stable method of order w satisfies $C\left[\frac{w-1}{2}\right]$ and $D\left[\frac{w-1}{2}\right]$.*

A consequence of these results is that the only algebraically stable collocation methods are those of order $2s-1$. Furthermore it is easily seen from Theorem 3.6.3 that the A-stability properties of SIRKs and SDIRKs are equivalent, but their nonlinear stabilities are different. In particular, Hairer (1980) has shown that the maximum order of an algebraically stable SDIRK is 4.

In a similar fashion Burrage and Butcher (1980) generalized the concept of G-stability (for one-leg methods) and B-stability (for Runge-Kutta methods) to multivalue methods. Using the concept of a G-norm, a multivalue method given by (2.8.1) is said to be monotonic if there exists $G \geq 0$ such that

$$\| y^{(n+1)} \|_G \leq \| y^{(n)} \|_G \, .$$

Burrage and Butcher (1980) have shown that if there exists $G \geq 0$ and a diagonal matrix $D \geq 0$ such that the matrix $M \geq 0$ where

$$M = \begin{pmatrix} G - A_2^\top G A_2 & A_1^\top D - A_2^\top G B_2 \\ D A_1 - B_2^\top G A_2 & B_1^\top D + D B_1 - B_2^\top G B_2 \end{pmatrix}$$

then the method is algebraically stable. Many of the nonlinear stability results that have been developed for Runge-Kutta methods can be extended to multivalue methods. In particular, Butcher (1987a) has shown that if the c_j are distinct then AN-stability is equivalent to algebraic stability. Furthermore, Burrage (1987) has generalized the result that Runge-Kutta methods of order $2s$ are algebraically stable to show that there exist families of MRKs (with free parameters $\alpha_2, \ldots, \alpha_r$) of order $2s$ and stage order s that are algebraically stable with $G = \text{diag}(\alpha_1, \ldots, \alpha_r)$, $\alpha_j \geq 0$, $j = 1, \ldots, r$.

3.7 Stiff order and differential-algebraic equations

It was first noted by Prothero and Robinson (1974) that when certain classes of methods are applied to (2.2.1) the observed order is not the classical order of consistency but, rather, more like the stage order.

Frank et al. (1981) generalized this work to nonlinear problems satisfying a one-sided Lipschitz constant by introducing the concepts of B-consistency and B-convergence, which are the analogues of consistency and convergence but applied to nonlinear stiff problems satisfying a one-sided Lipschitz constant. Thus a Runge-Kutta method is said to be B-**consistent** of order w for problems satisfying (2.2.5) if the local error is bounded such that

$$\| l_{n+1} \| \leq C_1 h^{w+1}, \quad \forall h \in (0, h_1], \tag{3.7.1}$$

where C and h_1 are independent of stiffness but can depend on ν and on bounds for certain derivatives of the exact solution over the range of integration.

Similarly, a Runge-Kutta method is said to be B-**convergent** of order w if the global error E_N satisfies

$$\| E_N \| \leq C_2 h^w, \quad \forall h \in (0, h_2] \tag{3.7.2}$$

where once again h_2 is a constant determined only by ν and C is determined by ν and bounds for certain derivatives of the exact solution (optimal B-convergence).

The crucial point here is that unlike the classical case, the bounds must not depend on bounds for the derivatives of f which may be large for stiff problems. In the case of variable stepsizes, h_1 and h_2 in (3.7.1) and (3.7.2) may also depend linearly on the maximum stepsize taken.

Frank et al. (1985) have shown that certain high-order, algebraically stable, methods such as Gauss, Radau IA and Radau IIA are B-convergent with B-convergence order equal to the stage order. This result is not optimal however. Burrage et al. (1986) have shown that for semi-linear problems, a number of classes of methods can have B-convergence order one more than the stage order. Kraaijevanger (1985) and Stetter (in an unpublished note) have shown that the midpoint rule is B-convergent of order 2 on dissipative problems if restrictions are imposed on the possible stepsize sequence (such as constancy or monotonicity).

This last point is an important one, because written as a Runge-Kutta method the trapezoidal rule is not algebraically stable, but can be considered to be algebraically stable (in a different sense) as a multivalue method (see Burrage (1990a) for further exposition). More recently, Dekker et al. (1990) have shown that for an irreducible method with distinct abscissae lying in [0,1] B-convergence implies algebraic stability. There is no discrepancy with this result and the previous one because of the imposition of certain restrictions on the stepsizes allowed for the trapezoidal method.

More recently, Hundsdorfer (1991) and Chartier (1993) have used some of the ideas in Burrage et al. (1986) and obtained global error bounds for multivalue methods when applied to stiff linear systems which are based only on bounds on the derivatives of y. This approach actually allows the similar analyses previously developed for Runge-Kutta methods to be treated as a special case. Thus rather than treat this special case, a brief discussion of the general multivalue approach will now be given.

Without loss of generality only the one-dimensional version of (2.2.4) will be considered in this analysis. By introducing the global error defined by

$$\epsilon_n = z(x_n, h) - y^{(n)},$$

where $z(x, h)$ denotes the correct value function, Hundsdorfer (1991) obtains a recurrence relation given by

$$\epsilon_{n+1} = M(z)\epsilon_n + l_n. \tag{3.7.3}$$

Here $M(z)$ is as in (3.3.1) and l_n is the local error expansion of the multivalue method (2.8.1). If this method has order w and stage order q with $w \geq q+1$ then

$$l_n = \sum_{p \geq q+1} (d_1^{(p)} + zB_2(I - zB_1)^{-1} d_2^{(p)}) h^p y^p(x_n)/p!,$$

where $d_1^{(p)}$ and $d_2^{(p)}$ can be interpreted as the stage order errors and are given by

$$d_1^{(p)} = \sum_{j=0}^{p} \binom{p}{j} q(\tau_j) - A_2 q(\tau^p) - pB_2 c^{p-1}$$

and

$$d_2^{(p)} = c^p - pB_1 c^{p-1} - A_1 q(\tau^p).$$

By introducing the vector-valued function

$$\psi(z) = (I - M(z))^{-1}(d_1^{(q+1)} + zB_2(I - zB_1)^{-1} d_2^{(q+1)}), \tag{3.7.4}$$

Hundsdorfer (1991) has proved the following result:

Theorem 3.7.1. *If an A-stable multivalue method of order $w \geq q+1$ and stage order q is applied to the dissipative linear problem given by (2.2.4) then either $\| \epsilon_n \| \leq C(y)x_n h^q$ or $\| \epsilon_n \| \leq C(y)x_n h^{q+1}$ if and only if $\| \psi(z) \|$ is uniformly bounded on $\mathbb{C}^{-1}$.*

Remarks.

1. The results in Theorem 3.7.1 assume that there exist β such that

$$\| zB_2(I - zB_1)^{-1} \| \leq \beta, \quad \forall z \in \mathbb{C}^{-1}.$$

2. $C(y)$ denotes some positive constant whose value is determined only by bounds on the derivative of y.

3. Theorem 3.7.1 is a generalization of a similar result given in Burrage et al. (1986) for Runge-Kutta methods.

Hundsdorfer (1991) (for multivalue methods) and Burrage and Hundsdorfer (1987) (for Runge-Kutta methods) have considered under which conditions methods can have a global order of one more than the stage order when applied to dissipative nonlinear problems. Of course a necessary condition is algebraic stability, but the remaining conditions seem to be somewhat restrictive and in the case of Runge-Kutta methods the maximum classical order of such methods is three.

It is important to note that order reduction does not occur on all classes of problems. Consider, for example, the class of singularly perturbed problems given by

$$y' = f(y, z), \quad \epsilon z' = g(y, z), \quad 0 < \epsilon \ll 1, \quad y \in \mathbb{R}^{m_1}, \quad z \in \mathbb{R}^{m_2}. \tag{3.7.5}$$

By letting $\epsilon \to 0$ (so that, in some sense, the stiffness becomes infinite) an index-one DAE in semi-explicit form is obtained. Note that in the case of index-one equations it is usually assumed that there are consistent initial values ($g(y_0, z_0) = 0$), that f and g are sufficiently smooth and $g_z(y, z)$ is nonsingular and of bounded inverse in a neighbourhood of the solution of (3.7.5).

The index-two formulation can be written in the form

$$y' = f(y, z), \quad 0 = g(y), \quad y \in \mathbb{R}^{m_1}, \quad z \in \mathbb{R}^{m_2}. \tag{3.7.6}$$

In this case it is usually assumed that there are consistent initial values ($g(y_0) = 0$, $g_y(y_0)f(y_0, z_0) = 0$), that f and g are sufficiently smooth and $g_y(y)f_z(y, z)$ is nonsingular in a neighbourhood of the solution.

In the case of singularly perturbed problems, Roche (1989) has shown that stiffly accurate Runge-Kutta methods suffer no order reduction. Hairer et al. (1988, 1989) have extended this theory to DAEs of indices two and three and then applied these results to singularly perturbed problems.

The important result here is that the solution to (3.7.5) can be written as a series in ϵ in which the coefficients of the powers of ϵ are solutions to DAEs of ever-increasing index. Using this result, Hairer et al. (1989) obtain formulas for the global errors in the y and z components in terms of the global errors of the Runge-Kutta method applied to DAEs of increasing index as the power of ϵ increases. Thus the three-stage, L-stable, stiffly accurate DIRK of order 3 given in (3.5.8) is an example of a method that suffers no order reduction when applied to singularly perturbed problems.

A similar analysis can be extended to multivalue methods in general, but this requires the concept of stiffly accurate multivalue methods. This has been given in Chartier (1993) through the stability matrix at infinity $M(\infty)$ and his analysis is briefly summarized here.

Thus the application of (2.8.1) to (3.7.5) written in tensor product form gives

$$\begin{aligned} Y &= (A_1 \otimes I_{m_1})y^{(n)} + h(B_1 \otimes I_{m_1})F(Y,Z) \\ \epsilon Z &= \epsilon(A_1 \otimes I_{m_1})z^{(n)} + h(B_1 \otimes I_{m_1})G(Y,Z) \\ y^{(n+1)} &= (A_2 \otimes I_{m_2})y^{(n)} + h(B_2 \otimes I_{m_1})F(Y,Z) \\ \epsilon z^{(n+1)} &= \epsilon(A_2 \otimes I_{m_2})z^{(n)} + h(B_2 \otimes I_{m_2})G(Y,Z), \end{aligned}$$

where F and G have the usual interpretation. Under the assumption that B_1 is nonsingular, substitution of the second equation for G in the fourth equation defines $z^{(n+1)}$ independently of ϵ. Now letting $\epsilon = 0$ gives

$$\begin{aligned} Y &= (A_1 \otimes I_{m_1})y^{(n)} + h(B_1 \otimes I_{m_1})F(Y,Z) \\ 0 &= G(Y,Z) \\ y^{(n+1)} &= (A_2 \otimes I_{m_2})y^{(n)} + h(B_2 \otimes I_{m_1})F(Y,Z) \\ z^{(n+1)} &= (M(\infty) \otimes I_{m_2})z^{(n)} + (B_2 B_1^{(-1)} \otimes I_{m_2})Z. \end{aligned}$$

Now if for all $l \in \{1,\ldots,r\}$, there exists $k \in \{1,\ldots,s\}$ such that

$$\begin{aligned} a_{lj}^{(2)} &= a_{kj}^{(1)}, \quad j = 1,\ldots,r \\ b_{lj}^{(2)} &= b_{kj}^{(1)}, \quad j = 1,\ldots,s \end{aligned} \tag{3.7.7}$$

then $M(\infty) = 0$ and

$$\begin{aligned} z_j^{(n+1)} &= Z_k, \quad j = 1,\ldots,r \\ y_j^{(n+1)} &= Y_k, \quad j = 1,\ldots,s. \end{aligned}$$

In particular,

$$g(y_j^{(n+1)}, z_j^{(n+1)}) = 0, \quad j = 1,\ldots,r$$

and so the numerical approximation lies on the manifold specified by $g(y,z) = 0$.

Thus the following definition can be given.

Definition 3.7.2. *A multivalue method is* **stiffly accurate** *if (3.7.7) holds.*

Note that this definition is entirely consistent with that for linear multistep and Runge-Kutta methods. Using this definition of stiff accuracy, Chartier (1993) has been able to prove a number of results on the global order of multivalue methods as applied to index-one and index-two problems. In particular, Chartier (1993) has shown:

Theorem 3.7.3. *If a zero-stable, multivalue method with consistent initial conditions and appropriate starting procedures, order w and stage order q is applied to an index-one problem satisfying the assumptions described earlier, then the global order will be w in the y-component and r in the z-component (whenever $nh \leq C$) with*

$r = w$, if the method is stiffly accurate.
$r = \min(w, q+1)$, if the method is stable at infinity and $1 \notin \sigma(M(\infty))$.
$r = \min(w-1, q)$, if the method is stable at infinity and $1 \in \sigma(M(\infty))$.

Theorem 3.7.4. *If a zero-stable, stiffly accurate multivalue method with consistent initial conditions and order $w \geq 2$ and stage order $q \geq 1$ is applied to an index-two problem satisfying the assumptions described earlier, then the global order will be* $\min(w, q+1)$ *in both the y-component and the z-component whenever $nh \leq C$.*

3.8 Variable-stepsize implementations

In order to be efficient, any implementation of a numerical method must allow for variable stepsizes. One way of doing this is to try to estimate the local error at each step and control it by taking a stepsize such that the local error is less than some prescribed tolerance. The hope is that the global error will also be controlled.

There are a number of techniques for estimating the local truncation error and hence providing a means for a variable-stepsize implementation. Just two will be discussed here: extrapolation and embedding. Although these approaches were originally developed for linear multistep and Runge-Kutta methods first, they can be understood from the multivalue method framework as will now be shown.

From (3.2.4) and (3.2.5) the local truncation error associated with a multivalue method of order w is given by

$$l_{n+1} = z_{n+1} - y^{(n+1)} = \psi(z_n)h^{w+1} + O(h^{w+2}), \tag{3.8.1}$$

where $z_n = z(x_n, h)$ and $z(x, h)$ is the correct value function and $\psi(z_n)$ is a vector function of elementary differentials of order $w+1$ evaluated at z_n.

If the same method is repeated with stepsize of $2h$ commencing from x_{n-1} then

$$\begin{aligned} \hat{l}_{n+1} &= z_{n+1} - \hat{y}^{n+1} = \psi(z_{n-1})(2h)^{w+1} + O(h^{w+2}) \\ &= \psi(z_n)(2h)^{w+1} + O(h^{w+2}), \end{aligned} \tag{3.8.2}$$

on expanding z_{n-1} about x_n.

Subtracting (3.8.2) from (3.8.1) gives

$$T_n = \hat{y}^{n+1} - y^{n+1} = \psi(z_n)(1 - 2^{w+1})h^{w+1} + O(h^{w+2}) \tag{3.8.3}$$

so that

$$P_n = \frac{\| T_n \|}{2^{w+1} - 1} \tag{3.8.4}$$

is an estimate of the principal local truncation error.

However, it is important to understand at this stage that not all the r components of a multivalue method necessarily contribute to the local error in the same way. This is readily seen in the case of k-step linear multistep methods in which $r = k$ but for which there is only one update component. In fact only those components which have a unit eigenvalue associated with A_2 contribute to the local error. Consequently, it is the spectral projector E of $A_2 - I$ satisfying

$$E^2 = E, \quad E(A_2 - I) = 0$$

which is significant. Note that since A_2 has, from consistency, always at least one unit eigenvalue, E is always at least rank 1. In the case that there is only one unit eigenvalue, E can be interpreted as a left eigenvalue of $A_2 - I$. Thus T_n should be replaced by $(E \otimes I)T_n$ in (3.8.4) to obtain a true estimate of the principal local truncation error.

Consequently, if TOL is the desired local error tolerance, then a new stepsize kh for the next step can be chosen with

$$k = c\left(\frac{(2^{w+1} - 1)\text{TOL}}{\| (E \otimes I)T_n \|}\right)^{\frac{1}{w+1}}, \tag{3.8.5}$$

where restrictions may be placed on k which limit the amount by which the stepsize can be changed in any given step and where c is a safety factor which is often set to be 0.9.

One of the drawbacks of Richardson extrapolation is that it is relatively expensive to implement. One way of overcoming this is by the process of embedding in which an s-stage, r-value multivalue method of order w is embedded in an $(s + 1)$-stage r-value multivalue method of order $w + 1$. This technique was first considered by Sarafyan (1966), Fehlberg (1968) and England (1969) for Runge-Kutta methods (see section 3.4).

In the case of a multivalue embedding the first s rows of the matrices A_1 and B_1 should be the same for the embedded method and the error estimating method, while the matrix A_2 should be the same for both. Only the update matrices associated with the derivative components are allowed to vary. Thus if $Y = (Y_1^\top, \ldots, Y_s^\top)^\top$ and $\hat{Y} = (Y^\top, Y_{s+1}^\top)^\top$ represent the internal components and B_2 and $(\hat{B}_2, b)$ represent the $r \times s$ and $r \times (s+1)$ update matrices for the derivative components, then T_n in (3.8.3) can be written as

$$T_n = h(((B_2 - \hat{B}_2) \otimes I)F(Y) - (b \otimes I)f(Y_{s+1})). \tag{3.8.6}$$

This approach particularizes to the standard embedding approach described in section 3.4 for Runge-Kutta methods and to predictor-corrector implementations.

In the case of a predictor-corrector implementation in variable-stepsize mode, the local truncation error can be controlled by Milne's device (see Lambert (1991), for example). The advantage for linear multistep methods is that the local error no longer consists of complicated expressions involving a myriad of elementary derivatives but rather terms involving high-order derivatives of y.

Thus, for example, if both a predictor and corrector have order p then the principal local truncation errors of the predictor and corrector, assuming the same initial values, are given by

$$\begin{aligned} y(x_{n+1}) - y_{n+1}^{(0)} &= C_{p+1}^* h^{p+1} y^{(p+1)}(x_{n+1}) + O(h^{p+2}) \\ y(x_{n+1}) - y_{n+1}^{(r)} &= C_{p+1} h^{p+1} y^{(p+1)}(x_{n+1}) + O(h^{p+2}) \end{aligned}$$

where r is the rth correction in $P(EC)^r E$ mode. Then, rearranging terms, an estimate for the local truncation error is

$$T = C_{p+1} \left(\frac{y_{n+1}^{(r)} - y_{n+1}^{(0)}}{C_{p+1}^* - C_{p+1}} \right).$$

It has been noted that the stepsize strategy based on (3.8.5) can often lead to highly oscillatory behaviour in the stepsize with frequent step rejections. In particular, Gustaffson et al. (1988) have used control theory concepts to estimate kh for the next Runge-Kutta step based on

$$k = c \left(\frac{\text{TOL}}{\| T_n \|} \right)^{\frac{1}{w+1}} \left(\frac{\| T_{n-1} \|}{\text{TOL}} \right)^p. \tag{3.8.7}$$

Appropriate values for p can be found by using a scheme of Shampine and Watts (1971) or an approach described in Higham and Hall (1990) who attempt to maximize stability intervals which lead to stable stepsize control mechanisms. In this approach Higham and Hall (1990) analyse the stability of certain fixed-points of a dynamical system of dimension two.

This concludes the presentation of material on sequential methods for ordinary differential equations. An attempt has been made to show that many numerical methods can be studied from the general framework of multivalue methods. In many cases it has only been through very recent work (Burrage (1988), Hundsdorfer (1991), Chartier (1993), Schneider (1993) for example) that this approach has been possible. However, the various properties of multivalue methods are now sufficiently well understood to make this approach a feasible and informative one.

4

PARALLEL LINEAR ALGEBRA

Perhaps one of the most crucial aspects in the development of parallel numerical methods for the solution of both initial value and boundary value problems is the incorporation of efficient parallel linear solution algorithms. These linear algebra aspects arise at the most fundamental of levels, when performing matrix-vector or matrix-matrix computations. They also arise when implicit methods are implemented, because the ensuing nonlinear equations are usually solved by a modified Newton technique which results in systems of linear equations. Thus a thorough understanding of the many facets associated with the solution of systems of equations is crucial to the reading of this monograph. Since only one chapter is devoted to these aspects, such a discussion cannot, by its very nature, be comprehensive and the reader should consult other books on this subject, such as Golub and van Loan (1989) and Dongarra et al. (1991) for a more thorough overview of numerical linear algebra.

In this chapter both general parallel algorithms for solving linear systems and specific algorithms for certain structured problems, such as banded and block-banded problems, will be considered, along with parallel direct and iterative methods for linear systems as well as multigrid techniques. This chapter will also discuss in depth some new work on solving linear systems based on deflation techniques presented in Burrage et al. (1994a, 1994b) and Erhel et al. (1994), which also have some direct applications to solving differential systems. In particular, this chapter will cover the following material:

- section 4.1: a brief introduction to how parallel computing can be exploited when solving ODEs with particular reference to the parallel solution of linear systems of equations;
- sections 4.2: the application of cyclic reduction techniques to linear systems with special structures;
- section 4.3: an introduction to the concept of tearing techniques via Wang's algorithm;
- section 4.4: direct techniques for solving dense, banded and sparse linear systems;

- section 4.5: iterative techniques and their associated convergence properties based on the concepts of M-matrices and general splittings;
- section 4.6: the use of multigrid techniques to accelerate the convergence of iterative methods;
- section 4.7: the use of deflation techniques to accelerate the convergence of iterative methods. This new technique works by approximating the eigenspace ($\mathbb{P}$) corresponding to the unstable or slowly converging modes using the iterates of the fixed-point iteration and then performing a coupled iteration process with Newton iteration on $\mathbb{P}$ and fixed-point iteration on $\mathbb{Q}$ (the orthogonal complement of $\mathbb{P}$);
- section 4.8: an introduction to Krylov subspace methods and, in particular, conjugate gradients and GMRES methods, with an extension to the concept of variable preconditioning (Erhel et al. (1994));
- section 4.9: a discussion on how iterative techniques can be parallelized including multicolouring, asynchronous methods and multisplitting algorithms.

4.1 Approaches to parallel programming

Parallel algorithms for solving scientific problems can be developed in several ways. One approach is to develop new and novel algorithms from scratch and to adapt to a particular architecture and memory management environment. A second approach is to focus on novel parallel development in a widely available and cost-effective environment such as a network of transputers. Another approach is that epitomized by LAPACK (Anderson et al. (1990)) in which existing software is used as building blocks. It is unlikely that this third approach is optimal in terms of performance but it is no doubt the simplest. In addition, the approach adopted is architecture dependent. For example, in the case of vector machines there are compilers which attempt to parallelize codes based on an appropriate ordering of the loop variables (called inner product, middle product and outer product orderings (see Ortega (1988), for example)) and this can lead to many different ways for developing parallel linear algebra routines.

As a specific example consider the numerical solution of a large, stiff IVP by some implicit method embedded in a sequential code. For such a problem there are two main areas where parallelism can be exploited:

(i) in the partitioning of the problem;

(ii) in the solution of the nonlinear equations by a linearization technique (such as the modified Newton method).

There are, of course, other areas where parallelism within an IVP solver can be exploited such as in the development of robust error control (see Enright and Higham (1990), for example), but such matters will be addressed later.

The partitioning can be solved by the development of standard communication protocols while (ii) could involve using some of the linear algebra routines (such as an *LU* factorization routine) from LAPACK which are written in terms of BLAS. Since, in this case, the data can be considered to be predistributed, there is likely to be an improved performance in using this approach than if data had to be distributed from the master processor. More recently, a scalable linear algebra library for distributed-memory concurrent computers has become available (SCALAPACK (Choi et al. (1992)).

Just as in the sequential case where there are different ODE codes depending on whether the problem to be solved is full, sparse or banded (ODEPACK has, for example, at least five different routines depending on the nature of the problem) so it is to be expected that a similar situation will exist in a parallel environment. In fact it seems likely that there will be an even greater multitude of parallel differential equation solvers as parallel techniques for solving full, sparse, banded, tridiagonal and block-banded systems can be quite disparate.

In addition to standard linear algebra packages available in LAPACK and SCALAPACK there are other standard basic parallel algorithms which are universally useful. These include direct techniques based on recursive doubling, cyclic reduction and partitioning as well as iterative techniques such as Krylov subspace techniques. It must also be remarked at this stage that the efficacy of a parallel algorithm cannot be judged only in terms of efficiency considerations, for parallelism often induces another complexity – namely numerical instability. Some parallel algorithms that have been proposed, while being seductively efficient, can fail to be stable.

As will be seen in Chapter 6 when solving linear initial value problems (LIVPs) and linear boundary value problems (LBVPs) by one-step methods a certain regular structure in the ensuing linear systems of equations arises. This structure is often block-bidiagonal or almost block-bidiagonal or block-tridiagonal. Thus many of the linear algebra techniques discussed in this chapter will focus on these structures.

4.2 Cyclic reduction

The technique of cyclic reduction was first applied to the solution of tridiagonal linear systems (see Lambiotte and Voight (1974), for example) and then extended to block-tridiagonal linear systems (Heller (1978)). Consider now the block-bidiagonal system ($Qy = b$)

$$\begin{pmatrix} I & & & \\ P_2 & Q_2 & & \\ & \ddots & \ddots & \\ & & P_N & Q_N \end{pmatrix} \begin{pmatrix} y_1 \\ y_2 \\ \vdots \\ y_N \end{pmatrix} = \begin{pmatrix} b_1 \\ \vdots \\ b_N \end{pmatrix} \tag{4.2.1}$$

of dimension mN.

In the simplest case, when $m = 1$, the maximum number of processors that can be gainfully employed when using cyclic reduction is $\frac{N}{2}$ and the system can be solved in $O(\log_2 N)$ parallel stages. If the system is normalized so that the diagonal elements are unity (this is only done if $m = 1$) then the first step of the cyclic reduction technique involves eliminating y_1 from the second equation, y_3 from the fourth, and so on. The process can then be repeated on the second, fourth, sixth equations, and so on.

In order to illustrate this process, consider the following example of order 6 where the vector b will be appended to the matrix Q.

$$\begin{pmatrix} 1 & & & & & \\ p_2 & 1 & & & & \\ & p_3 & 1 & & & \\ & & p_4 & 1 & & \\ & & & p_5 & 1 & \\ & & & & p_6 & 1 \end{pmatrix} \begin{pmatrix} y_1 \\ y_2 \\ y_3 \\ y_4 \\ y_5 \\ y_6 \end{pmatrix} = \begin{pmatrix} b_1 \\ b_2 \\ b_3 \\ b_4 \\ b_5 \\ b_6 \end{pmatrix}.$$

The first step of the cyclic reduction process eliminates the coupling between the odd and even rows and can be performed in parallel using $\frac{N}{2}$ processors (in this case 3) to give

$$\begin{pmatrix} 1 & & & & & & b_1 \\ 0 & 1 & & & & & b_2 - p_2 b_1 \\ & p_3 & 1 & & & & b_3 \\ & -p_3 p_4 & 0 & 1 & & & b_4 - p_4 b_3 \\ & & & p_5 & 1 & & b_5 \\ & & & -p_5 p_6 & 0 & 1 & b_6 - p_6 b_5 \end{pmatrix}.$$

The second stage of the reduction eliminates the coupling between all strict multiples of 2 (2, 6, 10, etc.) and multiples of 4. This would give in this example

$$\begin{pmatrix} 1 & & & & & & b_1 \\ 0 & 1 & & & & & b_2 - p_2 b_1 \\ 0 & p_3 & 1 & & & & b_3 \\ 0 & -p_3 p_4 & 0 & 1 & & & b_5 - p_5 b_4' \\ 0 & -p_5 p_4' & 0 & 0 & 1 & & b_5 \\ 0 & -p_4' p_6' & 0 & 0 & 0 & 1 & b_6' - p_6' b_4' \end{pmatrix},$$

where

$$b'_4 = b_4 - p_4 b_3, \quad p'_4 = -p_3 p_4, \quad p'_6 = -p_5 p_6, \quad b'_6 = b_6 - p_6 b_5.$$

The final step consists of a simple solve since y_2 is now known explicitly.

The case in which $m = 1$ masks some nontrivial behaviour when the matrix Q is block-bidiagonal. Consequently, a discussion for the case $m > 1$ will also be given. The first important point to note is that the system should not be normalized so that the diagonal blocks become the identity matrices. Rather it is assumed that the factorizations of the inverses of the appropriate Q_i matrices are all performed at the outset. Thus consider the previous example but with scalars replaced by blocks

$$\begin{pmatrix} Q_1 & & & & & \\ P_2 & Q_2 & & & & \\ & P_3 & Q_3 & & & \\ & & P_4 & Q_4 & & \\ & & & P_5 & Q_5 & \\ & & & & P_6 & Q_6 \end{pmatrix}.$$

The first step is now equivalent to performing Gaussian elimination to remove the blocks P_2, P_4 and P_6 and gives

$$\begin{pmatrix} Q_1 & & & & & \\ 0 & Q_2 & & & & \\ & P_3 & Q_3 & & & \\ & X_1 & 0 & Q_4 & & \\ & & & P_5 & Q_5 & \\ & & & X_2 & 0 & Q_6 \end{pmatrix}$$

$$X_1 = -P_4 Q_3^{-1} P_3, \quad X_2 = -P_6 Q_5^{-1} P_5;$$

while the second step gives

$$\begin{pmatrix} Q_1 & & & & & \\ 0 & Q_2 & & & & \\ & P_3 & Q_3 & & & \\ & X_1 & 0 & Q_4 & & \\ & X_3 & 0 & 0 & Q_5 & \\ & X_4 & 0 & 0 & 0 & Q_6 \end{pmatrix}$$

$$X_3 = -P_5 Q_4^{-1} X_1, \quad X_4 = -X_2 Q_4^{-1} X_1.$$

Of course one of the difficulties with cyclic reduction is that as the algorithm proceeds, fewer processors are fully utilized. Furthermore, if only

p processors are available, it is not clear how the work should be decomposed. Addison in a private communication suggests a general scheme in which the system can be partitioned into p independent main blocks which are coupled together by p separator equations in which the last block of equations is included with the other separators. The system is then solved in three phases:

(i) within each block the subdiagonal blocks are eliminated in parallel, leaving fill-in around the separator equations;

(ii) the separator equations are solved;

(iii) back substitution.

This technique is essentially a modification of the approach of Chen et al. (1978) for the parallel solution of banded triangular systems and of Wang (1981) for the parallel solution of tridiagonal systems. Wang's method is called the partition method and has been modified by van der Vorst and Dekker (1989) for bidiagonal systems.

4.3 Partitioning

The concept of partitioning or tearing is a very important one when solving certain types of large stiff differential equation problems which have a block-banded structure. The presentation given here is based on Wang's partitioning for tridiagonal systems but applies equally well to block-tridiagonal systems.

Given p processors and a tridiagonal system of dimension $n = pk$ then Q is partitioned into a $p \times p$ block-tridiagonal matrix in which each block is of order $k = n|p$. Each diagonal block is tridiagonal while each subdiagonal and superdiagonal block is the null matrix except for one nonzero element. Thus, for example, consider the case in which $n = 12, k = 3$ and $p = 4$.

$$\begin{pmatrix} a_1 & d_1 & & & & & & & & & & \\ c_2 & a_2 & d_2 & & & & & & & & & \\ & c_3 & a_3 & d_3 & & & & & & & & \\ & & c_4 & a_4 & d_4 & & & & & & & \\ & & & c_5 & a_5 & d_5 & & & & & & \\ & & & & c_6 & a_6 & d_6 & & & & & \\ & & & & & c_7 & a_7 & d_7 & & & & \\ & & & & & & c_8 & a_8 & d_8 & & & \\ & & & & & & & c_9 & a_9 & d_9 & & \\ & & & & & & & & c_{10} & a_{10} & d_{10} & \\ & & & & & & & & & c_{11} & a_{11} & d_{11} \\ & & & & & & & & & & c_{12} & a_{12} \end{pmatrix} . \qquad (4.3.1)$$

The system is now solved as outlined below, where Q after the second step is given by

$$
\begin{pmatrix}
a_1 & & q_1 & & & & & & & & & \\
& a_2 & d_2 & & & & & & & & & \\
& & a_3 & & & q_3 & & & & & & \\
& & c_4 & a_4 & & q_2 & & & & & & \\
& & p_1 & & a_5 & d_5 & & & & & & \\
& & p_2 & & & a_6 & & & q_5 & & & \\
& & & & & c_7 & a_7 & & q_4 & & & \\
& & & & & p_3 & & a_8 & d_7 & & & \\
& & & & & p_4 & & & a_9 & & & q_7 \\
& & & & & & & & c_{10} & a_{10} & & q_6 \\
& & & & & & & & p_5 & & a_{11} & d_{11} \\
& & & & & & & & p_6 & & & a_{12}
\end{pmatrix}.
$$

(i) Row transformations are applied on the p diagonal blocks simultaneously removing the subdiagonal elements but creating a fill-in in the last column of each subdiagonal block.

(ii) A similar process is applied to remove the superdiagonal elements within the diagonal blocks, creating a fill-in in the last column of each diagonal block and one nonzero element in the bottom right-hand corner of the superdiagonal block.

(iii) This matrix can now be triangularized by eliminating the subdiagonal block and then diagonalized by the elimination of the superdiagonal block, without any additional fill-in.

Wang (1981) notes that this technique is stable for diagonally dominant problems and that pivoting is not easily incorporated into the algorithm.

4.4 Direct methods

Full systems

A general numerical algorithm for solving the problem

$$Qy = b, \quad y \in \mathbb{R}^m, \quad b \in \mathbb{R}^m \tag{4.4.1}$$

is LU factorization, in which a unit lower-triangular matrix L (with ones on the diagonal), an upper-triangular matrix U and a permutation matrix P are found such that

$$PQ = LU. \tag{4.4.2}$$

In this case the solution of (4.4.2) can be found by a forward and back substitution of the triangular systems

$$Lz = Pb, \quad Uy = z. \tag{4.4.3}$$

The factorization given in (4.4.2) can be constructed by Gaussian elimination with partial pivoting, in which the element of largest magnitude in each column is used as the pivot for that particular column within the Gaussian elimination stage. For general systems of the form (4.4.1) this pivoting is necessary in order to produce stable algorithms, but in the case of problems which have certain structures, pivoting may not be necessary (as is the case for matrices which are both **symmetric** ($A = A^\top$) and **positive definite** ($x^\top Ax > 0, \forall x \neq 0$)).

There have been many attempts to adapt LU factorization to various parallel architectures (see, for example, Saad (1986) who has considered *LU* algorithms for bus and ring topologies with distributed memory). This adaptation can be based on a fine-grain or coarse-grain approach.

In particular, Ortega (1988) has considered a fine-grain algorithm based on a data flow approach which is suitable for a large SIMD machine with m^2 processors arrayed in a square which can compute and communicate simultaneously. At each step one more multiplier is computed and communicated to the right, with the matrix elements updating when they have the necessary data. The computation takes the form of a diagonal wavefront and Ortega (1988) demonstrates a speed-up of $O(m^2)$ if the number of processors is large. He also suggests a lumping of elements to reduce the number of processors when m is large.

A medium-grain approach can be adopted by using the rows or columns as the basic building blocks. In this case the ordering is important, so that if $m = kp$ then the blocked approach places the first k rows on processor one, the second k rows on processor two, etc, while the interleaved approach places rows $k \times p + i$, $k = 0, 1, \ldots,$ on processor i. This latter approach is a more suitable one in terms of reducing idle time.

As Bisseling and van der Vorst (1989) have remarked, the efficiency of parallel LU decomposition algorithms based on row or column oriented algorithms decreases rapidly with increasing p because of increasing communication costs relative to computation costs. Fox et al. (1988) and Bisseling and van der Vorst (1989) have achieved better performance by distributing rows and columns cyclically to the processors which gives a more efficient load balancing and lower communication complexity.

These approaches are particular instances of a more general blocked algorithm approach which produces matrix-matrix algorithms rather than matrix-vector algorithms (blocksize of 1). The block approach introduces parallelism at the Level 3 BLAS layer which offers greater scope than the

Level 1 or 2 BLAS for exploiting parallelism. However, the optimum value for the blocksize is very much architecture dependent.

A different approach from LU factorization is to develop factorizations which can exploit parallelism, such as $Q = WZ$ where

$$W = \begin{pmatrix} 1 & & & & 0 \\ & \ddots & 0 & \iddots & \\ & X & 1 & X & \\ & \iddots & 0 & \ddots & \\ 0 & & & & 1 \end{pmatrix}, \quad Z = \begin{pmatrix} X & \cdots & \cdots & \cdots & X \\ & \ddots & X & \iddots & \\ & 0 & \ddots & 0 & \\ & \iddots & X & \ddots & \\ X & \cdots & \cdots & \cdots & X \end{pmatrix}.$$

Here X and 0 denote either nonzero or zero elements, respectively, or alternatively, areas of nonzero and zero elements separated by partitioning lines (see Kaps and Schlegl (1987), for example).

Another factorization ($Q = MN$) is called the twisted factorization, where

$$M = \begin{pmatrix} X & X & & & & \\ & \ddots & \ddots & & & \\ & & X & X & & \\ & & & X & & \\ & & & X & X & \\ & & & & \ddots & \ddots \\ & & & & & X \quad X \end{pmatrix},$$

$$N = \begin{pmatrix} X & & & & & \\ X & X & & & & \\ & \ddots & \ddots & & & \\ & & X & X & X & \\ & & & & \ddots & \ddots \\ & & & & & X \quad X \\ & & & & & \quad\; X \end{pmatrix},$$

first proposed in Joubert and Cloeth (1984) who showed that the twisted factorization exists if Q is symmetric and positive definite, or diagonally dominant or an M-matrix. The advantage of this approach is that the decomposition and back substitutions can be done in parallel.

Banded systems

Banded systems can often arise when some modelling process has a local property. The problem (4.4.1) is said to have **semi-bandwidth** θ and bandwidth $2\theta - 1$ if

$$q_{ij} = 0, \quad |j - i| \geq \theta.$$

For symmetric banded problems with semi-bandwidth θ, parallelism can have a significant effect at the very lowest of levels – such as storage. In the case of serial machines, a banded matrix is usually stored by diagonals but this is not necessarily the right approach on a vector machine and Ortega (1988) suggests storing the nonzero elements row-wise in a single vector.

Ortega (1988) suggests a generalization of full LU decomposition to banded systems. At the first step multipliers of the first row are subtracted from rows 2 to $\theta + 1$. At the second step only elements from rows and columns 2 to $\theta + 2$ are involved. This is continued until a full submatrix of order θ remains and this is decomposed by the full matrix approach. Unfortunately, if all of the processors are to be used then necessarily $\theta \geq p$ and even then it is difficult to get an efficient load balancing. Thus this approach is particularly inefficient if θ is small and so many specialized algorithms have been developed for problems with small bandwidth (such as Wang's algorithm (1981)).

It is also worth noting here that invariably parallel algorithms require substantially more work than traditional sequential methods running on one processor. For example, Bondeli (1989) gives the following sequential operation count for the tridiagonal problem:

$$\begin{aligned} \text{Gaussian elimination} &: \quad 8m - 1 \\ \text{Cyclic reduction} &: \quad 17m - 12p - 4 \quad (m = 2^p - 1) \\ \text{Wang's algorithm} &: \quad 21m - 12p - 8k - 4 \quad (m = pk). \end{aligned}$$

Dongarra and Johnsson (1987) have analysed parallel algorithms for the solution of large, dense, banded systems on a variety of architectures and networks. The characteristic of these algorithms is that

(i) each partition is factored independently;

(ii) these factors are used to decouple the solution, but this produces fill-in;

(iii) a reduced decoupled system is solved by some parallel technique such as cyclic reduction and this stage involves some form of nearest neighbour communication between processors;

(iv) the complete solution is formed by back substitution.

Dongarra and Johnsson (1987) note that in the case of problems with a small bandwidth the potential for parallelism is via the independence of the operations in eliminating the different variables. However the parallel efficiency of this process is usually less than one because issues arising from communication, storage, and routing conflicts cannot be ignored. On the other hand for problems with a relatively large bandwidth, potential parallelism is via the independence of the operations in eliminating a single variable and this can have a parallel efficiency of one.

In another paper Dongarra and Sameh (1984) consider a parallel algorithm for solving block-tridiagonal systems where each of the diagonal blocks is assumed to have a diagonal dominance property. This process consists of four stages assuming p diagonal blocks of order q and p processors.

(i) The LU factors of each of the p diagonal blocks are found in parallel. This requires no interprocessor communication.

(ii) The system is converted to a form in which the diagonal blocks become the identity matrices. This can again be performed with no processor communication by the solution of $3p$ systems of linear equations in LU factored form.

(iii) The system is partitioned. If θ represents the half-bandwidth of the system then the θ equations on either side of each of the $p-1$ partitioning lines form a diagonally dominant system of dimension $2\theta(p-1)$. This solution step requires interprocessor communication.

(iv) Back substitutions for the remaining components.

Similar techniques can be applied when a reordering of the unknowns is performed. Thus consider, for example, the problem given in (4.4.1), where $Q = (-1, 2, -1)$. This system can be partitioned into p blocks of order q and a set of $p-1$ separator points (one between each block) if $m = pq + p - 1$. If the variables are then reordered so that the separator points are ordered last then the ensuing system can be written as

$$\hat{Q}\hat{y} = \hat{b}, \tag{4.4.4}$$

where

$$\hat{Q} = \begin{pmatrix} Q_1 & & & P_1 \\ & \ddots & & \vdots \\ & & Q_p & P_p \\ P_1^{\mathrm{T}} & \dots & P_p^{\mathrm{T}} & \tilde{Q} \end{pmatrix}. \tag{4.4.5}$$

Here the Q_i are tridiagonal, $\tilde{Q}$ is diagonal and the $P_i \in \mathbb{R}^{q\times(p-1)}$ and $\hat{Q}$ is said to be an **arrowhead matrix**. (4.4.5) can now be solved in a similar way to the decomposition techniques outlined previously based on an initial LU factorization of the Q_i.

Sparse systems

In the case that Q in (4.4.1) is large and sparse, different efficiency and implementation considerations arise. In particular, storage may be a limiting factor and the data handling problem is significant with floating point calculations possibly forming only a small proportion of the overall algorithm. These differences between the sparse and the dense cases have significant ramifications when parallelism is considered.

Dongarra et al. (1991) consider three different classes of direct methods for sparse systems: general, frontal and multifrontal techniques (see also Duff et al. (1986)). These are now briefly described.

An important factor when constructing algorithms for sparse systems is that the LU factors of Q will generally be denser than the original matrix Q, due to the appearance of fill-in. In addition the effect of ordering of the components in (4.4.1) can have a dramatic effect on sparsity preservation. However, Markowitz (1957) suggested an adaptive strategy for maintaining the sparsity within Gaussian elimination by choosing as the pivot the element of the remaining reduced submatrix which minimizes the product of the number of nonzero elements in the associated row and column. This approach has to be adapted to preserve numerical stability.

In this general approach, sparse data structures are used throughout the algorithm and this has significant impact on vectorization and parallelization as Dongarra et al. (1991) discuss in some detail. These data structures allow the matrix to be stored as a collection of packed vectors which involves storing a pointer to its start and the number of entries. The reason for this approach, of course, is to reduce the $O(m^2)$ flops at each Gaussian elimination step to $O(\tau) + O(m)$ for a matrix of order m with τ nonzeros.

Parallelism can be exploited by partitioning and the use of block methods and by performing full Level 3 BLAS at the innermost loops. In the case of symmetric problems, sparse techniques can be implemented based only on a diagonal pivoting strategy which avoids the difficulty of explicitly updating the sparsity pattern at each Gaussian elimination stage. This scheme is known as the **minimum degree algorithm** and was introduced by Tinney and Walker (1967).

In the case of sparse problems with a special structure, such as banded problems, frontal methods (and their generalization to multifrontal methods) can be used (see Dongarra et al. (1991), for example). Thus in the case of a banded problem with semi-bandwidth θ, a frontal matrix of dimension $\theta \times (2\theta - 1)$ is held in main storage and continually updated as the elimination progresses. By allowing for larger frontal matrices it may even be possible to perform several pivot steps within the same frontal matrix. Since in this case all elimination operations are performed within a full matrix, standard techniques for vectorization and blocking parallel techniques can be utilized. Unfortunately, unless θ is small, frontal methods can require more storage and many more floating point operations than a general sparse scheme based on Gaussian elimination. It is for this reason that multifrontal methods (see Duff et al. (1986)) have been introduced.

In the case of multifrontal methods, an elimination tree is used to define a precedence order within the factorization in which the factorization starts at the leaves of the tree and proceeds towards the root. This approach allows for considerable flexibility in the parallel environment since, for example, any computation at a leaf node can take place simultaneously while computation at nodes not on the same direct path from the root to a leaf are independent. Some of these parallel implementational issues have been explored by Duff (1986).

4.5 Stationary iterative techniques

A very important class of initial value problems arises from the semi-discretization of the spatial variables in a time-dependent parabolic partial differential equation. The ensuing IVPs are usually characterized as being large, stiff and with eigenvalues widely scattered along the negative real axis. Furthermore, the Jacobian usually has a block-banded structure with the size of the block bandwidth dependent on the dimension of the original partial differential equation.

Thus consider the solution of the r-dimensional inhomogeneous diffusion convection problem

$$u_t(x,t) = D\nabla^2 u(x,t) - Cu_x(x,t) + f(x,t), \quad t > 0, \tag{4.5.1}$$

where $x \in \mathbb{R}^r$ and ∇^2 denotes the elliptic operator. Note that in the case $C = 0$, (4.5.1) is purely a diffusion problem and f can be considered as representing some internal heat source or sink. The one-dimensional case has been discussed in Example 2.3.1, but consider now the two-dimensional problem in which Ω is a square region $[0,(m+1)h] \times [0,(m+1)h]$; then if the problem is of Dirichlet type and a five-point stencil is used with ordering from left to right and bottom to top then a system of IVPs of the form

$$y'(x) = Qy(x) + g(x), \quad y \in \mathbb{R}^{m^2} \tag{4.5.2}$$

arises. Here Q is a block tridiagonal matrix of the form $\theta(I_m, T_m, I_m)$ where T_m is the tridiagonal matrix $(1,-4,1)$ and $\theta = D/h^2$.

In the case of a stationary (elliptic) problem, linear systems of the form

$$Qy = b, \quad Q = (I_m, T_m, I_m), \tag{4.5.3}$$

where $y = (y_{11}, \ldots, y_{1m}, \ldots, y_{m1}, \ldots, y_{mm})^{\mathrm{T}}$, have to be solved.

In order to take into account the fact that differing grid sizes may be used for the different spatial dimensions, and that there may be convective terms and mixtures of Dirichlet and Neumann boundary conditions, a more general form for Q in (4.5.3) is

$$\begin{pmatrix} Q_1 & P_1 & & \\ R_1 & Q_2 & \ddots & \\ & \ddots & \ddots & P_{N-1} \\ & & R_{N-1} & Q_N \end{pmatrix}. \tag{4.5.4}$$

Here Q is often symmetric and positive definite. Indeed Q may even have a stronger property related to the idea of diagonal dominance.

Definition 4.5.1. *A matrix $Q \in \mathbb{R}^{m\times m}$ is said to be, respectively,* **diagonally dominant** *(DD) or* **strictly diagonally dominant** *(SDD) if*

$$|q_{ii}| \;\geq\; \sum_{\substack{j=1\\ j\neq 1}}^{m} |q_{ij}|, \quad i = 1, \ldots, m$$

$$|q_{ii}| \;>\; \sum_{\substack{j=1\\ j\neq 1}}^{m} |q_{ij}|, \quad i = 1, \ldots, m.$$

Definition 4.5.2. *A matrix $Q \in \mathbb{R}^{m\times m}$ is said to be* **connectedly diagonally dominant** *(CDD) if for each $i \in \{1, \ldots, m\}$ either*

$$|q_{ii}| > \sum_{\substack{j=1\\ j\neq 1}}^{n} |q_{ij}| \tag{4.5.5}$$

or i is connected to some $k \in \{1, \ldots, m\}$ by a sequence of nonzero matrix entries $q_{ik_1}, q_{k_1k_2}, \ldots, q_{k_rk}$ (where $k_1, \ldots, k_r \in \{1, \ldots, m\}$) for which (4.5.5) holds.

The property of CDD was introduced by Varga (1962) to cover systems such as (4.5.3) which are not SDD but are CDD.

In order to obtain more general convergence results for schemes such as (4.5.4), it is necessary to impose additional structure on the matrix Q. This leads to the concept of H-matrices, M-matrices and consistently ordered matrices.

Definition 4.5.3. *A matrix Q is said to be* **nonnegative** $(Q \geq 0)$ *if and only if*

$$q_{ij} \geq 0, \quad \forall i, j.$$

Definition 4.5.4. *A real matrix Q is an* M**-matrix** *if it is nonsingular and*

$$\textit{diag}\,(Q) \geq 0, \quad Q^{-1} > 0.$$

An equivalent definition of an M-matrix (see Berman and Plemmons (1979)) is

Definition 4.5.5. *A real matrix $Q = dI - C$ is an* M**-matrix** *if*

$$d > 0, \quad C \geq 0, \quad \rho(C) \leq d.$$

Definition 4.5.6. *The* **comparison matrix**, C, *of a complex matrix Q is characterized by*

$$c_{ii} = |q_{ii}|, \quad c_{ij} = -|q_{ij}|, \quad i \neq j.$$

Definition 4.5.7. *A complex matrix Q is an* H**-matrix** *if its comparison matrix is an M-matrix.*

Definition 4.5.8. *An (M, N) splitting $Q = M - N$ is said to be an* M**-splitting** *if M is an M-matrix and $N \geq 0$.*

Definition 4.5.9. *An (M, N) splitting is said to be* **regular (weak regular)** *if $M^{-1} \geq 0$, $N \geq 0$, or $M^{-1} \geq 0$, $M^{-1}N \geq 0$, respectively.*

Definition 4.5.10. *A matrix Q of order m is* **consistently ordered** *if for some t there exist disjoint subsets $S_1, \ldots, S_t$ with $\cup_{i=1}^{t} S_i = \{1, \ldots, m\}$ in which if $i \in S_k$ and j are such that $|q_{ij}| + |q_{ji}| > 0$ then $j \in S_{k+1}$, if $j > i$ and $j \in S_{k-1}$ if $j < i$.*

An example of a consistently ordered matrix is a block-tridiagonal matrix whose diagonal blocks are themselves diagonal matrices (see Young and Gregory (1972), for example).

Large linear systems which possess the DD property can be efficiently solved by iterative techniques rather than by direct methods such as Gaussian elimination or LU factorization because there is no fill-in. The standard classical way to solve the general linear system of equations given by (4.4.1) is to split the system in such a way that

$$\omega Q = M - N, \tag{4.5.6}$$

where M is nonsingular and ω is some appropriate scalar value, and then perform the iterations

$$My_{k+1} = Ny_k + \omega b. \tag{4.5.7}$$

Clearly iteration schemes such as (4.5.7) will converge to the solution of (4.4.1) if

$$\rho(M^{-1}N) = \rho(I - \omega M^{-1}Q) < 1.$$

The matrix $H = M^{-1}N$ will henceforth be called the **iteration matrix** associated with (4.5.7).

Traditional iterative methods such as Jacobi, Gauss-Seidel, Richardson and SOR fit into this format with, respectively,

$$\begin{aligned} M &= D, \quad \omega = 1 \\ M &= D + \omega L, \quad \omega = 1 \\ M &= I \\ M &= D + \omega L, \quad 0 < \omega < 2, \end{aligned}$$

where D represents the diagonal part of Q and L the strictly lower-triangular part of Q.

It can be shown (Varga (1962)) that Jacobi, Gauss-Seidel and SOR will converge linearly for problems with the CDD property and that, in general, Gauss-Seidel converges approximately twice as fast as Jacobi. For problems of the form (4.5.4) which are symmetric and have the CDD property there are formulas for optimally choosing ω in SOR which dramatically improve the convergence behaviour over Gauss-Seidel. Alternatively, the ω in (4.5.7) can be allowed to vary from step to step, giving an iteration scheme of the form

$$My_{k+1} = (M - \omega_k Q)y_k + \omega_k b, \tag{4.5.8}$$

although in this case it is often difficult to determine appropriate values for the ω_k.

With these definitions it is now possible to prove two results (see Ortega (1988) and Frommer (1990), respectively).

Theorem 4.5.11. *If Q is SDD or CDD, with positive diagonal elements and negative nondiagonal elements then it is an M-matrix.*

Theorem 4.5.12. *If Q is an M-matrix then the Jacobi and Gauss-Seidel methods will converge.*

Theorem 4.5.13. *For real, symmetric, positive definite, consistently ordered matrices or consistently ordered matrices with positive constant diagonal elements*

$$\rho(G) = \rho(J)^2,$$

where $\rho(G)$ and $\rho(J)$ represent the spectral radii of the iteration matrices associated with the Gauss-Seidel and Jacobi schemes, respectively.

The following two results appear in Varga (1962) and will be of some significance when considering the use of overlapping subsystems with respect to waveform relaxation algorithms in Chapters 7, 8 and 9.

Theorem 4.5.14. *If $Q = M - N$ is a regular splitting of Q with $Q^{-1} \geq 0$ then $\rho(M^{-1}N) < 1$.*

Theorem 4.5.15. *Let $Q = M_1 - N_1 = M_2 - N_2$ be two regular splittings of Q where $Q^{-1} > 0$. Then if $N_2 \geq N_1 \geq 0$ with $N_2 \neq N_1$, $N_1 \neq 0$*

$$1 > \rho(M_2^{-1}N_2) > \rho(M_1^{-1}N_1) > 0.$$

It is also possible to perform more than one iteration per step, such as in the Peaceman-Rachford method which performs a double iteration per step (see Young and Gregory (1972), for example). This technique can be extended to the so-called generalized alternating direction methods of the form

$$\begin{aligned} M_1 y_{k+1/2} &= N_1 y_k + b \\ M_2 y_{k+1} &= N_2 y_{k+1/2} + b, \end{aligned}$$

where $Q = M_1 - N_1 = M_2 - N_2$ (see Conrad and Wallach (1979), for example). If

$$\begin{aligned} M_1 &= \alpha I + M, \quad N_1 = \alpha I - N \\ M_2 &= \alpha I + N, \quad N_2 = \alpha I - M \\ Q &= M + N \end{aligned}$$

and Q, M and N are symmetric, positive definite then Ortega (1988) has proved that the iterations will converge for any initial guess if $\alpha > 0$.

In the case of both the one and two dimensional elliptic problems, defined on a square, it is possible to derive exact expressions for $\rho(G)$, $\rho(J)$, $\rho(J_B)$ (line Jacobi method), $\rho(J(\omega))$ (weighted Jacobi method) and

$\rho(G(\omega))$ (SOR). Thus if m discretization points are used in each space direction with an equidistant mesh length of $h = \frac{1}{m+1}$ then it can be shown (see Varga (1962)) that

$$\begin{array}{rcccc} \rho(J) & = & \cos h\pi & \approx & 1-(\pi h)^2/2 \\ \rho(G) & = & (\cos h\pi)^2 & \approx & 1-(\pi h)^2 \\ \rho(J_B) & = & \frac{\cos h\pi}{2-\cos h\pi} & \approx & 1-(\pi h)^2 \\ \rho(J(\omega)) & = & 1-\omega+\omega\cos h\pi & \approx & 1-\omega(\pi h)^2/2 \\ \rho(G(\omega_{\mathrm{opt}})) & \approx & 1-2\pi h+O(h^2), \quad \omega_{\mathrm{opt}} & \approx & 2-2\pi h. \end{array} \tag{4.5.9}$$

A consequence of this is that as the spatial mesh becomes finer and finer, which is often necessary with parabolic and elliptic problems if greater accuracy is required, then the ensuing linear systems become increasingly more difficult to solve by traditional iterative schemes such as Jacobi and Gauss-Seidel schemes since they require many more iterations to guarantee satisfactory convergence. These schemes can be accelerated by techniques such as multigrid and deflation which are described in the next two sections.

4.6 Multigrid techniques

It was seen in (4.5.9) that if the weighted Jacobi method is applied to (4.4.1) with $Q = (1, -2, 1)$ then

$$\rho(J(\omega)) = 1-\omega+\omega\cos h\pi, \quad h = \frac{1}{m+1}. \tag{4.6.1}$$

It can also be shown that the eigenvalues and eigenvectors of $J(\omega)$ are given by

$$\begin{array}{rcl} \lambda_j & = & 1-\omega+\omega\cos jh\pi, \quad j=1,\ldots,m \\ v_j & = & (\sin jh\pi,\ldots,\sin jmh\pi)^{\mathsf{T}}, \quad j=1,\ldots,m. \end{array} \tag{4.6.2}$$

Let the modes associated with small values of j (in particular, $1 \le j < [\frac{m+2}{2}]$) be called the **smooth modes** and those associated with large values of j (in particular, $[\frac{m+2}{2}] \le j \le m$) be called the **oscillatory modes**; then it is clear from (4.6.1) and (4.6.2) that decreasing the grid spacing, h, only worsens the convergence of the smooth components of the error. Furthermore, if the convergence behaviour is monitored then it can be seen (see Briggs (1987), for example) that the errors initially decrease rapidly due to the quick elimination of the oscillatory components of the error but then become stalled by the presence of the smooth components which are left relatively unchanged by the iteration process. This behaviour is known as the **smoothing property**.

This effect is important in trying to determine an appropriate value for the relaxation parameter, ω. It can be seen from (4.6.2) that $\rho(J(\omega))$ will be minimized if $\omega = 1$ (given that the weighted Jacobi method converges). However, if ω is chosen to provide the best damping of the oscillatory components by setting

$$\lambda_{[\frac{m+2}{2}]} = \lambda_{m+1} := 2\omega - 1,$$

then an optimal value for ω is $\omega = 2/3$ when m is odd, in which case it can be shown (see Briggs (1987)) that the oscillatory components are reduced by a **smoothing factor** of at least $1/3$ after each relaxation. This factor is independent of the grid spacing h.

Now, let Ω_h denote the set of grid points based on a grid spacing of h, and Ω_{2h} denote the grid points of the coarse grid which is obtained from the fine grid by choosing the even-numbered grid points. For $k \in [1, [\frac{m+2}{2}])$ the kth mode on Ω_h is also the kth mode on Ω_{2h}, so that a mode becomes more oscillatory when moving from a fine grid to a coarse grid, or equivalently, smooth modes on a fine grid become less smooth on a coarse grid. This observation is the basis of multigrid techniques in which as iterations slow, due to the dominance of the smooth components, the relaxations can be moved to a coarser grid where these smooth components appear more oscillatory. This procedure can be continued by further coarsening and relaxing on the residual equation on each grid. Of course these results have to be returned to the original grid and this process leads to the concept of **restriction** and **interpolation** operators.

In the case of the one-dimensional elliptic problem, Briggs (1987) defines the interpolation operator I_{2h}^h which acts on coarse grid vectors to give fine grid vectors by

$$I_{2h}^h u^{2h} = u^h$$

where

$$\begin{aligned} u_{2j}^h &= u_j^{2h} \\ u_{2j+1}^h &= \frac{1}{2}(u_j^{2h} + u_{j+1}^{2h}). \end{aligned}$$

This represents a direct transfer of even-numbered points and an average of adjacent coarse grid values for odd-numbered points.

In a similar way, the restriction operator I_h^{2h} which acts on fine-grid vectors to give coarse-grid vectors can be defined by

$$I_h^{2h} u^h = u^{2h}$$

where

$$u_j^{2h} = u_{2j}^h,$$

or by a full weighting with

$$u_j^{2h} = \frac{1}{4}(u_{2j-1}^h + 2u_{2j}^h + u_{2j+1}^h).$$

There are natural generalizations to higher-dimensional elliptic problems.

It is possible to relax an arbitrary number of times at each level. The following is a compact recursive definition of a V-cycle implementation ($V_h(u_h, b_h)$) for solving

$$Q_h y_h = b_h$$

(see Briggs (1987)), given an initial guess of u_h.

ALGORITHM : $u_h \leftarrow V_h(u_h, b_h)$

1. Relax ν_1 times on $Q_h y_h = b_h$, $\quad y^{(0)} = u_h$;
2. **If** Ω_h = coarsest grid
   go to 3
   else
   $b_{2h} \leftarrow I_h^{2h}(b_h - Q_h u_h)$;
   $u_{2h} \leftarrow 0$;
   $u_{2h} \leftarrow V_{2h}(u_{2h}, b_{2h})$;
   endif
3. $u_h \leftarrow u_h + I_{2h}^h u_{2h}$;
4. Relax ν_2 times on $Q_h y_h = b_h$, $\quad y^{(0)} = u_h$.

There are many different implementations of the multigrid process based, for example, on V-cycles, W-cycles and FM-cycles. These can be represented by the following patterns on $\Omega_h, \Omega_{2h}, \Omega_{4h}, \Omega_{8h}$.

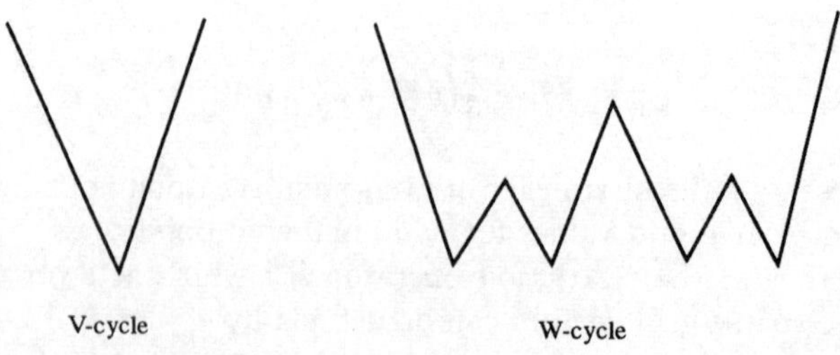

FIG. 4.1. V and W cycles over four levels

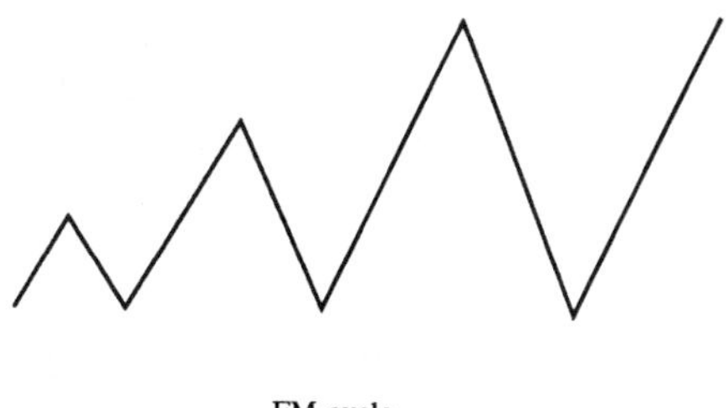

FIG. 4.2. full multigrid cycle

In fact multigrid techniques are particular elements of a more general class of level methods known as multilevel methods in which the concept of a computational grid has been replaced by more general computational levels (see Xu (1992), for example).

The study of the convergence properties of multigrid techniques is fairly technical and depends on the underlying relaxation algorithm (see Hackbusch and Trottenberg (1982), for example), and will not be given in any great detail here.

However, it can be shown (see Briggs (1987), for example) that the coarse grid corrector operator CG over two levels is given by

$$CG = (I - I_{2h}^{h}(I_{h}^{2h}Q_hI_{2h}^{h})^{-1}I_{h}^{2h}Q_h)H^{\nu} \tag{4.6.3}$$

assuming that ν relaxations take place on Ω_h with the scheme whose amplification matrix is H. Here it should be noted that I_{2h}^{h} and I_{h}^{2h} are linear operators from, respectively, $\mathbb{R}^{[\frac{m+2}{2}]-1}$ to $\mathbb{R}^m$ and $\mathbb{R}^m$ to $\mathbb{R}^{[\frac{m+2}{2}]-1}$.

Let $\{v_j\}$ be the modes of Q_h and consider the application of the CG operator, defined above, on the complementary modes v_j, v_{m+1-j}. Then it can be shown that

$$CG\begin{pmatrix} v_j \\ v_L \end{pmatrix} = \begin{pmatrix} \lambda_j^{\nu} & 0 \\ 0 & \lambda_L^{\nu} \end{pmatrix}\begin{pmatrix} s_j & s_j \\ c_j & c_j \end{pmatrix}\begin{pmatrix} v_j \\ v_L \end{pmatrix} \tag{4.6.4}$$

where

$$c_j = \cos^2\frac{j\pi}{2}h, \quad s_j = \sin^2\frac{j\pi}{2}h, \quad L = m + 1 - j,$$

and where it is assumed that the relaxation scheme with iteration matrix H has the same eigenvectors as Q_h, and that λ_j is an eigenvalue of H associated with v_j.

As Briggs (1987) points out, (4.6.4) implies considerable damping of both smooth and oscillatory error components since both λ_L^{ν} and s_j are small and provides the theoretical justification of the efficacy of the multigrid approach. This analysis will be continued further when multigrid waveform relaxation algorithms for differential systems are studied in Chapter 7.

4.7 Deflation techniques for systems of equations

Iterative methods for solving linear systems of equations can be very efficient in a sequential or parallel computing environment if the structure of the coefficient matrix can be exploited to accelerate the convergence of the iterative process. However, for classes of problems for which suitable preconditioners cannot be found or for which the iteration scheme does not converge or converges slowly, iterative techniques are inappropriate.

In order to overcome some of these difficulties associated with iterative schemes, a completely general technique for deflating those eigenvalues of the iteration matrix, which either slow or cause divergence, is presented. This process takes place automatically while the iterations are proceeding.

The process of accelerating the convergence of iterative methods by a deflation process which progressively extracts the largest eigenvalues (in magnitude) associated with the Jacobian of the problem has been studied by Jarausch and Mackens (1987) and Shroff and Keller (1993) for nonlinear systems. It was applied by Shroff and Keller (1993) to the numerical solution of nonlinear parameter-dependent problems of the form

$$y = F(y, \lambda), \quad F : \mathbb{R}^m \times \mathbb{R} \to \mathbb{R}^m$$

by a coupled iteration process which forces or accelerates the convergence of a fixed-point iteration scheme and represents an extension of the technique proposed by Jarausch and Mackens (1987) for solving symmetric nonlinear problems.

Jarausch (1993) has considered a different approach which uses singular subspaces for splitting the fixed-point equation associated with systems of parabolic partial differential equations. Jarausch (1993) claims that this approach is an efficient one since the systems are effectively decoupled by the construction of right singular subspaces associated with the Jacobian of the problem. In addition, the use of invariant subspaces is avoided by transforming the system by a so-called rotator matrix. This approach is called the "ideal normal equation approach" in that it avoids the squaring of the singular values by the usual approach of normalizing the equations. In this case the singular values of the transformed problem have the same singular values as the original problem. In spite of considerable applications of these projection techniques to nonlinear parameter-dependent problems, little appears to have been done in applying these techniques computationally to linear systems of equations. However, Burrage et al. (1994a, 1994b) and Erhel et al. (1994) have considered the applications of these techniques to linear systems.

A brief review of some of these techniques will now be given. The notation that will be used is the notation used by Shroff and Keller (1993) which is very similar to the notation used by Jarausch and Mackens (1987) and Jarausch (1993). Furthermore, only the techniques used by Shroff and Keller (1993) will be considered, although the approach of Jarausch (1993) also seems a fruitful one.

The approach of Shroff and Keller (1993), known as the **recursive projection method**, is based on the fact that divergence or slow convergence of the fixed-point iteration scheme

$$y^{(k+1)} = F(y^{(k)}, \lambda)$$

is due to the eigenvalues of F_{y^*} (the Jacobian of F evaluated at the fixed-point y^*) approaching or leaving the unit disk. The recursive projection method recursively approximates the eigenspace ($\mathbb{P}$) corresponding to the unstable or slowly converging modes using the iterates of the fixed-point iteration. A coupled iteration process takes place by performing Newton iteration on $\mathbb{P}$ and fixed-point iteration on $\mathbb{Q}$ (the orthogonal complement of $\mathbb{P}$) where fast convergence is assured. The scheme will be particularly effective if the dimension of $\mathbb{P}$ is small.

Assume now that the problem to be solved is parameter independent and can be written as

$$y = F(y), \quad y \in \mathbb{R}^m. \tag{4.7.1}$$

Defining

$$\left.\begin{array}{rcl} y & = & p+q, \quad p = Py, \quad q = Qy = (I-P)y \\ f(p,q) & = & PF(p+q), \quad g(p,q) = QF(p+q) \\ f_p(p,q) & = & PF_y(y), \end{array}\right\} \tag{4.7.2}$$

where P and Q are the orthogonal projections of $\mathbb{R}^m$ onto $\mathbb{P}$ and $\mathbb{Q}$, respectively, and letting

$$h(p,q) = p + (I - f_p(p,q))^{-1}(f(p,q) - p);$$

then the coupled iteration scheme of Shroff and Keller (1993) is given by

$$\left.\begin{array}{rcl} p^{(k+1)} & = & h(p^{(k)}, q^{(k)}), \quad k = 0, \ldots, N-1 \\ q^{(k+1)} & = & g(p^{(k)}, q^{(k)}), \quad k = 0, \ldots, N-1 \\ y & = & p^{(N)} + q^{(N)}. \end{array}\right\} \tag{4.7.3}$$

The overall iteration represents a Jacobi-type process in which the iterations are coupled by a Newton iteration and a fixed-point iteration.

Clearly r, the dimension of $f_p(p,q)$, should be kept as small as possible in order to minimize the linear algebra costs. Ultimately, however, the size of r depends on how quickly convergence takes place. Here it should be noted that the fixed-point solution (p^*,q^*) of (4.7.3) satisfies

$$f(p^*,q^*)=p^*, \quad g(p^*,q^*)=q^*, \quad h(p^*,q^*)=p^*,$$

while F^* will denote the Jacobian of F evaluated at p^*,q^*.

The projectors P and Q can be computed by observing that if $Z \in \mathbb{R}^{m\times r}$ is an orthonormal basis for $\mathbb{P}$ then

$$P = ZZ^\top, \quad Q = I - ZZ^\top, \quad Z^\top Z = I_r.$$

The matrix Z can be recursively updated by noting from (4.7.3) that

$$\begin{aligned}\Delta q^{(k)} &= q^{(k+1)} - q^{(k)} = g(p^{(k)},q^{(k)}) - g(p^{(k-1)},q^{(k-1)}) \\ &= g(p^{(k-1)} + \Delta p^{(k-1)}, q^{(k-1)} + \Delta q^{(k-1)}) - g(p^{(k-1)},q^{(k-1)}) \\ &= g_p^*\Delta p^{(k-1)} + g_q^*\Delta q^{(k-1)} + O(\varepsilon^2). \end{aligned} \tag{4.7.4}$$

Here ε represents terms that are hopefully negligible compared with the linear terms in the expansion and

$$\begin{aligned} g_p^* &= g_p(p^*,q^*) &= QF^*P \\ g_q^* &= g_q(p^*,q^*) &= QF^*Q. \end{aligned} \tag{4.7.5}$$

This implies that in the case of invariant subspaces $g_p^* = 0$ holds. Thus (4.7.4) and (4.7.5) imply that (4.7.4) is the power method

$$\Delta q^{(k)} = g_q^*\Delta q^{(k-1)} \tag{4.7.6}$$

and asymptotically $\{\Delta q^{(k)}\}$ will lie in the dominant eigenspace of g_q^* (assuming $\Delta q^{(0)}$ has a nonzero component in this direction). (4.7.6) can now be used to approximate the dominant eigenspace of g_q^* by forming a window of t difference vectors $S = \{\Delta q^{(k)}\}_{k-t+1}^{k}$ as the fixed-point iterations proceed and then computing an orthogonal basis U for span(S) by the modified Gram-Schmidt process. The eigenspace $B = U^\top F^* U$ is then formed and the eigenvectors along with the Schur vectors of B are computed. For efficiency reasons Shroff and Keller (1993) therefore suggest taking $t = 2$ in which case S is factored as $S = \hat{S}T$, where $\hat{S}$ is orthogonal of dimension $m \times 2$ and T is upper triangular of dimension 2.

If $T_{11} \geq 10^3 T_{22}$ then just the first column of $\hat{S}$ is appended to the basis Z. Alternatively, if convergence is deemed to be slow due to a complex conjugate pair the first two columns of $\hat{S}$ are appended to Z.

As more eigenvalues are removed the basis Z will become increasingly inaccurate due to the loss of orthogonality in the Gram-Schmidt process and $QF^*P \neq 0$. Hence Shroff and Keller (1993) suggest performing a subspace iteration on the columns of Z every so often. This takes the form

$$Z \to \text{orth}(F^*Z),$$

where orth(F^*Z) denotes computing an orthonormal basis for the columns of F^*Z by the Gram-Schmidt process. Of course F^* is not computed directly, but can be computed by

$$F^*Z_i \approx \frac{1}{\varepsilon}(F(y+\varepsilon Z_i) - F(y)), \quad i = 1, \ldots, r.$$

The convergence properties of this approach can be analysed by examining the Jacobian of (4.7.3) evaluated at (p^*, q^*). It is given by

$$J = \begin{pmatrix} h_p^* & h_q^* \\ g_p^* & g_q^* \end{pmatrix}, \tag{4.7.7}$$

where

$$\begin{aligned} g_p^* &= QF^*P, \quad g_q^* = QF^*Q, \quad h_p^* = 0 \\ f_p^* &= PF^*P, \quad f_q^* = PF^*Q, \quad h_q^* = (I - f_p^*)^{-1} f_q^*. \end{aligned} \tag{4.7.8}$$

If the orthonormal basis is computed exactly then $g_p^* = 0$ and

$$\sigma(J) = \{0, \sigma(g_q^*)\},$$

and so the convergence of (4.7.3) is governed by the spectral norm of g_q^* which is progressively made smaller by the deflation process described above. On the other hand it should be noted that if a modified Newton process is used to compute p then $h_p^* \neq 0$, but is nevertheless small. In fact up to order $O(\epsilon^2)$ in (4.7.4), the global error for (4.7.3) can be written as

$$e^{(k+1)} = Je^{(k)}, \quad e^{(k)} = (p^{(k)^\top} - p^\top, q^{(k)^\top} - p^\top)^\top, \tag{4.7.9}$$

so that the error behaviour is determined by the behaviour of the power method.

It is possible to modify the Jacobi-type iteration scheme described in (4.7.3) to produce Gauss-Seidel or SOR schemes. This generalization has been described in Burrage et al. (1994a). In the former case there are two possible iterations, a Gauss-Seidel and Reverse Gauss-Seidel iteration, of the form

$$\begin{aligned} p^{(k+1)} &= h(p^{(k)}, q^{(k)}) \\ q^{(k+1)} &= g(p^{(k+1)}, q^{(k)}) \end{aligned} \tag{4.7.10}$$

and

$$\begin{aligned} q^{(k+1)} &= g(p^{(k)}, q^{(k)}) \\ p^{(k+1)} &= h(p^{(k)}, q^{(k+1)}), \end{aligned} \tag{4.7.11}$$

respectively.

Clearly these two processes should be very similar since they both compute the same sequence but with different starting and finishing procedures, and this is borne out by the following analysis of the spectra of the Jacobians associated with these schemes.

It is easy to show that the Jacobians associated with (4.7.10) and (4.7.11), denoted by J_G and J_R, are given by

$$J_G = \begin{pmatrix} h_p^* & h_q^* \\ g_p^* h_p^* & g_p^* h_q^* + g_q^* \end{pmatrix}$$

and

$$J_R = \begin{pmatrix} g_q^* & g_p^* \\ h_q^* g_q^* & h_p^* + h_q^* g_p^* \end{pmatrix} = \begin{pmatrix} I \\ h_q^* \end{pmatrix} \begin{pmatrix} g_q^* & g_p^* \end{pmatrix} + \begin{pmatrix} 0 & 0 \\ 0 & h_p^* \end{pmatrix}.$$

Under the assumption that $h_p^* = 0$ (so that a Newton iteration is used for solving for p), the spectra of the associated Jacobian matrices are, respectively, given by

$$\sigma(J_G) = \{0, \sigma(g_q^* + g_p^* h_q^*)\} = \sigma(J_R) \tag{4.7.12}$$

while in the case of (4.7.3) the eigenvalues of J satisfy

$$\det(\lambda^2 I - \lambda g_q^* + g_p^* h_q^*) = 0. \tag{4.7.13}$$

If the orthonormal basis is computed exactly, the spectral radii of all three schemes are exactly the same but in the presence of inaccuracies in Z, Gauss-Seidel and reverse Gauss-Seidel and Jacobi behave differently, as was shown in Burrage et al. (1994a).

Linear systems

The extension and numerical study of these nonlinear techniques to linear systems of equations has been given in Burrage et al. (1994a), Erhel et al. (1994) and Burrage et al. (1994b). These schemes can in fact be applied with very general iteration schemes such as conjugate gradient or GMRES techniques. GMRES was introduced by Saad and Schultz (1986) and is commonly used to solve large nonsymmetric linear systems, and Erhel et al. (1994) have adapted these deflation techniques to produce a new variable preconditioning based on an invariant subspace approximation for the restarted GMRES algorithm. However, in this section only stationary iteration schemes of the form

$$My^{(k+1)} = Ny^{(k)} + b \tag{4.7.14}$$

will be considered.

In this case the fixed-point formulation of (4.7.1) is given by

$$y = F(y), \quad F(y) = M^{-1}Ny + M^{-1}b. \tag{4.7.15}$$

Thus let $\mathbb{P}$ be the invariant subspace of dimension r for

$$H = M^{-1}N,$$

I_r be the identity matrix of order r, and Z the orthogonal basis of $\mathbb{P}$. Thus with

$$Q = I - ZZ^\top, \quad P = ZZ^\top, \quad I_r = Z^\top Z, \quad QP = 0$$

and writing

$$y = (P+Q)y = Py + q = Zu + q, \quad u = Z^\top y,$$

then (4.7.1) and (4.7.15) imply

$$\begin{aligned} (I_r - Z^\top HZ)u &= Z^\top M^{-1}b + Z^\top Hq \\ q &= Q(M^{-1}b + Hq + HZu) \\ y &= Zu + q. \end{aligned} \tag{4.7.16}$$

The Jacobi, Gauss-Seidel and reverse Gauss-Seidel schemes can then be written in the general iterative form

$$\begin{aligned} u^{(k+1)} &= (I_r - Z^\top HZ)^{-1}Z^\top(M^{-1}b + Hq^{(i)}) \\ q^{(k+1)} &= (I - ZZ^\top)(M^{-1}b + Hq^{(k)} + HZu^{(j)}) \end{aligned} \tag{4.7.17}$$

where the relationships between i, j and the method is given by

i	j	method
k	k	Jacobi
k	$k+1$	Gauss-Seidel
$k+1$	k	reverse Gauss-Seidel.

In the last case it is understood that the q iteration is performed first.

It can be seen from (4.7.17) that Gauss-Seidel and reverse Gauss-Seidel have very similar properties in that they both compute the same sequence but with different starting and finishing values. The spectra of these iteration schemes are again given by (4.7.11) and (4.7.12) with

$$\begin{aligned} g_q^* &= (I - ZZ^\top)H(I - ZZ^\top) \\ g_p^* &= (I - ZZ^\top)HZ \\ h_q^* &= Z(I_r - Z^\top HZ)^{-1}Z^\top H(I - ZZ^\top). \end{aligned}$$

In the case that $\mathbb{P}$ is invariant then the spectral norm of all three iteration schemes is $\rho((I - ZZ^\top)H)$. Recall here that H is the iteration matrix of the underlying iteration scheme, and this underlying scheme can be chosen depending on both the problem and the computer architecture. In the case of a parallel environment a Jacobi or block Jacobi iteration may be appropriate in which case M will be diagonal or block diagonal; while in a sequential environment Gauss-Seidel or block Gauss-Seidel or SOR schemes may be more appropriate as this will lead to faster convergence but less parallelism.

The number of eigenvectors that are appended to Z depend on the nature of the desired convergence properties, and this involves the development of a cost function which can be interrogated every so often to see if it is worthwhile to increase the dimension of Z (see Burrage et al. (1994a)).

In order to see the efficacy of this approach a series of linear systems is solved arising from the fitting of surfaces to a set of sparsely scattered meteorological stations in Australia (Burrage et al. (1994b), Williams and Burrage (1994a)). The problem size can vary from a few hundred to almost 20,000. Here just three test sets are chosen of dimension 550, 1025 and 1500. These matrices are symmetric positive definite and dense. The condition number can be varied by the addition of a positive scalar to the diagonal elements of the influence matrix. These systems were solved on a Cray YMP-2D sited at the University of Queensland using the reverse Gauss-Seidel deflation technique with an underlying Jacobi iteration and compared with Cray routines for LU factorization and backward and forward substitution. The timings in Table 4.1 are presented in seconds and were obtained by extracting three eigenvalues every five iterations. The systems here are only mildly conditioned.

Table 4.1 *comparison timings*

data points	550	1025	1500
LU solve	0.377	2.354	7.443
deflation	0.097	0.386	0.809
speed-up	3.88	6.1	9.2

It can be seen that the performance of the deflation approach compares very favourably with a highly optimized Cray library routine for doing a linear solve based on *LU* factorization especially in view of the fact that the problem is dense. In order to investigate the effect of conditioning on the deflation process a problem of dimension 1025 was solved for different conditionings. These results are presented in Figure 4.3.

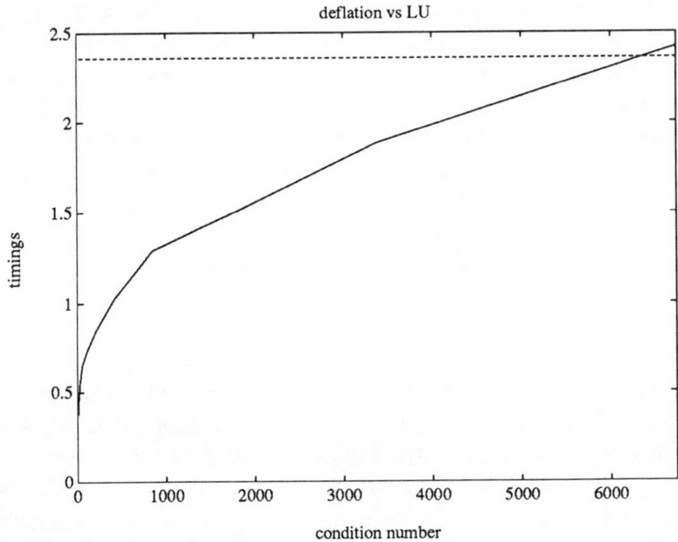

FIG. 4.3. comparison timings

The constant dotted line denotes the time based on *LU* factorization. In the deflation implementation three eigenvalues were deflated every five iterations. For these values the deflation process slows as the conditioning worsens. However, these timings can be improved by using different extraction protocols. An adaptive implementation is currently under development which adapts the frequency with which differing numbers of eigenvalues are deflated to the convergence rate of the deflation process.

Some additional testing has been performed on problems which have multiple eigenvalues or clustered eigenvalues. The convergence behaviour of the deflation process is described in (4.7.9), and because it is essentially the power method it is important to realize that the deflation process works differently in both the case of multiple eigenvalues and clustered eigenvalues.

Thus for example, on a problem of dimension 100 with 99 eigenvalues clustered at 0.95 and one eigenvalue at 0.2 the deflation process will converge in a very few iterations by deflating only one eigenvalue. For this example the deflation process behaves as if the problem is of dimension 2 with one eigenvalue at 0.99 and another at 0.2. This is entirely consistent with the way the power method behaves on problems with multiple eigenvalues (and distinct eigenvalues).

For problems which have clusters of eigenvalues, the behaviour of the deflation process depends on how finely the eigenvalues are clustered within each cluster. If the clustering within each cluster is very fine then the inherent errors in the deflation process will cause all the eigenvalues within the cluster to be treated the same and the process will behave as in the multiple eigenvalue case (without any noticeable loss of accuracy). As the clustering becomes less fine then the deflation process will extract a small number of eigenvalues from each cluster and consider the rest in that cluster as being multiple eigenvalues. Finally, at a moderate clustering the process may well attempt to extract nearly all the eigenvalues from the clusters which are near the unit disk.

As an example of this, a problem of dimension 100 with 10 clustered sets with 10 eigenvalues in each cluster was solved. In the two clusters closest to the unit disk the eigenvalues were equally spaced in the interval (0.990, 0.9987) and (0.892, 0.901). For this problem the Jacobi method could not converge in 5000 iterations, while the reverse Gauss-Seidel deflation technique with an underlying Jacobi iteration converged in 81 iterations. In this case up to four eigenvalues could be extracted every five iterations and 68 eigenvalues were extracted in total, with nine eigenvalues being extracted from each of the first two clusters and seven eigenvalues from the next cluster.

In conclusion, it can be seen that the deflation process is a remarkably robust procedure working on a variety of pathological cases. The number of eigenvalues that have to be extracted is usually modest and the speed-up in time can be impressive. Not only does the deflation process represent a very efficient sequential method but there is considerable scope for parallelism. The deflation process is rich in Level 2 and Level 3 BLAS and the small system of linear equations defined on the P-space can be solved on one processor using a standard sequential solver based, for example, on LU factorization. These parallel investigations are currently being studied by Williams and Burrage (1994b). A preliminary implementation gives a speed-up of about a factor of 5 using 16 processors over a sequential implementation on problems of dimension 1500. However, this implementation represents a synchronized approach so that while the stiff system is being updated on the master processor, the other processors are in wait mode. This suggests that an asynchronous approach will be needed in order to avoid these large times of inactivity.

Finally, it should be mentioned that this deflation process may also prove to be significant in the differential equations arena, not only because of its applications to linear systems but because it can be viewed as an automatic adaptive partitioning procedure for splitting large differential systems into stiff and nonstiff components. This is discussed in more detail in Chapter 6.

4.8 Krylov subspace techniques

An apparently different approach, but one which can be placed in a similar format to (4.5.8) is the conjugate gradient (CG) method which is based on the idea of minimizing the function

$$F(y) = \frac{1}{2}y^\top Qy - y^\top b \tag{4.8.1}$$

where Q is assumed to be symmetric, positive definite. Since the minimum of this function occurs at $y = Q^{-1}b$, minimizing F and solving (4.4.1) are equivalent.

The standard iterative approach to minimizing (4.8.1) is to form the following iterates

$$y_{k+1} = y_k - \alpha_k s_k, \tag{4.8.2}$$

where the s_k are a set of direction vectors. One possible choice for these direction vectors is by forming the gradient of F. Thus

$$s_k = \nabla F(y_k) = Qy_k - b$$

and a Richardson-type iteration of the form

$$y_{k+1} = y_k - \alpha_k(Qy_k - b)$$

is obtained.

Unfortunately, the convergence of this scheme can be slow. Hestenes and Stiefel (1952) introduced the conjugate gradient method which works by the successive minimization of F over the span of a preceding set of search vectors $\{s_1, \ldots, s_k\}$. Thus the iteration takes the form given in (4.8.2) where

$$\alpha_k = s_k^\top r_k / r_k^\top Q r_k, \quad r_k = Qy_k - b, \tag{4.8.3}$$

and where the s_k satisfy $s_k^\top Q s_k = 0$, $s_k^\top r_k \neq 0$. Various options are considered for computing the s_k in Golub and van Loan (1989) and it can be shown that this technique performs well on linear systems which are well conditioned or have clustered eigenvalues. The following results on the performance of the CG method are given in Ortega (1988):

Theorem 4.8.1. *If Q is an $m \times m$ real, symmetric, positive definite matrix then the CG method converges in at most m iterations.*

Theorem 4.8.2. *If Q is symmetric, positive definite with r distinct eigenvalues then the CG method converges in at most r iterations.*

Theorem 4.8.3. *If Q is symmetric, positive definite and $K=$ cond(A) in the L_2 norm then with $e_k = y_k - y$,*

$$\|e_k\|_2 \leq 2\sqrt{K}\left(\frac{\sqrt{K}-1}{\sqrt{K}+1}\right)^k \|e_0\|_2.$$

Preconditioning

A consequence of this last result is that if the condition number of a matrix is large then the CG method converges very slowly. In fact the performance of the CG method and other iterative techniques can often be dramatically improved by the technique of preconditioning, where (4.6.1) is transformed to

$$\hat{Q}\hat{y} = \hat{b}, \quad \hat{Q} = L^{-1}QL, \quad \hat{y} = L^{-1}y, \quad \hat{b} = L^{-1}b. \tag{4.8.4}$$

In the case of the CG method, L is chosen so that $\hat{Q}$ is well conditioned or has clustered eigenvalues. Rather than apply the CG method to this transformed system it can be modified directly so that systems of the form

$$Pz = r, \quad P = LL^\top$$

have to be solved and P is called the **preconditioning matrix**.

There are many possible preconditioning approaches, with one of the most popular being polynomial preconditioning in which (4.6.1) is replaced by

$$p(Q)Qy = p(Q)b,$$

where p is a **preconditioning polynomial**. The main advantage of this approach is its suitability for vector and parallel architectures since there are only two intrinsic operations: matrix-vector multiplication and vector addition. It is also very suited to matrix-free computations when the matrix Q is known only implicitly.

Clearly a suitable approach for determining $p(z)$ is to require that $p(Q)$ approximates Q^{-1} in some way. There is no optimal polynomial since the choice of p will depend on the eigenvalue distribution of Q which is usually not known a priori. However, if it is known that $\sigma(Q) \subseteq [a,b]$, $b > a > 0$, then a suitable polynomial p is the one which minimizes s where

$$s = \min \parallel 1 - p(\lambda)\lambda \parallel, \quad \lambda \in [a,b]. \tag{4.8.5}$$

In the case of the uniform norm, the Chebyshev polynomial minimax property implies

$$p(\lambda)\lambda = 1 - \frac{T_k(\frac{a+b-2\lambda}{a-b})}{T_k(\frac{a+b}{a-b})},$$

where $T_k(x)$ is the Chebyshev polynomial of the first kind of degree k.

In the case of a least squares approach with a Jacobi weight function $\omega(\alpha,\beta,\lambda) = (b-\lambda)^\alpha(\lambda-a)^\beta$, $\alpha,\beta > -1$, Saad (1985) has extended the work of Johnson et al. (1983) and has shown that

$$p(\lambda)\lambda = 1 - \frac{J_k(\frac{b+2\lambda}{b})}{J_k(1)},$$

where $J_k(x)$ is the Jacobi polynomial of degree k, orthogonal on [-1,1] with respect to $\omega(\alpha+1,\beta,\lambda)$.

Both these approaches can be implemented efficiently using a three-term recurrence relation (see Ashby et al. (1992), for example). Chebyshev preconditioning seems to be suited to matrices whose eigenvalues are dense and nearly uniformly distributed on $[a,b]$, while the least squares approach is superior if there is a clustering of eigenvalues at b.

Of course, there still remains the difficulty in knowing how to choose a and b and the degree k of the preconditioning polynomial. Ashby (1987, 1989) and Ashby et al. (1992) consider a number of adaptive approaches based on the fact that for symmetric, positive definite systems there is an intimate relationship between the CG algorithm and the Lanczos algorithm, which quickly determines the extremal eigenvalues of $p(Q)Q$, and hence of Q.

It should also be noted that these approaches can be extended to the case of symmetric, indefinite systems. In this case, it is necessary to assume that $\sigma(Q) \subset [a_1,b_1] \cup [a_2,b_2]$, where $a_1 < b_1 < 0 < a_2 < b_2$ and s in (4.8.5) is replaced by

$$s = \min \parallel 1 - p(\lambda)\lambda \parallel, \quad \lambda \in [a_1,b_1] \cup [a_2,b_2], \quad (p(\lambda)\lambda)'_{\lambda=0} = 0$$

which leads to the concept of **Grcar polynomial preconditioning** (Ashby et al. (1989)).

Another important preconditioning technique is based on the idea of incomplete preconditioning. Thus if Q is symmetric, a preconditioner can be chosen as $LL^{\top}$ where L is an approximation to the Cholesky factor of Q. The sparsity structure can be preserved by letting l_{ij} be the appropriate Cholesky element if q_{ij} is nonzero otherwise if $q_{ij} = 0$ then $l_{ij} = 0$. It can be shown that this factorization does not necessarily exist even for symmetric, positive definite matrices, although Ortega (1988) has shown that the factorization exists if Q is an H-matrix. Van der Vorst (1987) has proposed a similar preconditioning technique in conjunction with twisted factorization for block systems of equations. Unfortunately, however, the incomplete preconditioning approach is difficult to vectorize or parallelize efficiently.

Some of these approaches have been applied to block structured matrices (see Concus et al. (1985), for example). Thus if Q is a block tridiagonal matrix of the form

$$\begin{pmatrix} Q_1 & P_1^{\top} & & \\ P_1 & Q_2 & \ddots & \\ & \ddots & \ddots & P_{N-1}^{\top} \\ & & P_{N-1} & Q_N \end{pmatrix}$$

it can be shown that a complete block decomposition has the form

$$Q = (D + L)D^{-1}(D + L^{\top})$$

where

$$D = \begin{pmatrix} D_1 & & \\ & \ddots & \\ & & D_N \end{pmatrix}, \quad L = \begin{pmatrix} 0 & & & \\ P_1 & 0 & & \\ & \ddots & \ddots & \\ & & P_{N-1} & 0 \end{pmatrix}$$

and

$$D_1 = Q_1, \quad D_i = Q_i - P_{i-1}D_{i-1}^{-1}P_{i-1}^{\top}, \quad i = 2, \ldots, N.$$

If the Q_i are tridiagonal and the P_i are diagonal, Ortega (1988) suggests an incomplete factorization of the form

$$D_1 = Q_1, \quad D_i = Q_i - P_{i-1}C_{i-1}P_{i-1}^{\top}, \quad i = 2, \ldots, N$$

where C_{i-1} is a tridiagonal approximation to D_{i-1}^{-1} which can be shown to exist if Q is an M-matrix and has the CDD property.

For nonsymmetric problems, the CG scheme cannot be applied directly since orthogonality amongst the residual vectors is not possible through the three-term recurrence relation which is the basis of the CG approach. Of course CG methods can be applied to the system

$$Q^\top Qy = Q^\top b$$

at the cost of squaring the condition number and increasing the number of iterations needed to attain the necessary convergence. These difficulties have led to the concept of GMRES methods first introduced by Saad and Schultz (1986).

GMRES methods

GMRES and CG-type methods are in fact elements of Krylov subspace methods in which problems are solved in a smaller-dimensional projection subspace (the Krylov subspace). By noting in (4.5.7), with $\omega = 1$, that

$$y_{k+1} = y_k + M^{-1}(b - Qy_k)$$

then with $r_0 = b - Qy_0$,

$$y_k = y_0 + \alpha_0 M^{-1} r_0 + \alpha_1 M^{-1} Q M^{-1} r_0 + \ldots + \alpha_{k-1}(M^{-1}Q)^{k-1} M^{-1} r_0.$$

Consequently, the y_k can be written as $y_0 + x$ where x is an element of the k-dimensional space

$$K_k = \text{span}\{u, Gu, \ldots, G^{k-1}u\}, \quad u = M^{-1}r_0, \quad G = M^{-1}Q,$$

and K_k is known as the k-dimensional **Krylov subspace** corresponding to u and G.

In the case of symmetric systems, the projected system onto the Krylov subspace is formed by a three-term recurrence relation (the Lanczos algorithm), while in the nonsymmetric case, the matrix of the projected system is an upper Hessenberg matrix (the Arnoldi algorithm). Thus the GMRES algorithm uses the Arnoldi process to construct an orthonormal basis $V_k = [v_1, \ldots, v_k]$ for $K_k(Q, r_0)$. In preconditioned GMRES the y_k are the vectors that minimize $\| b - Qx \|_2$ over all x with $x - y_0$ in $K_k(M^{-1}Q, r_0)$, so that the correction of y_k with respect to y_0 minimizes the residual over this Krylov subspace.

As with CG methods, it is easy to see that GMRES converges in at most m iterations. However, a disadvantage is that all residual vectors (up until convergence is attained) have to be stored and the construction of the projected system becomes complicated as the iterations increase. For this reason GMRES is often restarted after every k iterations and this is known as GMRES(k). There are additional difficulties, however, in knowing how to select the parameter k. Thus the restarted GMRES method restricts the Krylov subspace dimension to a fixed value k and restarts the Arnoldi process using the last iterate y_k as an initial guess. Below is an algorithmic description of the restarted GMRES for solving $Qy = b$:

```
ALGORITHM : GMRES(k)
ε is the tolerance for the residual norm ;
convergence:= false ;
choose y_0 ;
until convergence do
    r_0 = b − Q y_0 ;
    β = ||r_0|| ;
    v_1 := r_0/β ;
    for j = 1, ···, k do
    |   p := Q v_j ;
    |   for i = 1, ··· j do
    |   |   h_ij := v_i^T p ;
    |   |   p := p − h_ij v_i ;
    |   endfor
    |   h_{j+1,j} := ||p||_2 ;
    |   v_{j+1} := p/h_{j+1,j} ;
    |   if ||b − Q y_j|| < ε then
    |       solve min_{x_j ∈ R^j} ||β e_1 − H̄_j x_j|| ;
    |       y_j := x_0 + V_j x_j ;
    |       convergence := true ;
    |   endif;
    endfor ;
    solve min_{x_k ∈ R^k} ||β e_1 − H̄_k x_k|| ;
    y_k := y_0 + V_k x_k ;
    if ||b − Q y_k|| < ε then
        convergence : = true ;
        else y_0 := y_k ;
    endif;
    y_0 = y_k ;
enddo
```

Here the matrix $\overline{H_l} = (h_{ij})$, $l = 1, \ldots, k$, is an upper Hessenberg matrix of order $(l+1) \times l$ and this gives the fundamental relation

$$QV_l \quad = \quad V_{l+1}\overline{H_l}.$$

Thus the GMRES algorithm computes $y = y_0 + V_l x_l$ where x_l solves the least squares problem $\min_{x_l \in \mathbb{R}^l} \|\beta e_1 - \overline{H_l} x_l\|$. Usually a QR factorization of $\overline{H_l}$ using Givens rotations is used to solve this least squares problem.

The convergence behaviour of GMRES has been analysed by van der Vorst and Vuik (1993) and superlinear convergence has been related to the convergence of Ritz values. The full GMRES version behaves as if the smallest eigenvalues were removed after some iterations. It has been observed that the convergence of the restarted algorithm depends heavily on the dimension of the Krylov subspace and may be slower than in the full case (Huang and van der Vorst (1989)). It appears as if the restarting procedure loses information on the smallest Ritz values. An adaptive procedure is proposed in Joubert (1994) to choose the restart frequency according to the convergence and work requirements.

The concept of polynomial preconditioning for CG algorithms is closely related to hybrid methods which combine, for example, a GMRES algorithm with a Richardson iteration. The idea is to use first GMRES to approximate both the solution and eigenvalues and then to use Richardson iteration using a polynomial derived from the estimated eigenvalues. A survey of hybrid methods which rely on eigenvalue estimations can be found in Nachtigal et al. (1992). These estimations are usually done by the power method or by the Arnoldi technique but they can also be computed from modified moments (see Calvetti et al. (1994)). Other hybrid solutions do not rely on eigenvalue estimations but use directly a polynomial generated by GMRES itself (Nachtigal et al. (1992)). An alternative approach discussed in van der Vorst and Vuik (1991) and Saad (1993) is to build a preconditioner based on the application of GMRES.

Erhel et al. (1994) have therefore developed a preconditioning technique which aims at keeping the information when restarting. The idea is to estimate the invariant subspace corresponding to the smallest eigenvalues and represents a generalization of the work of Burrage et al. (1994a, 1994b) on deflation techniques. Thus after each restart, a new preconditioner is built and the flexible GMRES method can be used (see Saad (1993)). At each restart, new eigenvectors are estimated in order to increase the invariant subspace. The preconditioner is almost equal to the matrix on the approximated invariant subspace and is taken as the identity on the orthogonal subspace.

Numerical results presented in Erhel et al. (1994) show that for many systems this technique can converge much faster than the restarted version and almost as fast as the full scheme. Moreover, the preconditioned scheme can be faster in CPU time thanks to a lower complexity, and also requires substantially less memory than a full version. Furthermore, this method is readily parallelizable by means of Level 2 dense BLAS operations, and since the algorithm only requires a matrix-vector product it can be applied to so-called matrix-free versions of GMRES where the matrix is not stored. This has important ramifications when using the new version of VODE known as VODEPK.

Assume now that all the eigenvalues of Q are nondefective, and let $|\lambda_1| \leq |\lambda_2| \leq \ldots \leq |\lambda_m|$ be the eigenvalues of Q. Let P be an invariant subspace of dimension r corresponding to the smallest r eigenvalues of A, and let $Z = (U, W)$ be an orthonormal basis of $\mathbb{R}^m$ where U is an orthonormal basis of P. In this basis, Q is similar to a matrix $\tilde{Q}$ which is written in the Schur form as

$$\tilde{Q} = \begin{pmatrix} T & Q_{12} \\ 0 & Q_{22} \end{pmatrix}, \tag{4.8.6}$$

where $T = U^\top QU$ is the restriction of Q onto the subspace P.

Let M be a matrix defined by

$$M = Z \begin{pmatrix} T/|\lambda_m| & 0 \\ 0 & I_{m-r} \end{pmatrix} Z^\top. \tag{4.8.7}$$

M is nonsingular and its inverse is easily computed by

$$M^{-1} = Z \begin{pmatrix} |\lambda_m| T^{-1} & 0 \\ 0 & I_{m-r} \end{pmatrix} Z^\top. \tag{4.8.8}$$

In practice, the basis W is unknown so that this matrix form of M^{-1} cannot be used. Therefore M^{-1} can be written as

$$\begin{aligned} M^{-1} &= |\lambda_m| U T^{-1} U^\top + (I_m - UU^\top), \\ &= I_m + U(|\lambda_m| T^{-1} - I_r) U^\top, \end{aligned} \tag{4.8.9}$$

which is cheap to compute.

The preconditioned matrix $M^{-1}Q$ is therefore similar in this basis to

$$E = \begin{pmatrix} |\lambda_m| I_r & |\lambda_m| T^{-1} Q_{12} \\ 0 & Q_{22} \end{pmatrix},$$

so that its eigenvalues are $|\lambda_m|$ and the remaining eigenvalues of Q, that is to say the eigenvalues of Q_{22}. In particular, the condition number is now $|\lambda_m|/|\lambda_{r+1}|$ instead of $|\lambda_m|/|\lambda_1|$ in some appropriate norm. Hence the GMRES algorithm applied to $M^{-1}Q$ converges faster than when applied to Q.

However, in general the invariant subspace P is only an approximation $\tilde{P}$. Using an orthonormal basis $(\tilde{U}, \tilde{W})$ the preconditioner M now has a nonzero block in the (2,1) position of the matrices $\tilde{Q}$ and E:

$$\tilde{Q} = \begin{pmatrix} T & Q_{12} \\ Q_{21} & Q_{22} \end{pmatrix} \quad E = \begin{pmatrix} |\lambda_m| I_r & |\lambda_m| T^{-1} Q_{12} \\ Q_{21} & Q_{22} \end{pmatrix}, \tag{4.8.10}$$

where $T = \tilde{U}^\top Q \tilde{U}$. This gives a perturbed matrix where the perturbation is given by the block Q_{21}. If this block is small enough, the eigenvalues of $M^{-1}Q$ are close to $|\lambda_m|$ and to the eigenvalues of Q_{22} (recall that the eigenvalues are supposed to be nondefective), and an improved convergence rate for this preconditioned GMRES will be obtained.

After each restart the new Ritz values, which approximate the eigenvalues of $M^{-1}Q$ which in turn approximate the eigenvalues of Q_{22} and hence the remaining eigenvalues of Q, are estimated. The size of the invariant subspace can be increased in order to obtain a more powerful preconditioner by adding new Schur vectors. In order to avoid loss of orthogonality, these vectors are orthogonalized against the previous basis U.

The procedure outlined represents a variable preconditioning on the left. Naturally a right preconditioning approach is also valid. In fact it is well known that a right preconditioning guarantees a decreasing residual while this is not always true for a left preconditioner. Figure 4.4 shows an increase in the residual at the start of each restart using a left preconditioner.

Thus a flexibly-preconditioned-restarted GMRES version is built. In some sense, this algorithm recovers the superlinear convergence of the full GMRES version which behaves as if the smallest eigenvalues were removed. This approach has some merit when dealing with a restarted version. In this case, the preconditioner keeps the information on the smallest Ritz values which would be lost by the restart. Moreover, this preconditioner is cheap and easily parallelizable, so that it can be considerably faster in CPU time than a full scheme.

Below is an algorithmic description of the new flexibly-preconditioned-restarted GMRES based on a left preconditioning. A right-preconditioning approach can also be constructed. Currently a fixed number of eigenvalues, r, are extracted after each restart.

Erhel et al. (1994) have compared the performance of this approach with full GMRES and restarted GMRES using Matlab and the template of GMRES provided in Barret et al. (1993) on a number of test problems. In these test problems one or two eigenvalues were deflated at each restart and restarts occurred every 10 iterations. As a particular example, consider (4.4.1) where $Q = SDS^{-1}$ with $Q, S, D \in \mathbb{R}^{100\times 100}$, $S = (1, \beta)$ a bidiagonal matrix with 1 on the diagonal and β on the upper subdiagonal, $b = (1, \ldots, 1)^\top$ and $D = \mathrm{diag}(-10, -9, \ldots, -1, 1, 2, \ldots, 90)$. The following results were given in Erhel et al. (1994).

ALGORITHM : FLEXGMRES(k, r)

ϵ is the tolerance for the residual norm ;
convergence:= false ;
choose y_0 ;
$M := I_m$;
$U := \{\}$;
until convergence **do**
  $r_0 = M^{-1}(b - Qy_0)$;
  Arnoldi process applied to $M^{-1}Q$ to compute V_k ;
  solve $\min_{x_k \in \mathbb{R}^k} \|\beta e_1 - \overline{H_k} x_k\|$;
  $y_k := y_0 + V_k x_k$;
  if $\|M^{-1}(b - Qy_k)\| < \epsilon$ convergence := true ;
    $y_0 = y_k$;
  else
    compute r Schur vectors of H_k denoted by S_r ;
    orthogonalize $V_k S_r$ against U ;
    increase U by $V_k S_r$;
    $T := U^\top Q U$;
    $M^{-1} := I_m + U(|\lambda_m| T^{-1} - I)U^\top$;
  endif
enddo

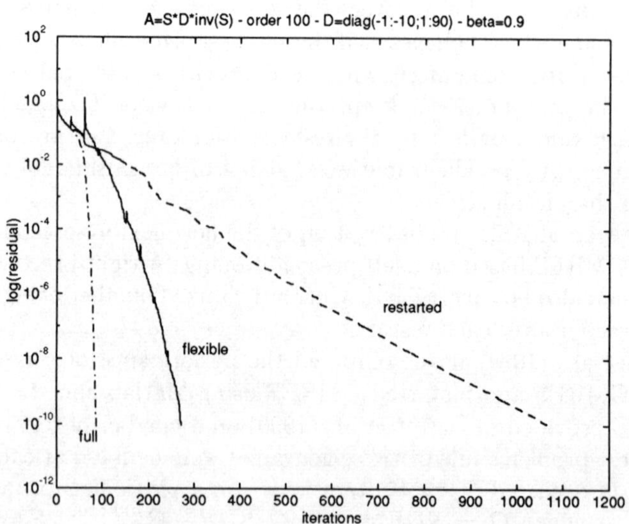

FIG. 4.4. convergence results

Table 4.2 *iterations and timings*

method	restart	eigs deflated.	iterations	CPU time
full	100	0	88	32.6667
restarted	10	0	∞	*
flexible	10	7	∞	*
flexible	10	8	788	20.9500
flexible	10	17	179	6.6167

This problem demonstrates the importance of deflating a sufficient number of eigenvalues. If only seven eigenvalues are deflated, for example, the flexible scheme fails to converge as does the restarted scheme. However, if about 20% of the eigenvalues are deflated then Table 4.2 shows dramatic improvement in timings over the full GMRES scheme. Thus this flexible approach appears to be a very robust and efficient algorithm.

4.9 Aspects of parallelization

The implementational differences of polynomial preconditioning versus incomplete *LU* preconditioning in a parallel environment as well as the apparently non-parallel aspects of the Gauss-Seidel method have already been remarked upon in section 4.5. However, although polynomial preconditioning would appear to offer some advantages in a parallel environment, especially for sparse problems, the limitations of existing architectures and compilers often mitigate against these advantages. For example, Dyke (1995) has investigated the efficiency of multigrid techniques with a Gauss-Seidel smoother versus polynomial preconditioned and non-preconditioned CG methods on a MasPar MP1 using MPFortran. These results show that of these three techniques non-preconditioned CG is in fact the fastest on the MP1. The reason for this appears to be due to the implementation of the CSHIFT operation within MPFortran which is slow compared with other matrix-vector operations. In the case of polynomial preconditioning a number of CSHIFT operations are needed per iteration and hence the inefficiency. No doubt this state of affairs will change as communication hardware and software improves and parallel compilers become more efficient.

Stationary iterative methods

A naturally parallelizable algorithm is the Jacobi iteration which if applied to (4.5.3) can be written as

$$y_{ij}^{(k+1)} = \frac{1}{4}(y_{i+1,j}^{(k)} + y_{i-1,j}^{(k)} + y_{i,j+1}^{(k)} + y_{i,j-1}^{(k)} + b_{ij}), \quad i,j = 1,\ldots,m.$$

However, despite this natural parallelization, a number of implementational questions have to be addressed when implementing this scheme in a parallel environment, with probably the most important one being how to distribute the computations efficiently amongst the available processors. For example, if each processor computes just one approximation to one component of the unknown vector then the implementation is dominated by communication considerations and may have serious bottlenecks. On the other hand suppose that the computational grid is split into blocks of $q \times q$ grids (where it is assumed that $pq^2 = N$, $N = m^2$); then the points on the boundaries of these grids must, at each step, be transmitted to the neighbouring processors. If the processor network is also a square then for the internal blocks there are $4q$ numbers to be transmitted at each iteration while there are q^2 computations per iteration. For a fixed number of processors p, as N grows communication grows linearly while computation grows quadratically and a balance has to be found between the two. The fact that this is very much architecture dependent complicates the issue.

A natural generalization of the Jacobi method for block systems of the form

$$\begin{pmatrix} Q_{11} & \cdots & Q_{1m} \\ \vdots & & \vdots \\ Q_{m1} & \cdots & Q_{mm} \end{pmatrix} \begin{pmatrix} y_1 \\ \vdots \\ y_m \end{pmatrix} = \begin{pmatrix} b_1 \\ \vdots \\ b_m \end{pmatrix}$$

is called the block Jacobi method and is given by

$$Q_{ii} y_i^{(k+1)} = - \sum_{j=1, j \neq i}^{m} Q_{ij} y_j^{(k)} + b_i, \quad i = 1, \ldots, m, \tag{4.9.1}$$

where the Q_{ii} are assumed to be nonsingular and can be factored into LU factors in parallel at the outset. Here the y_j and b_j are themselves vectors.

In the case of (4.5.3) the block Jacobi method takes the form

$$T y_i^{(k+1)} = -y_{i+1}^{(k)} - y_{i-1}^{(k)} + b_i, \quad i = 1, \ldots, m. \tag{4.9.2}$$

Assuming that there are m processors each processor should form its own LU factorization of the tridiagonal matrix T and at the end of each iteration each processor i communicates to the neighbouring processors to the left and the right its updated value of y. This is a very efficient implementation in that all communication is local. However, for dense problems in which no structure can be exploited as in (4.9.1), an appropriate implementation is one based on the **master-slave model** in which, at the end of each iteration, each processor must communicate to the master its updated value. The master then collects these values and redistributes them to all the processors in preparation for the next iteration.

Gauss-Seidel and SOR also have a blocked form, but require the solution of a lower-triangular system of equations at each iteration. In this latter case there is no obvious parallelism if the equations are ordered in the standard form from left to right and bottom to top, but if the grid points are renumbered then the situation changes dramatically. One successful way of reordering the equations is based on the so-called chequerboard effect using two colourings (red and black). In order to see how this works consider the case of the central differencing problem applied to the two-dimensional Laplacian problem on a square. At any interior point the red (black) points are coupled with the four neighbouring black (red) points. Thus if the system is rewritten with all red unknowns (x_R) ordered first followed by all the black points (x_B) then for this problem the following system is obtained

$$\begin{pmatrix} 4I_R & P \\ P^\top & 4I_B \end{pmatrix} \begin{pmatrix} y_R \\ y_B \end{pmatrix} = \begin{pmatrix} b_R \\ b_B \end{pmatrix},$$

where I_R and I_B are the identity matrices of appropriate order. Block Gauss-Seidel now gives

$$\begin{aligned} 4I_R y_R^{(k+1)} &= b_R - P y_B^{(k)} \\ 4I_B y_B^{(k+1)} &= b_B - P^\top y_R^{(k+1)}, \end{aligned}$$

and this technique now consists of Jacobi-like operations.

If $m^2 = 2qp$ then q black and q red components can be assigned to each processor, while if this equation does not hold then some other load balancing approximate to this one must be performed. It should be noted that in the case of a two-colouring, the maximum number of processors that can be gainfully employed is $m^2/2$. This technique can of course be generalized to SOR and there is again flexibility in the types of communication protocols adopted.

In the case of large-dimensional partial differential equations or if larger finite differencing stencils are used some of the nearest neighbours will have the same colour as the centre point if only a two-colouring is used. Thus the two-colouring technique can be generalized to an arbitrary number of colourings (c, say), in which case q can be chosen as $q = m^2/cp$ (assuming integer division). Thus at each step, each processor updates all the components of one colour and then communicates the necessary data to the neighbouring processors with the components of the next colour being updated, and so on, until c sweeps of all the colours have been completed. In some architectures, such as transputer arrays, it is possible to overlap the computation of the next colour and the communication of the previous colour (see, for example, Missirlis (1987) and Reed and Patrick (1985)).

It should be noted that as well as a chequerboard-type ordering it is possible to do a line ordering in which the components are coloured by lines. The ordering can again have some effect on the rates of convergence and this has been explored by Adams and Jordan (1985). These iterative techniques have been used with considerable success in conjunction with multigrid methods (see Brandt (1977), for example) in which iterations are moved back and forth between fine and coarse grids by interpolation and injection.

Asynchronous methods

The parallel implementations described so far have all been in terms of a synchronized approach. As Bertsekas and Tsitsiklis (1989) note, in a synchronous algorithm the algorithm is divided into **phases** and all processor interaction takes place at the end of each phase. Synchronization can be performed **globally** in which each processor sends (and receives from) all the other processors a phase termination message before proceeding to the computations in the next phase, or **locally** in which the phase termination messages are sent to and received from a known subset of processors. Clearly, there are overheads associated with synchronous algorithms due to waiting times while the message passing is underway and due also to poor load balancing, in which the execution time depends on the time taken by the slowest processor to complete its computations. One way of reducing these overheads in a parallel environment is to use asynchronous algorithms so that processors can carry on computing without having to worry about a specified sequence of data communication. The danger of performing these extra computations is, of course, that the extra work may be of no use or, even worse, counterproductive.

Thus in an asynchronous iterative implementation, updates are performed based on some past update of every other processor but not necessarily the most recent update. This has of course considerable impact on the convergence properties of the algorithm (see Shao and Kang (1987) and Bertsekas and Tsitsiklis (1989), for example).

Bertsekas and Tsitsiklis (1989) have analysed the convergence properties of many different types of asynchronous iterative schemes in a very general setting. Typically, asynchronous implementations converge faster than the equivalent synchronous implementations but since the number of message transmissions may increase significantly, communication delays can be substantial and unpredictable, in which case asynchronous implementations may be slower.

In order to study the convergence of asynchronous iteration schemes,

Bertsekas and Tsitsiklis (1989) differentiate between two types of schemes. Thus consider the fixed-point iteration of the nonlinear equation

$$y(x) = f(y(x))$$

by the asynchronous scheme

$$\begin{aligned} y_i(x+1) &= f_i(y_1(\tau_1^i(x)), \ldots, y_m(\tau_m^i(x))), \quad x \in T_i \\ y_i(x+1) &= y_i(x), \quad x \notin T_i. \end{aligned} \tag{4.9.3}$$

Here the T_i represent the set of points at which the y_i are updated and the $\tau_j^i(x)$ are the delays. If the T_i are infinite and $\{x_k\}$ is a sequence of elements of T_i tending to infinity such that

$$\lim_{k\to\infty} \tau_j^i(x_k) = \infty, \quad \forall j, \quad i = 1, \ldots, m$$

then (4.9.3) is said to be **totally asynchronous**.

Some of the consequences of a totally asynchronous algorithm are that each component can be updated infinitely often and that old information is eventually purged, although there can be arbitrarily large communication and computation delays. Thus given any x_1 there exists $x_2 > x_1$ such that

$$\tau_j^i(x) \geq x_1, \quad \forall i, j \quad \text{and} \quad x \geq x_2.$$

Bertsekas and Tsitsiklis obtain general results about the convergence of such schemes, but excessive asynchronism can be detrimental to convergence. Consequently, they introduce the concept of **partially asynchronous algorithms** in which communication delays are bounded. Thus there exists an integer θ such that

$$x - \theta \leq \tau_j^i(x) \leq x, \quad \forall i, j.$$

A consequence of this is that each processor performs an update at least once during any window of length θ and the information used by any processor becomes outdated after at most θ time units. Bertsekas and Tsitsiklis characterize two different types of partially asynchronous algorithms: those for which a totally asynchronous implementation does not converge but a partially asynchronous implementation will, if θ is finite, and those which converge if θ is small and diverge if θ is large.

There are many delicate aspects associated with asynchronous implementations with one of the most important being the termination problem. Because message arrival is uncertain, Bertsekas and Tsitsiklis suggest an approach whereby each processor monitors its own computations and decides when some local termination becomes valid. At this point no more messages are sent by that processor. Termination occurs after the local termination of all processors.

Multisplittings

The block approach described in (4.9.2) is very much a coarse-grained approach and this can be extended through the concept of multisplitting in which each subsystem is solved on a different processor. This process will now be described in some depth as it will be relevant to the discussion in Chapter 9 of a MIMD implementation of an ODE waveform relaxation algorithm.

O'Leary and White (1985) have considered **multisplittings** of Q in which

$$Q = M_l - N_l, \quad l = 1, \ldots, L, \tag{4.9.4}$$

where L is the number of splittings. Frommer and Mayer (1990) have considered the overlappings of subsystems. This approach has been extended by Jeltsch and Pohl (1991) to waveform relaxation methods for systems of ordinary differential equations.

Jeltsch and Pohl (1991) define a **multisplitting** of an order m matrix Q as in (4.9.4) and then the system of equations

$$M_l \hat{y}_l = N_l \hat{y}_l + b, \quad l = 1, \ldots, L$$

is solved either directly or indirectly. The solution y is then found from

$$y = \sum_{l=1}^{L} E_l \hat{y}_l, \tag{4.9.5}$$

where the E_l are nonnegative diagonal matrices satisfying

$$\sum_{l=1}^{L} E_l = I. \tag{4.9.6}$$

Jeltsch and Pohl (1991) also consider overlapping block decompositions. Thus if $S = \{1, \ldots, m\}$, and L subsets $S_1, \ldots, S_L$ of S are chosen so that $\cup_{l=1}^{L} S_l = S$, and if there exist at least one pair of indices $i \neq j$ with $i, j \in \{1, \ldots, L\}$ so that $S_i \cap S_j \neq \phi$ then the multisplitting is said to be **overlapped**, otherwise it is disjoint.

In the case of overlapping splitting, the splitting is performed not on Q but on $P_l Q$ so that

$$P_l Q = M_l - N_l, \quad l = 1, \ldots, L.$$

Here P_l is a diagonal matrix, with

$$P_l(i,i) \;=\; 0, \quad i \notin S_l$$

$$P_l(i,i) = 1, \quad i \in S_l.$$

The weighting matrices E_l now satisfy

$$\begin{aligned} E_l(i,i) &= 0, \quad i \notin S_l \\ E_l(i,i) &\neq 0, \quad i \in S_l. \end{aligned}$$

Example 4.9.1. Consider the tridiagonal system with

$$Qy = b, \quad y \in \mathbb{R}^5, \quad Q = (-1, 2, -1).$$

Then a splitting into the first two components and the last three components would give

$$M = \begin{pmatrix} 2 & -1 & 0 & 0 & 0 \\ -1 & 2 & 0 & 0 & 0 \\ 0 & 0 & 2 & -1 & 0 \\ 0 & 0 & -1 & 2 & -1 \\ 0 & 0 & 0 & -1 & 2 \end{pmatrix}, N = \begin{pmatrix} 0 & 0 & 0 & 0 & 0 \\ 0 & 0 & 1 & 0 & 0 \\ 0 & 1 & 0 & 0 & 0 \\ 0 & 0 & 0 & 0 & 0 \\ 0 & 0 & 0 & 0 & 0 \end{pmatrix}.$$

On the other hand a multisplitting based on overlapping components with $S_1 = \{1, 2, 3, 4\}$ and $S_2 = \{3, 4, 5\}$ would give

$$M_1 = \begin{pmatrix} 2 & -1 & 0 & 0 & 0 \\ -1 & 2 & -1 & 0 & 0 \\ 0 & -1 & 2 & -1 & 0 \\ 0 & 0 & -1 & 2 & 0 \\ 0 & 0 & 0 & 0 & 0 \end{pmatrix}, N_1 = \begin{pmatrix} 0 & 0 & 0 & 0 & 0 \\ 0 & 0 & 0 & 0 & 0 \\ 0 & 0 & 0 & 0 & 0 \\ 0 & 0 & 0 & 0 & 1 \\ 0 & 0 & 0 & 0 & 0 \end{pmatrix}$$

$$M_2 = \begin{pmatrix} 0 & 0 & 0 & 0 & 0 \\ 0 & 0 & 0 & 0 & 0 \\ 0 & 0 & 2 & -1 & 0 \\ 0 & 0 & -1 & 2 & -1 \\ 0 & 0 & 0 & -1 & 2 \end{pmatrix}, N_2 = \begin{pmatrix} 0 & 0 & 0 & 0 & 0 \\ 0 & 0 & 0 & 0 & 0 \\ 0 & 1 & 0 & 0 & 0 \\ 0 & 0 & 0 & 0 & 0 \\ 0 & 0 & 0 & 0 & 0 \end{pmatrix}$$

$$E_1 = \mathrm{diag}(1, 1, \theta_1, \theta_2, 0), \quad E_2 = \mathrm{diag}(0, 0, 1 - \theta_1, 1 - \theta_2, 1).$$

Overlapping can have a significant effect on the convergence of the underlying iteration scheme. For example, if the problem in Example 4.9.1

has $m = 3$ then a block Jacobi iteration based on the splitting $S_1 = \{1, 2\}$ and $S_2 = \{3\}$ would give

$$\rho(H_J) = \frac{1}{\sqrt{3}}$$

for the iteration matrix. On the other hand an overlapping splitting based on $S_1 = \{1, 2\}$ and $S_2 = \{2, 3\}$ would give

$$\rho(H_0) = \frac{1}{3}.$$

An overlapping splitting can always be rewritten as a non-overlapping splitting of a larger system of equations and this has been studied by Burrage et al. (1995). These differences in convergence rates between overlapping and non-overlapping iterations have also been studied in some depth by Burrage et al. (1995). The conclusion there is that the overlapping (of even a small number of components) of the block Jacobi scheme can be very effective on differential systems approximating the one-dimensional heat equation but for the two-dimensional heat equation, overlapping is less efficacious.

More recently, two-stage splittings of linear systems have been introduced (Lanzkron et al. (1991)). Thus if $Q = M - N$ is an **outer splitting** and $M = B - C$ is an **inner splitting** then the resulting method in which there are p inner iterations to every outer iteration gives the iteration scheme

$$y^{(k+1)} = (B^{-1}C)^p y^{(k)} + \sum_{j=0}^{p-1} (B^{-1}C)^j B^{-1}(Nx^{(k)} + b). \tag{4.9.7}$$

The iteration matrix for this scheme is

$$H_p = I - (I - (B^{-1}C)^p)(I - M^{-1}N) \tag{4.9.8}$$

and Lanzkron et al. (1991) have shown that if the outer splitting is regular and the inner splitting is weak regular then (4.9.7) will converge for any value p. More generally, Szyld and Jones (1992) have generalized this to give a two-stage multisplitting. Thus $(M_l, B_l, C_l, D_l, E_l)$ is said to be a **two-stage multisplitting** if

$$\begin{aligned} Q &= M_l - N_l, \quad l = 1, \ldots, L \\ M_l &= B_l - N_l, \quad l = 1, \ldots, L \\ \sum E_l &= I, \end{aligned}$$

where $E_l \geq 0$ are diagonal matrices. One outer iteration of the two stage multisplitting algorithm can be written as

ALGORITHM : MSPLIT(L)

do $l = 1$ to L
  $x_{l0} = y^{(k)}$
  do $j = 0$ to $p - 1$
    $B_l x_{l,j+1} = C_l x_{l,j} + N_l y^{(k)} + b$
enddo
$y^{(k+1)} = \sum_{l=1}^{L} E_l x_{l,p}$

The generalization of (4.9.8) to the multisplitting case leads to an iteration matrix given by

$$H_p = \sum_{l=1}^{L} E_l (B_l^{-1} C_l)^p + \sum_{l=1}^{L} E_l \sum_{j=0}^{p-1} (B_l^{-1} C_l)^j B_l^{-1} N_l. \tag{4.9.9}$$

Szyld and Jones (1992) have shown that if $Q^{-1} \geq 0$ and the outer splitting is regular and the inner multisplitting weak regular, then the two-stage multisplitting process will converge for any value of p.

Jones and Szyld (1994) have extended this work to allow for overlapping blocks and have shown that if Q is an M-matrix then the convergence of a two-stage overlapping block Jacobi multisplitting is asymptotically faster than that of a two-stage non-overlapping block Jacobi multisplitting if the same weighting matrices are used. On the other hand Szyld and Jones (1992) have shown that asymptotically the two-stage method is slower than the standard block approach with $p = 1$. However, for certain parallel architectures with nonuniform memory access times a gain in computational efficiency can accrue by the improved locality of operations in the two-stage process and this is borne out by some numerical testing in both Szyld and Jones (1992) and Jones and Szyld (1994).

Finally in this section some remarks about the parallelization of multigrid techniques are made. Here the implementation is very much architecture dependent. In a MIMD environment a very natural technique is to use domain decomposition but this means that artificial boundaries are created on each subdomain. This can have considerable impact on the convergence of the line smoothers as a spurious coupling is introduced. Various techniques have been proposed to overcome these convergence difficulties such as only coarsening in one direction but for problems with tight coupling a domain decomposition can degrade the convergence of the multigrid process. In a SIMD environment, on the other hand, there is the problem of redundant processors if the grid becomes too coarse, while there is an additional problem of matching the grid to the processor topology. For example, on a 4096-processor MasPar MP1 the time to perform a Level 2

BLAS on a system of dimension 4097 would be twice the time for a system of dimension 4096 or less using MPFortran. This is because the compiler automatically stacks the components onto two grids – one which maps the first 4096 elements to the 4096 processors and another grid with just a single element on it. Thus it is important to choose the dimension of the problem carefully if there is some choice allowed.

This concludes the material on methods for linear systems. A substantial body of material has been given here because the solution of linear systems is a very important component in stiff differential equation solvers, and recent codes such as VODEPK (Brown and Hindmarsh (1989)) and DASPK (Brown et al. (1993)) now use preconditioned GMRES iteration schemes in combination with an inexact Newton approach. In addition, other work in this area on matrix-free methods and iterative solvers has been carried out by Brown and Hindmarsh (1986), Chan and Jackson (1986) and Byrne (1992). Thus it is becoming increasingly important that researchers in the area of differential equations have a good grasp of the important issues in the field of the numerical solution of linear systems.

This latter remark becomes even more significant when considering parallel implementational issues, not only because the use of parallel linear solvers offers one hope of substantial speed-ups but because new waveform relaxation approaches based on multigrid techniques, and described in Chapters 7, 8 and 9, also appear very promising.

5

DIRECT METHODS FOR ODES

Much of the contents of the previous chapters have been necessary for an understanding of subsequent material on parallel differential equation techniques. This material on parallel techniques has essentially been divided into two sections. In Chapters 5 and 6 the various different forms that parallelism can take when solving differential equations are discussed with the main focus being the analysis and construction of direct and iterative methods for exploiting small-scale parallelism. The underlying concepts here are parallelism across the method and parallelism across the steps. Chapters 7, 8 and 9 will however focus on techniques that exploit parallelism across the system.

In this present chapter the focus will be on parallelism across the method. The basic tool that will be used here is the general framework of multivalue methods (introduced in Chapter 2). The chapter will cover the following material:

- section 5.1: an introduction to the types of parallelism that are possible when solving IVPs, namely parallelism across the method, parallelism across the system and parallelism across the steps;
- section 5.2: a discussion on how the structure of multistage methods can be exploited to give a limited form of parallelism for both nonstiff and stiff problems;
- section 5.3: the generalization of the concept of single implicitness to allow the eigenvalues of the method coefficient matrix to be real and distinct which decouples the nonlinear equations defining the internal stages;
- section 5.4: a brief historical description on the use of predictor-corrector methods to exploit a modest number of processors;
- section 5.5: the development of a general framework which allows nonstiff and stiff predictor-corrector methods to be written, respectively, in block explicit and block diagonally implicit multivalue form;
- section 5.6: the construction of parallel one-block methods based on a multivalue format;

- section 5.7: an analysis and discussion on the limits of parallelism that the predictor-corrector approach can allow;
- section 5.8: the development of a rigorous tool for studying and hence constructing efficient predictor-corrector methods (based on the trivial predictor) with small error coefficients;
- section 5.9: the extension of the error analysis tool developed in section 5.8 to the case of higher-order predictors based on interpolation;
- section 5.10: the development of stiff methods based on the diagonally implicit iteration of Runge-Kutta methods, which can be viewed as the splitting of the coefficient matrix by a diagonal matrix;
- section 5.11: the generalization of the predictor-corrector approach to a new splitted process, which can be viewed as a preconditioning, and the implementation in a SIMD environment.

5.1 Introduction

Parallel techniques for solving differential systems can arise at many levels. At the first level there is the parallelization of extant sequential code through the use of parallelizing compilers and the restructuring of DO loops, while at the second level, parallelization of serial algorithms can take place. Both these approaches may offer very little, as the number of instructions that can be executed in parallel is often not very large. Of course, if the system is stiff and large, then enhanced performance can also be gained through the use of parallel linear algebra techniques

A more realistic approach requires the modification and redesign of sequential algorithms in terms of the target parallel architecture (SIMD, MIMD, hybrid, etc.). At a very basic level this can involve exploiting concurrent function evaluations within a step or methods that compute blocks of values simultaneously. This latter approach can be done, for example, by the use of block predictor-corrector methods which can be used in both stiff or nonstiff mode. Such approaches are called by Gear (1986) **parallelism across the method.**

A very important coarse-grained technique is via the decomposition of a problem into subproblems which can then be solved in parallel with the processors communicating as appropriate. In a sequential environment, the analogue of this approach is the so-called multirate method in which a system is decomposed in such a way that subproblems can be solved by different methods with differing stepsize strategies. The decoupled computations have to be synchronized in such a way that the largest stepsize taken is an integral multiple of all smaller stepsizes. Thus, for example, if (2.1.2) is decomposed into two subproblems of the form

$$\begin{aligned} y_1' &= f_1(x, y_1, y_2) \\ y_2' &= f_2(x, y_1, y_2) \end{aligned}$$

then N steps with stepsize h can be performed on subsystem 1 using extrapolated values of y_2, followed by a single step with stepsize $H = Nh$ on subsystem 2.

This approach has an obvious inherent coarse-grained parallelism which can be exploited (see Skelboe (1991), for example), but it is also a forerunner of a much more general class of decomposition techniques such as waveform relaxation and these methods will be discussed in Chapters 7, 8 and 9. Such approaches are examples of **parallelism across the system.**

A third approach called **parallelism across the steps** includes techniques (such as the parallel solution of linear or nonlinear recurrences) which solve simultaneously over a large number of steps.

It is highly likely that efficient parallel algorithms may well take elements from all three of these categories, so that such algorithms will lie in a three-dimensional space as indicated in Figure 5.1.

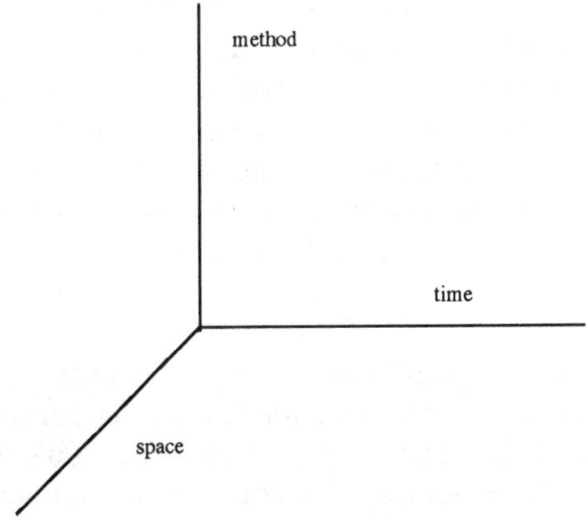

FIG. 5.1. the parallelism space

In designing and implementing parallel algorithms it is important to make the distinction between small-scale parallelism in which there may only be a dozen or so relatively powerful processors, or massive parallelism in which thousands of processors may be available. Thus as well as classifying parallelism by method, system and step it is also important to classify algorithms in terms of small-scale and large-scale parallelism, and to realize that techniques that exploit parallelism across the method, for example, are invariably examples of small-scale parallelism, while techniques that exploit parallelism across the system offer the possibility of massive parallelism if the differential systems are large enough.

It is also important to realize that in order to get effective performance from a parallel implementation, a number of conditions relating to the nature of the problem should hold. In particular, the problem should possess at least one of the following attributes:

- the function should be expensive to evaluate;
- the integration interval should be long;
- repeated integration, as in parameter fitting, is taking place;
- the dimension of the problem is large.

Of all these factors, perhaps the last is the most significant. There is no point in expecting reasonable parallel performance if the dimension of the problem is only of one or at most two orders of magnitude, since it is highly likely that for such problems communication time will dominate computation time. It is clearly also important to be able to match the algorithm to an appropriate architecture – a decomposition approach (such as a multirate technique) may be fine for a coarse-grained MIMD environment but be completely unsuited to a SIMD environment.

These remarks also suggest that the way algorithm development takes place may have to change when moving from the sequential to the parallel. In the sequential environment, many classes of efficient methods are fine tuned against a standard battery of test problems. This is especially true in the case of explicit Runge-Kutta methods which are often tuned against the problem set DETEST. However, this set of test problems is entirely unsuited to testing the efficacy of parallel algorithms, as most of the problems in DETEST do not satisfy the above-mentioned attributes. There seems to be no comparable set of test problems for fine-tuning parallel algorithms, and the fact that this is so should not be surprising. This is because any test problem set would be entirely architecture and implementation dependent. Some of these issues are further discussed in Burrage and Pohl (1993).

Gear in the 1980s was one of the first to investigate extensively the potential for parallelism in solving ordinary differential equations. In his first paper in this area (Gear (1986)), Gear discusses a number of application areas where parallelism can be exploited as well as major impediments to efficient parallel implementation. This discussion ranges from efficient parallel stepsize control based on delayed error computation to the importance of communication between stiff variables. Gear (1988) surveys parallel methods for ordinary differential equations and considers the possibility of exploiting large-scale parallelism across the method for solving stiff systems by performing as much as possible of the linear algebra associated with the implicit components asynchronously with the updating process. However, it is generally recognized now that method parallelism is suitable for low-degree parallelism only. Consequently, in later works, Gear (1991a) and Gear and Xuhai (1991) consider the possibility of massive parallelism across the system and across the steps, respectively. The basis of the work on parallelism across the system is through the analysis of the structure of the computational graph imposed by the differential equation.

The information flow in the solution of any IVP can be represented as in Figure 5.2,

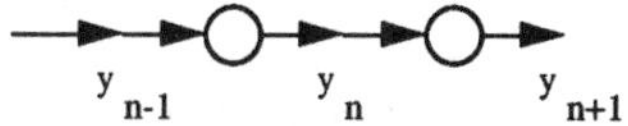

FIG. 5.2. information flow for an IVP

where the circles represent the computation necessary to advance the solution. Clearly there is no parallelism in this approach. However, as Gear (1991b) notes, these circles can be expanded as computational graphs and this can lead to method parallelism. Thus it is natural to restructure the computation graph to exploit parallelism.

For example, the computation graph for a $PE(CE)^2$ predictor-corrector method is shown below in Figure 5.3.

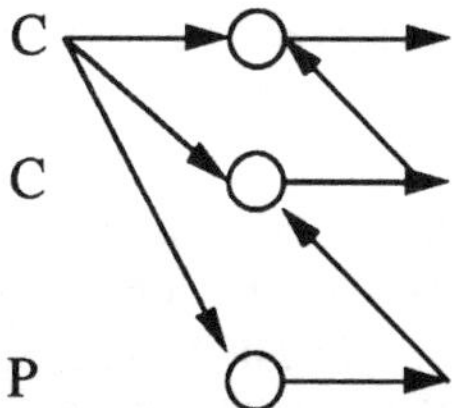

FIG. 5.3. $PE(CE)^2$ implementation

The use of the final corrector as the starting value for the predictor and subsequent corrections imparts an inherent sequentiality to the process. However, by modifying the order in which the prediction and correction at various step points are performed, a modest parallel performance is obtained through a frontal approach (Miranker and Liniger (1967)), as indicated in Figure 5.4.

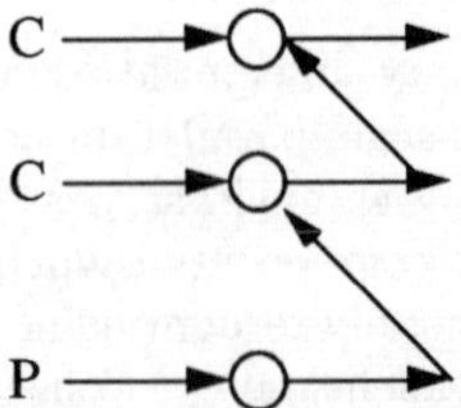

FIG. 5.4. a frontal approach

Although this computational graph approach has some uses in exploiting parallelism across the method it is of less use when trying to exploit parallelism across the system or across the steps. In an attempt to classify different approaches for exploiting parallelism across the system, Gear (1991a) identifies two types of systems: **homogeneous** and **heterogeneous**. An example of a homogeneous system is one arising from the semi-discretization of a partial differential equation in which the processing time per step is approximately the same for each subsystem so that each subsystem can be integrated on a different processor in a synchronous fashion. An example of a heterogeneous system is one derived from a lumping process such as that occuring in the modelling of VLSI circuits. In this case there is a lack of similarity between the subsystems. Thus there may be irregular communication between the subsystems, and poor load balancing due to some subsystems being able to use larger stepsizes than others. For this second class of problems a waveform relaxation approach may be appropriate, while in the case of homogeneous problems a straightforward parallelism across the system is a natural one especially for SIMD machines.

In exploiting parallelism across the system for time-dependent partial differential equations the main limitation is communication. For hyperbolic problems this is not such a problem since communication is limited to the speed along the characteristics. Thus as the system size increases, communication does not increase relative to computation. However, for parabolic problems information can travel much further so that communication apparently grows as $O(m^2)$. Nevertheless, Kuznetsov (1988) has noted that in order to obtain a given accuracy communication is only needed from mesh points a certain distance away and this limits communication growth. This implies that a SIMD implementation is a sensible approach in this case.

5.2 Direct methods

Direct methods attempt to exploit parallelism through concurrent function evaluations of multistage methods. Most of the work that has been done in this area has been applied to Runge-Kutta methods but the approach is equally applicable to more general multistage multivalue methods. The focus in this section will be on Runge-Kutta methods, and parallelism will be investigated based on an analysis of the sparsity pattern of the Runge-Kutta matrix. This approach has been formalized by Iserles and Nørsett (1990) who develop a theoretical framework for studying parallel Runge-Kutta methods based on diagraph analysis of the Runge-Kutta matrix. However, while the approach is satisfying from a theoretical viewpoint it really offers little additional insight that can be gained from constructing a few basic parallel structures, and it is this later approach that will be focused upon here.

As a first example consider a five-stage explicit Runge-Kutta method whose Runge-Kutta matrix has the sparsity pattern

$$\begin{pmatrix} 0 & & & & \\ \times & 0 & & & \\ \times & 0 & 0 & & \\ \times & \times & \times & 0 & \\ \times & \times & \times & 0 & 0 \end{pmatrix}$$

where the symbol $\times$ denotes a nonzero element. Then, clearly, stages 2 and 3, and stages 4 and 5 can each be computed concurrently, but at the cost of losing an order of accuracy that can usually be achieved with a five stage explicit method.

In attempting to construct direct methods which exploit some form of parallelism it is convenient to define the concept of a **parallel stage**. Thus a group of function evaluations is a parallel stage if all the arguments are independent and if all the function evaluations within the parallel stage r only use information from the first $r-1$ parallel stages.

Jackson and Nørsett (1990) consider block explicit Runge-Kutta methods in which the Runge-Kutta matrix, A, consists of $p \times p$ blocks with each block of at most dimension q and with A strictly block lower triangular. Such a method is called an **s-stage, p-parallel, q-processor** explicit method and the parallelism arises from the fact that up to q processors can be used to compute the stages within a block concurrently. The effective number of parallel stages in this case is p. Clearly in order to get some appropriate load balancing of the work the size of each block should be close to q (except for $A_{21}, \ldots, A_{p1}$ which typically have only one column each). However, because $A^p = 0$, the maximum order of such methods is at most p (independent of s and q) which is no improvement in order over the standard sequential case. Kiehl (1992) has generalized this result to other classes of one-step methods and has also shown that for ROW methods the minimum number of parallel stages needed to attain order p is $p-1$. Thus there is very limited potential for parallelism through the standard explicit approach, but for implicit methods there is greater flexibility.

Perhaps the simplest way to demand some form of parallelism in the implicit case is to require that the Runge-Kutta matrix be diagonal. This completely decouples the sm nonlinear systems of equations which define the internal stages into s independent nonlinear systems of dimension m. Unfortunately, the maximum order of such methods is two (although it can be greater than this for a restricted class of linear problems).

A natural way of generalizing this approach and overcoming this order barrier is to require that the Runge-Kutta matrix be block diagonal. In this case the blocks themselves can be chosen to give various efficiency properties. For example, Lie (1987) has constructed a four-stage A-stable method of order 3 suitable for a two-processor implementation in which the two diagonal blocks are themselves lower triangular with the same constant value on all four diagonal elements. By allowing each diagonal element in a given block to be the same, but these values to be different between the blocks, Iserles and Nørsett (1990) construct a two-parallel, two-processor, A-stable method of order 4 given by

$$\begin{array}{c|cccc}
\frac{1}{3} & \frac{1}{3} & & & \\
\frac{2}{3} & \frac{1}{3} & \frac{1}{3} & & \\
\frac{21+\sqrt{57}}{48} & 0 & 0 & \frac{21+\sqrt{57}}{48} & \\
\frac{27-\sqrt{57}}{48} & 0 & 0 & \frac{3-\sqrt{57}}{24} & \frac{21+\sqrt{57}}{48} \\
\hline
 & \frac{9+3\sqrt{57}}{16} & \frac{9+3\sqrt{57}}{16} & -\frac{(1+3\sqrt{57})}{16} & -\frac{(1+3\sqrt{57})}{16}
\end{array}$$

By assuming that these diagonal blocks are themselves full, Iserles and Nørsett (1990) have constructed a four-stage, two-parallel, two-processor method of order 4 which is L-stable given by

$$\begin{array}{c|cccc}
\frac{3-\sqrt{3}}{6} & \frac{5}{12} & \frac{1-2\sqrt{3}}{12} & & \\
\frac{3+\sqrt{3}}{6} & \frac{1+2\sqrt{3}}{12} & \frac{5}{12} & & \\
\frac{3-\sqrt{3}}{6} & 0 & 0 & \frac{1}{2} & \frac{-\sqrt{3}}{6} \\
\frac{3+\sqrt{3}}{6} & 0 & 0 & \frac{\sqrt{3}}{6} & \frac{1}{2} \\
\hline
 & \frac{3}{2} & \frac{3}{2} & -1 & -1
\end{array}$$

The cost here of a two-processor implementation is the same as a serial implementation of the two-stage Gauss method of order 4. The advantage is that this method is L-stable whereas the Gauss method is only A-stable.

The implicit examples given so far represent a complete decoupling of the blocks, but this is not necessary in order to extract some kind of parallelism. Jackson and Nørsett (1990) consider parallel diagonally implicit methods by allowing the Runge-Kutta matrix to be block lower triangular with diagonal blocks which are themselves diagonal. An example of such a method is the following A-stable method of order 4 suitable for implementation on two-processors:

$$\begin{array}{c|cccc} 1 & 1 & & & \\ \frac{3}{5} & 0 & \frac{3}{5} & & \\ 0 & \frac{171}{44} & -\frac{215}{44} & 1 & \\ \frac{2}{5} & -\frac{43}{20} & \frac{39}{20} & 0 & \frac{3}{5} \\ \hline & \frac{11}{72} & \frac{25}{72} & \frac{11}{72} & \frac{25}{72} \end{array}$$

Iserles and Nørsett (1990) have presented a similar method of order 4 which has the advantage that it is L-stable and has an order 3 method embedded within, which can be used for local error control. This method is given by

$$\begin{array}{c|cccc} \frac{1}{2} & \frac{1}{2} & & & \\ \frac{2}{3} & 0 & \frac{2}{3} & & \\ \frac{1}{2} & -\frac{5}{2} & \frac{5}{2} & \frac{1}{2} & \\ \frac{1}{3} & -\frac{5}{3} & \frac{4}{3} & 0 & \frac{2}{3} \\ \hline & -1 & \frac{3}{2} & -1 & \frac{3}{2} \end{array} \qquad (5.2.1)$$

For both of these methods, the parallelism arises from the fact that the first two stages can be computed concurrently, while the second set of two stages can then be calculated in parallel (using the same LU factors as for the first two stages) given that the first two stages have been computed sufficiently accurately.

In fact the effectiveness of this parallel diagonally implicit approach is limited by the following result which is due to Jackson and Nørsett (1990).

Theorem 5.2.1. *Let $\lambda_1, \ldots, \lambda_k$ be the distinct diagonal coefficients of A with respective multiplicities $m_1, \ldots, m_k$; then the maximum order w of any p-parallel, q-processor parallel diagonally implicit method satisfies*

$$w \leq 1 + \sum_{i=1}^{k} \min\{m_i, p\}.$$

This result is independent of s and q and can be proved by investigating the minimal polynomial of A and noting that it divides $\prod_{i=1}^{k}(x-\lambda_i)^{r_i}$, $r_i = \min\{m_i, p\}$. A consequence of this result is that the maximum order of any p-parallel, q-processor, singly diagonally implicit method is $p+1$ which is the same order bound for SDIRK methods when implemented serially.

Hence this order bound limits the effectiveness of parallel SDIRK methods However, if the diagonal elements are allowed to vary in a block, some advantages do arise as in method (5.2.1) which is L-stable of order 4.

As mentioned earlier, Iserles and Nørsett (1990) have developed a general framework to describe appropriate parallel sparsity patterns for the Runge-Kutta matrix, A. They construct a diagraph $G(A) = \{V, E\}$ where V denotes the set of vertices $\{1, 2, \ldots, s\}$ and E denotes a set of connections. Thus

$$(i,j) \in E \quad \text{if} \quad i,j \in V \quad \text{and} \quad a_{ij} \neq 0.$$

For any method, Iserles and Nørsett (1990) construct a characterizing triple denoted by (ν, μ, σ). Here σ denotes the length of the longest loop in the connections of E and can be interpreted as there being σ stages linked in an implicit irreducible block. Thus for a DIRK $\sigma = 1$ while for an explicit method $\sigma = 0$.

The parameter ν represents a processor partitioning of the method into ν disjoint nonempty sets $P_1, \ldots, P_\nu$, while for each P_k there is a certain level partitioning into μ disjoint but not necessarily empty sets $L_{k1}, \ldots, L_{k\mu}$ in which for each k and j all vertices in L_{kj} must belong to a loop and if $a_{ij} \neq 0$ then
(a) either there exist $p \in \{1, \ldots, \nu\}$ and $r, t \in \{1, \ldots, \mu\}$ such that

$$r \leq t, \quad i \in L_{pt}, \quad j \in L_{pr}$$

(b) or there exist $p, q \in \{1, \ldots, \nu\}$ and $r, t \in \{1, \ldots, \mu\}$ such that

$$p \neq q, \quad r < t, \quad i \in L_{qt}, \quad j \in L_{pr}.$$

This definition leads to a (ν, μ, σ) implementation of a method which is executed on ν processors in μ cycles with no more than σ copies of the derivative vector. For example, the method with the following sparsity pattern for the Runge-Kutta matrix and associated diagraph, given in Figure 5.5, has a (3,2,1) implementation.

$$\begin{pmatrix} \times & & & & & \\ & \times & & & & \\ & & \times & & & \\ \times & \times & \times & \times & & \\ \times & \times & \times & & \times & \\ \times & \times & \times & & & \times \end{pmatrix}$$

In the notation of Jackson and Nørsett (1990) this method would be called a six-stage, two-parallel, three-processor method. Since most of the

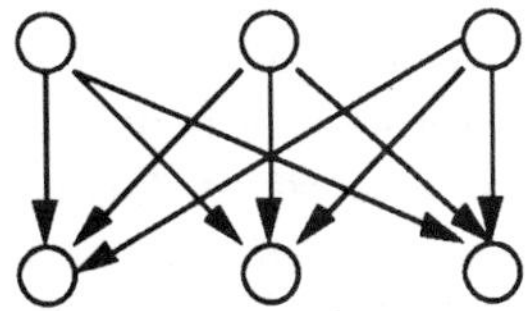

FIG. 5.5. diagraph – processors labelled row-wise

methods that are designed for a parallel implementation have a regular sparsity structure and a modest number of stages the simple approach of Jackson and Nørsett (1990) would appear to be preferable to the more complicated diagraph approach in characterizing the parallel properties of a method.

In order to see the likely performance of such methods, Kalvenes (1989) has performed some parallel implementations on the stiff DETEST package (Enright et al. (1975)) on a shared-memory machine (Alliant FX/8). In this case comparisons were made between SIMPLE, ODEPACK and PARK (a two-processor, block diagonally implicit A-stable Runge-Kutta method of order 4 due to Iserles and Nørsett (1990)). PARK was unable to solve the test problems from groups D–F and even though PARK runs on two-processors it was significantly slower than both SIMPLE and ODEPACK. In the case of stringent tolerances it was slower than ODEPACK by at least a factor of 5.

In conclusion, these approaches offer very limited parallelism. They can be viewed in some sense as a structural approach based on diagonal implicitness. Thus another way to perhaps gain some modest parallel performance is to base a structural approach on single implicitness but by removing the coupling in a transformed setting. This is done by assuming that A is similar to a diagonal matrix with real and distinct elements. This approach has been considered by Lie (1987), Nørsett and Simonsen (1987) and in considerable depth by Cooper (1991), and will be discussed in the next section.

5.3 Multiply implicit methods

In section 2.7 it was mentioned that of the family of implicit Runge-Kutta methods, those methods whose Runge-Kutta matrix (A) has a one-point spectrum are the most efficiently implementable on a sequential machine. These methods include SDIRKs in which the Runge-Kutta matrix is lower triangular with constant values on the diagonal and SIRKs in which A is similar to a single Jordan block.

SDIRKs, however, suffer from the drawback that their stage order is only one, whilst in the case of SIRKs, transforming and untransforming can be comparatively costly on small or moderate-sized problems.

In a parallel environment these methods may not necessarily be the most efficient implicit Runge-Kutta methods. Given an s-processor machine, for example, it is possible to decouple the sm nonlinear equations which define the internal stages into s independent blocks of dimension m and place each subsystem on a different processor. This can happen if A is similar to a diagonal matrix with distinct positive real entries, so that

$$A = T^{-1}DT, \quad D = \mathrm{diag}(\lambda_1, \ldots, \lambda_s). \tag{5.3.1}$$

Such methods are called by Keeling (1989) multiply implicit Runge-Kutta methods (MIRKs), and represent a natural generalization of singly implicit methods (Burrage (1978a)). Note that although the definition of Keeling is an obvious one, the acronym MIRK has been used by Cash (1975) to describe a class of methods known as mono-implicit Runge-Kutta methods. In order to avoid confusion, the class of methods characterized by (5.3.1) will be called parallel multiply implicit Runge-Kutta methods (PMIRKs).

Parallel implementations of PMIRKs have been considered by a number of authors including Lie (1987), Karakashian and Rust (1988) and Voss and Khaliq (1989). Using the notation of Chapter 2 and the tensor product notation described there, the nonlinear equations that describe the internal approximations $Y_1, \ldots, Y_s \in \mathbb{R}^m$ can be expressed as

$$\begin{aligned} G(Y): &= Y - e \otimes y_n - h(A \otimes I)F(Y) = 0 \\ Y &= (Y_1^\top, \ldots, Y_s^\top)^\top \in \mathbb{R}^{sm}. \end{aligned} \tag{5.3.2}$$

(5.3.2) can now be solved using a modified Newton iteration technique

$$\begin{aligned} M\Delta &= G(Y^{(k)}), \quad Y^{(k+1)} = Y^{(k)} - \Delta, \\ M &= I_s \otimes I_m - hA \otimes J, \end{aligned} \tag{5.3.3}$$

where J is the Jacobian matrix evaluated at one of the components of $Y^{(k)}$.

Using (5.3.1) and defining

$$\bar{\Delta} := (\bar{\Delta}_1^\top, \ldots, \bar{\Delta}_s^\top)^\top = T \otimes \Delta$$

$$\bar{G} := (\bar{G}_1^\top, \ldots, \bar{G}_s^\top)^\top = T \otimes G(Y),$$

then (5.3.3) can be split into s independent subsystems

$$(I_m - h\lambda_i)\bar{\Delta}_i = \bar{G}_i, \quad i = 1, \ldots, s. \tag{5.3.4}$$

The system is now decoupled and can be solved simultaneously in parallel independently of one another.

Collocation

If V is the $s \times s$ vandermonde matrix whose (i,j) element is c_i^{j-1} then the approach adopted by Burrage (1978b) in constructing the family of SIRKs with stage order $s-1$ and order s can be applied to the class of PMIRKs. Thus the general family of Runge-Kutta methods satisfying $C(s-1)$ and $B(s)$ is given by

$$A = V\bar{A}V^{-1}, \quad b^{\mathsf{T}} = (1, \ldots, 1/s)V^{-1}$$

$$\bar{A} = \begin{pmatrix} 0 & 0 & \cdots & 0 & \alpha_s \\ 1 & 0 & \cdots & 0 & \alpha_{s-1} \\ 0 & 1/2 & & & \\ \vdots & & \ddots & & \vdots \\ 0 & & & \frac{1}{s-1} & \alpha_1 \end{pmatrix}, \quad \alpha_j \in \mathbb{R}, \quad j = 1, \ldots, s. \quad (5.3.5)$$

Let $\lambda_1, \ldots, \lambda_s$ denote the real distinct eigenvalues of A and let $p_s(x, \lambda) = \prod_{j=1}^{s}(x - \lambda_j)$. Then

$$p_s(x, \lambda) = x^s - X_1(\lambda)x^{s-1} + X_2(\lambda)x^{s-2} + \cdots + (-1)^s X_s(\lambda), \quad (5.3.6)$$

where the $X_j(\lambda)$ are the **elementary symmetric functions** associated with p_s. Hence

$$X_1(\lambda) = \sum \lambda_i, \quad X_2(\lambda) = \sum_{i<j} \lambda_i\lambda_j, \quad \ldots, \quad X_s(\lambda) = \prod_{j=1}^{s} \lambda_j. \quad (5.3.7)$$

Then $\bar{A}$ will have eigenvalues $\lambda_1, \ldots, \lambda_s$ if and only if the characteristic polynomial of $\bar{A}$ given by

$$c(x) = x^s - \frac{1}{(s-1)!}\sum_{j=1}^{s}(s-j)!\alpha_j x^{s-j}$$

is identically equal to $p_s(x, \lambda)$. Hence

$$(s-j)!\alpha_j = (-1)^{j-1}(s-1)!X_j(\lambda). \quad (5.3.8)$$

If $\lambda_1 = \lambda_2 = \ldots = \lambda_s = \lambda$ then the family of singly implicit methods given in Burrage (1978b) is derived.

Just as in the case of SIRKs where collocation methods of stage order s can be constructed if and only if

$$L_s(c_j/\lambda) = 0, \quad j = 1, \ldots, s,$$

where the $L_k(x)$ are the **Laguerre polynomials** of degree k, orthogonal on $[0, \infty)$ with weight function e^{-x}, given by

$$L_k(x) = \sum_{j=0}^{k} (-1)^j \binom{k}{j} \frac{x^j}{j!}, \tag{5.3.9}$$

so collocatory PMIRKs can be constructed.

Noting that Runge-Kutta methods given by (5.3.5) are collocatory if and only if

$$\alpha = (\alpha_s, \ldots, \alpha_1)^\top = \frac{1}{s} V^{-1} c^s$$

or

$$\alpha_j = (-1)^{j-1} X_j(c)/s, \tag{5.3.10}$$

then from (5.3.8), necessarily

$$X_j(c) = \frac{s!}{(s-j)!} X_j(\lambda), \quad j = 1, \ldots, s. \tag{5.3.11}$$

Thus, given eigenvalues $\lambda_1, \ldots, \lambda_s$, the collocatory polynomial is (from (5.3.11)) given by

$$q_s(x) := \prod_{j=1}^{s} (x - c_j) = x^s - sX_1(\lambda)x^{s-1} + \ldots (-1)^s s! X_s(\lambda). \tag{5.3.12}$$

Example. The family of two-stage collocatory PMIRKs whose Runge-Kutta matrix has real eigenvalues λ_1 and λ_2 is given by (5.3.5), (5.3.10) and (5.3.11). It is characterized by

$$A = \frac{1}{c_2 - c_1} \begin{pmatrix} c_1 c_2 - c_1^2/2 & -c_1^2/2 \\ c_2^2/2 & -c_1 c_2 + c_2^2/2 \end{pmatrix}$$

$$b^\top = \frac{1}{c_2 - c_1} (c_2 - \frac{1}{2}, \frac{1}{2} - c_1),$$

where

$$c_1, c_2 = \lambda_1 + \lambda_2 \pm \sqrt{\lambda_1^2 + \lambda_2^2},$$

(see Karakashian and Rust (1988), for example). If, in addition,

$$\lambda_1 \lambda_2 - \frac{1}{2}(\lambda_1 + \lambda_2) + \frac{1}{6} = 0$$

these methods are of order 3. □

It is trivial to see from (5.3.12) that

$$q_s(x) = \prod_{j=1}^{s}(1 - \lambda_j \frac{d}{dx})x^s \tag{5.3.13}$$

so that

$$q_s'(x) = sq_{s-1}(x) - s\lambda_s q_{s-1}'(x). \tag{5.3.14}$$

Now given a set $\lambda_1, \ldots, \lambda_s$ it is important to be able to determine whether the collocatory polynomial $q_s(x)$ has real zeros or not. In the case of SIRKs it is easy to see from (5.3.9) and (5.3.12) that

$$q_s(x) = (-1)^s s! \lambda^s L_s(x/\lambda)$$

and so $\frac{c_1}{\lambda}, \ldots, \frac{c_s}{\lambda}$ must be the zeros of $L_s(x)$ which are known from orthogonal polynomial theory to be real, distinct and lie in $(0, \infty)$. The generalization of this result to PMIRKs is based on an elegant theory of biorthogonality developed by Iserles and Nørsett (1985, 1988).

The theory of biorthogonal polynomials is a generalization of orthogonality and is a useful technique in being able to prove results about sets of polynomials defined by some parameter set. In its generality a polynomial of degree n, $p_n(x, \alpha_1, \ldots, \alpha_n)$, is said to be biorthogonal on an interval (a, b) with respect to a set of distinct points $\alpha_1, \ldots, \alpha_n$ if

$$\int_a^b p_n(x, \alpha_1, \ldots, \alpha_n) d\phi(x, \alpha_j) = 0, \quad j = 1, \ldots, n, \tag{5.3.15}$$

where $\phi(x, \alpha)$, for every real α in a set $\mathcal{A}$, is a distribution function in x.

Denoting $p_n(x, \alpha_1, \ldots, \alpha_n)$ by $p_n(x, \alpha)$ and setting $p_0(x) = 1$ then the set $\{p_n\}_{n=0}^{\infty}$, given that it exists, is called by Iserles and Nørsett (1988) a biorthogonal polynomial system with respect to ϕ. Iserles and Nørsett (1988) study a particular subclass of biorthogonal systems where

$$d\phi(x, \alpha) = w(x, \alpha) d\sigma(x), \tag{5.3.16}$$

and σ is a distribution independent of α.

Many of the concepts and techniques used in the theory of orthogonal polynomials carry across to the theory of biorthogonality. In particular, Iserles and Nørsett (1988) state that w has the **interpolation property** if, with distinct $\alpha_1, \ldots, \alpha_n \in (a, b)$ and $y_1, \ldots, y_n \in \mathbb{R}$,

$$\sum_{j=1}^{n} w(x_k, \alpha_j) b_j = y_k, \quad k = 1, \ldots, n, \quad n \geq 1$$

has a solution, and hence prove

Theorem 5.3.1. *If w is C^1 in α and has the interpolation property on (a,b) then $p_n(x,\alpha)$ has n distinct, real zeros in (a,b).*

Iserles and Nørsett (1988) also show that there exist Rodrigues-type formulas for biorthogonal polynomial systems if, for example,

I: $\phi(x,\alpha) = x^\alpha/\alpha, \quad x \in (0,1), \quad \mathcal{A} = (0,\infty)$
II: $\phi(x,\alpha) = -e^{-x/\alpha}, \quad x \in (0,\infty), \quad \mathcal{A} = (0,\infty)$.

Furthermore,

$$\prod_{j=1}^{s}(1-\alpha_j\frac{d}{dx})x^s \tag{5.3.17}$$

is a Rodrigues-type formula in case II.

Comparing (5.3.13) and (5.3.14) it is immediately apparent that $q_s(x)$ is a biorthogonal polynomial with distribution $\phi(x,\lambda_k) = e^{-x/\lambda_k}$. Consequently, the following result holds:

Theorem 5.3.2. *The zeros of the collocating polynomial $q_s(x)$ are real, distinct and lie in $(0,\infty)$.*

It should be noted that Theorem 5.3.2 can be proven without using biorthogonality theory. Thus defining

$$\begin{aligned} Q(z) &= \sum_{j=0}^{s} q_j z^{s-j} = \prod_{j=1}^{s}(1-\lambda_j z) \\ N(z) &= \frac{1}{s!}\prod_{j=1}^{s}(z-c_j), \end{aligned}$$

Nørsett and Wanner (1979) have shown that for any collocation method

$$N(z) = \sum_{j=0}^{s} q_j \frac{z^j}{j!},$$

and hence that if Q has only real zeros then N must also have only real zeros. Bales et al. (1988) and Orel (1991) have generalized this result to show that if the zeros of Q are positive, real and distinct then the zeros of N have the same property.

The cost to implement an s-stage PMIRK on s processors is essentially one LU factorization plus one forward and back substitution or

$$m^3/3 + O(m^2) \text{ flops}.$$

This compares with

$$m^3/3 + sO(m^2) \text{ flops}$$

for the implementation of an s-stage SIRK in a sequential environment. If m is of moderate size there is no appreciable saving in costs by moving to a parallel environment for these methods. Thus, in order for PMIRKs to have any advantages they should have superior order, error or stability properties to SIRKs. Unfortunately, this does not appear to be the case.

Using the theory of C-polynomials, Nørsett and Wolfbrandt (1977) and Siemieniuch (1976) have shown that the maximum order of a PMIRK whose stability function has degree r in the numerator is $r+1$ and this is precisely the same order bound as for SIRKs. Hence weakening the eigenvalue condition of the Runge-Kutta matrix to allow for both real and distinct eigenvalues leads to no increase in order. Furthermore, in some sense, this weakening leads to more inefficient methods as the following results indicate.

Definition 5.3.3. *For any Runge-Kutta method of order w with stability function R, the exponential error constant, E_{w+1}, is defined as $e^z - R(z) = E_{w+1}z^{w+1} + O(z^{w+2})$.*

Now, since for any method of order w

$$R(z) = 1 + b^\top z(I - Az)^{-1}e = 1 + z + \cdots + \frac{z^w}{w!} + z^{w+1}b^\top A^w e + O(z^{w+2})$$

then

$$E_{w+1} = \frac{1}{(w+1)!} - b^\top A^w e. \qquad (5.3.18)$$

Using this fact, Nørsett and Wolfbrandt (1977) and Orel (1991) have shown

Theorem 5.3.4. *For any PMIRK of order w whose stability function has degree w in the numerator, the absolute value of the exponential error constant is minimized if all the eigenvalues are equal.*

Stability

The stability properties of PMIRKs have been investigated by Keeling (1989) and Bales et al. (1988) and again PMIRKs appear to offer little improvement over SIRKs in terms of A-stability. For example, it was shown by Wanner et al. (1978) that if an A-stable s-stage SIRK has order $s+1$ then necessarily s=1, 2, 3 or 5 and Keeling (1989) has shown that exactly the same result holds for PMIRKs.

By weakening the stability analysis to allow for A_0-stability (the stability region includes the negative real axis) or I-stability (the stability region includes the imaginary axis), Keeling (1989) has shown

Theorem 5.3.5. *For every s there exists an s-stage, A_0-stable PMIRK of order $s+1$, while for every even s there exists an s-stage, I-stable PMIRK of order s.*

Along similar lines Bales et al. (1988) have shown

Theorem 5.3.6. *The s-stage collocation PMIRKs of order s are A_0-stable if $\lambda_j \geq 1/2,\ j = 1, \ldots, s$.*

Implementation

Thus apparently, PMIRKs are particularly suited for problems arising from parabolic partial differential equations by the method of lines. A number of authors including Karahashian and Rust (1988) and Voss and Khaliq (1989) have implemented low-stage PMIRKs on a vector machine with a small number of processors. In particular, Karahashian and Rust (1988) report on the behaviour of a two-stage PMIRK of order three on (2.2.4) on a two-processor Cray, and compare this with the same implementation on one processor. They report that for $m < 50$, the two-processor implementation is less efficient while for $m = 25000$ a speed-up of 1.94 is obtained. The authors claim that "a high degree of parallelism of our schemes ... is indeed achievable on modern multiprocessor computers".

In a similar light, Voss and Khaliq (1989) have constructed a family of three-stage, L-stable, multiply implicit methods of order 3 and demonstrated an almost linear speed-up on three processors over the same method implemented on one processor.

However, these results need careful scrutiny and two remarks can be made.

1. Firstly, it makes no sense from an efficiency viewpoint to compare the speed-up of a two-stage, two-processor implementation with the same implementation on a one-processor machine since the two-stage PMIRK is not efficient in a sequential environment. The two-processor implementation should in fact be compared with an equivalent implementation of the A-stable, two-stage SIRK of order 3 which is efficient in a sequential environment.
2. Secondly, if this were done then it would seem likely that the two-processor implementation would in fact be slower than an efficient SIRK implementation.

Thus, in conclusion, although the theory of PMIRKs is aesthetically pleasing and often elegant, there would appear to be little future for these methods as practical and efficient solvers of stiff problems in a parallel environment. It should also be noted that the concept of PMIRKs can be extended to multivalue methods, but this approach would also suffer from the drawbacks outlined above.

Cooper (1991) has adopted a different approach by attempting a parallel implementation of singly implicit methods. Using the notation of (5.3.2) this gives

$$\begin{aligned} I_s \otimes (I - h\lambda J))\Delta &= (B \otimes I)G(Y^{(k)}) \\ Y^{(k+1)} &= Y^{(k)} - \Delta, \end{aligned} \tag{5.3.19}$$

where B is to be determined, and this allows all the stages to be computed in parallel.

Thus if Y represents the limit of the sequence computed in (5.3.19) and defining $V^{(k)} = Y - Y^{(k)}$, Cooper (1991) shows that

$$V^{(k+1)} = RV^{(k)} + W^{(k)},$$

where

$$R = ((I_s - B) \otimes I + \bar{h}(B\bar{A} - I_s) \otimes J)(I_s \otimes (I - \bar{h}J)^{-1})$$

and

$$\bar{h} = \lambda h, \quad A = \lambda \bar{A}.$$

Cooper (1991) then chooses B so that the spectral radius of R is zero for arbitrary J. This is achieved by letting

$$B = (1+\alpha)(\bar{A} + \alpha I_s)^{-1}, \quad \alpha \neq -1.$$

Cooper notes that an appropriate value for α is $\alpha = 1$ and also notes that for linear systems, the iteration scheme is guaranteed to terminate after s iterations since R is nilpotent for this choice of B.

In a parallel environment there is a clear advantage in using (5.3.19) over the modified Newton approach irrespective of the dimension of the problem. With this new scheme, the s sets of linear equations can be solved in parallel while the computation of G on the right-hand side of (5.3.19) and the transforming and the untransforming can be carried out using $3s$ sets of sm parallel multiplications. Thus given s processors, one iteration requires $3sm + m^2$ sets of multiplications performed in sequence.

In the modified Newton process, however, although the transformations can be performed using parallel multiplications, the s sets of linear equations have to be solved sequentially. Hence, $3sm + sm^2$ sets of multiplications have to be performed in sequence. Thus for m large, a speed-up of about s is available. However, since s is usually at most 5 or so, the speed-up is again modest. Nor does this take into account that the convergence rate of the new scheme will, in general, be slower than that of the modified Newton process. Nevertheless, this technique does offer genuine small-scale speed-up, unlike other methods that have been discussed in this section. Cooper and Vignesvaran (1993) have briefly investigated the efficiency of these schemes on a six-processor Sequent Symmetry machine using multitasking.

5.4 Prediction-correction: a background

Predictor-corrector methods have a long history in the numerical solution of IVPs. They have been proposed by Moulton (1926) in conjunction with Adams methods, by Rosser (1967) in conjunction with Runge-Kutta methods and by Shampine and Watts (1969) in terms of block methods.

The underlying feature of the traditional block method is that each application of the method generates a collection of equally spaced approximations to the solution. Block methods can be one-step in nature in which the values in a new block depend only on the last values in the preceding block or have a multiblock format in which the values depend on information from the previous blocks.

The advantage of using a block method in a parallel environment is that individual processors can compute, independently of one another, the approximation values to the solution at the different grid points. Generalizations of the traditional block method can be obtained by allowing the spacing of the grid points to be variable (see Burrage (1989), for example). This has considerable impact on both the stability and accuracy of the method, as was demonstrated in Burrage (1989).

Thus if Y_{n+1} represents a vector of solution values $y_{n+c_1}, \ldots, y_{n+c_s}$ and $F(Y_{n+1})$ represents the vector of derivatives $f(y_{n+c_1}), \ldots, f(y_{n+c_s})$, where c_s may equal 1 if this last value is used as an output point, then a s-processor implementation would hope to gain a speed-up of up to a factor of s by partitioning the problem and the method so that at each correction each processor simultaneously computes a subset of approximations.

An example of this approach is the two-point, fourth-order block method of Shampine and Watts (1969) given by

$$
\begin{aligned}
y_{n+1}^p &= \frac{1}{3}(y_{n-2}^c + y_{n-1}^c + y_n^c) + \frac{h}{6}(3f_{n-2}^c - 4f_{n-1}^c + 13f_n^c) \\
y_{n+2}^p &= \frac{1}{3}(y_{n-2}^c + y_{n-1}^c + y_p^c) + \frac{h}{12}(29f_{n-2}^c - 72f_{n-1}^c + 79f_n^c) \\
y_{n+1}^c &= y_n^c + \frac{h}{12}(5f_n^c + 8f_{n+1}^p - f_{n+2}^p) \\
y_{n+2}^c &= y_n^c + \frac{h}{12}(f_n^c + 4f_{n+1}^p + f_{n+2}^p),
\end{aligned}
$$

in which the two predicted steps are computed in parallel, followed by the concurrent evaluation of the two corrections. Note that in this example the solution values are computed at equidistant points (which, in general, need not be the case).

Similar approaches have been considered by Franklin (1978), Birta and

Abou-Rabia (1987) who have constructed block methods based on equidistant points, while Mouney, Authie and Gayraud (1991) have implemented an eight-point method of Birta and Abou-Rabia on a network of 32 transputers. Furthermore, Chu and Hamilton (1987) have constructed a two-block PECE method of order 4. However, its stability properties are inferior to those associated with the standard PECE approach based on the Adams methods of order 4 and 5.

A different approach was adopted by Miranker and Liniger (1967) and later refined by Katz, Franklin and Sen (1977) in which predictions and higher corrections at past points move forward simultaneously in a diagonal wavefront. An example of this technique is the following method

$$\begin{aligned} y^p_{n+1} &= y^c_{n-1} + 2hf^p_n \\ y^c_n &= y^c_{n-1} + \frac{h}{2}(f^p_n + f^c_{n-1}) \end{aligned}$$

in which y^p_{n+1} and y^c_n are computed in parallel.

Miranker and Liniger (1967) also considered the use of explicit Runge-Kutta methods in a diagonal wavefront and gave an example of a three-processor order 3 method given by

$$\begin{aligned} Y^1_{n+2} &= y^1_{n+2} \\ Y^2_{n+1} &= y^2_{n+1} + hc_1 f(Y^1_{n+1}) \\ Y^3_n &= y^3_n + h(c_2 - \frac{1}{6c_1})f(Y^1_n) + h\frac{1}{6c_1}f(Y^2_n) \\ y^1_{n+3} &= y^1_{n+2} + hf(Y^1_{n+2}) \\ y^2_{n+2} &= y^2_{n+1} + h((1 - \frac{1}{2c_1})f(Y^1_{n+1}) + \frac{1}{2c_1}f(Y^2_{n+1})) \\ y^3_{n+1} &= y^3_n + h((\frac{2c_2 - 1}{2c_1})(f(Y^1_n) - f(Y^2_n)) + f(Y^3_n)), \end{aligned}$$

where

$$c_1 = \frac{2(1 - 3c_2^2)}{3(1 - 2c_2)}.$$

This diagonal frontal approach can be generalized so that the computation of the predictor at x_{n+1}, the first corrector at x_n, the second correction at x_{n-1}, etc. are performed in parallel in conjunction with more sophisticated underlying methods. Such a technique can be represented graphically as in Figure 5.4. However, since, in general, relatively few corrections are needed to get suitable behaviour this parallel wavefront approach appears to be of limited value, allowing at best only very moderate parallelism.

One of the drawbacks of the method of Miranker and Liniger is that it is weakly stable and the global errors grow linearly with the number of integration steps. Fei (1994) gives a new class of parallel explicit Runge-Kutta formulas, which strictly speaking are not Runge-Kutta methods but pseudo-Runge-Kutta methods. With these methods all internal approximations in an integration step can be computed in parallel but they rely on all the internal approximations from the previous step. The general class of methods can be written as

$$\begin{aligned} Y_{n+1}^i &= y_n + h\sum_{j=1}^{s} a_{ij} f(Y_n^j), \quad i = 1, \ldots, s \\ y_{n+1} &= y_n + h\sum_{j=1}^{s} b_j f(Y_{n+1}^j). \end{aligned}$$

Of course, such a method can be written as a multivalue method which carries $s+1$ pieces of information from step to step. Fei (1994) constructs two-stage and three-stage methods of orders 3 and 4, with respect to some appropriate starting procedure.

Enenkel (1988) derives a number of different types of parallel predictor-corrector implementations based on Runge-Kutta methods. These include a pipelined approach which can be represented as

$$\begin{aligned} Y_{n+1}^{(0)} &= e \otimes y_n^{(l)} + (A_0 \otimes I)Y_n^{(l)} + h(L_0 \otimes I)F(Y_n^{(l)}) \\ Y_{n+1}^{(k)} &= e \otimes y_n^{(k)} + h(L \otimes I)F(Y_{n+1}^{(k-1)}), \quad k = 1, \ldots, l \\ y_{n+1}^{(k)} &= y_n^{(k)} + h(b^\top \otimes I)F(Y_{n+1}^{(k-1)}), \quad k = 1, \ldots, l. \end{aligned}$$

In this approach, $Y_{n+1}^{(k)}$ and $y_{n+1}^{(k)}$ are calculated as soon as $Y_{n+1}^{(k-1)}$ and $y_n^{(k)}$ are available. For example, in the case of $l = 3$, the computations proceed as described by Table 5.1.

Table 5.1 *pipelining*

	y_1^1	y_2^1	y_3^1	y_4^1
y_0	Y_1^1	Y_2^1	Y_3^1	Y_4^1
	F_1^1	F_2^1	F_3^1	F_4^1
		y_1^2	y_2^2	y_3^2
		F_1^2	F_2^2	F_3^2
			y_1^3	y_2^3

All operations in this table that are aligned above one another are performed in parallel. Apart from a startup cost, a new solution value is generated at each step. Enenkel (1988) shows that the order of the pipelined method increases by one at each correction until the order of the underlying corrector is reached. However, an examination of the error coefficients of the local truncation error is very complicated and not well understood so that proposed strategies for stepsize control are inconclusive.

5.5 A general framework

In designing efficient methods for parallel architectures attention must be given to such properties as order, stability and error behaviour. In this section a very general class of predictor corrector methods will be introduced which is sufficiently general to cover many methods of interest, and it will be shown how this class of methods can be written in block form as a multivalue method. This allows the order, stability and error behaviour theories presented in Chapter 3 for multivalue methods to be used in the construction of efficient parallel methods.

In order to illustrate this approach consider the block method in which a set of s values are updated concurrently at equidistant points $x_{n+1}, \dots, x_{n+s}$ by using an explicit Euler predictor and then corrected twice by a trapezoidal corrector applied in a composition case. This method can be computed in three steps for each equidistant step point $(j = 1, \dots, s)$ as

$$\begin{aligned} y_{n+j}^{(0)} &= y_n + jh\, f(x_n, y_n) \\ y_{n+j}^{(1)} &= y_n + \tfrac{h}{2} f(x_n, y_n) + h\sum_{i=1}^{j-1} f(x_{n+i}, y_{n+i}^{(0)}) + \tfrac{h}{2} f(x_{n+j}, y_{n+j}^{(0)}) \\ y_{n+j}^{(2)} &= y_n + \tfrac{h}{2} f(x_n, y_n) + h\sum_{i=1}^{j-1} f(x_{n+i}, y_{n+i}^{(1)}) + \tfrac{h}{2} f(x_{n+j}, y_{n+j}^{(1)}). \end{aligned} \tag{5.5.1}$$

This method was proposed by Abbas and Delves (1989) who have shown that it is of order 2 with local discretization error

$$\frac{-sh^3}{12} y^{(3)}(x_n) + O(h^4).$$

It has the obvious apparent advantage that the principal error term is linear in s. If higher-order methods could be constructed with similar error behaviour then they could be used with massive parallelism in mind. However, as will be seen in section 5.8 this is not in fact possible.

Although this method is a very simple one it is illustrative of a much more general technique in which a block of s values, with components

$y_1^{(n)}, \ldots, y_s^{(n)}$, are computed concurrently from step to step based on a Hermite predictor of the form

$$Y^{(0)} = (A_0 \otimes I)Y_n + h(L_0 \otimes I)F(Y_n) \tag{5.5.2}$$

and an implicit corrector (with $Z_n = (A_1 \otimes I)Y_n + h(L_1 \otimes I)F(Y_n)$)

$$\begin{aligned} Y^{(j)} &= Z_n + h(L \otimes I)F(Y^{(j-1)}), \quad j = 1, \ldots, l \\ Y_{n+1} &= Y^{(l)}, \end{aligned} \tag{5.5.3}$$

where $F(Y_n)$ denotes the vector with components $f(y_1^{(n)}), \ldots, f(y_s^{(n)})$, and where A_0, A_1, L_0, L_1 and L are all matrices of dimension s. Note that the update stage has been removed by assuming $c_s = 1$ and using the approximation at this point as the update, although this need not be the case. If an update calculation is required then the update is given by

$$Y_{n+1} = (A \otimes I)Y_n + h(B_1 \otimes I)F(Y_n) + h(B \otimes I)F(Y^{(l)}). \tag{5.5.4}$$

In general, this class of methods requires a starting procedure to generate $Y^{(0)}$ from y_0, the most natural being based on the correction of y_0 a certain number of times by an s-stage Runge-Kutta method. It was noted in Tam (1989) and Burrage (1991) that if (5.5.3) is corrected an arbitrary number of times then it can be always written in a block explicit multivalue formulation. For example, if method (5.5.3) and (5.5.4) is corrected $l-2$ times then it can be written as the block explicit multivalue method

$$\begin{array}{c|cccccc} I & 0 & & & & & \\ A_0 & L_0 & 0 & & & & \\ A_1 & L_1 & L & 0 & & & \\ \vdots & \vdots & & \ddots & \ddots & & \\ A_1 & L_1 & 0 & \ldots & L & 0 & \\ \hline A & B_1 & 0 & \ldots & \ldots & 0 & B \end{array} \tag{5.5.5}$$

with sl stages.

As a particular subclass of (5.5.5), Tam (1989, 1992a, 1992b) constructs families of block methods in which the scaled stability regions do not deteriorate as the order increases. In particular, Tam constructs a family of one-correction methods in which the predictor has order equal to the block-size and the corrector has order one more than the predictor. The stability region of this method is the interior of the region $\{z \in \mathbb{C} : |1+z+z^2/2| = 1\}$. However, although the stability regions do not deteriorate with increasing order, the error coefficients do (see Burrage (1989) for a more detailed discussion). Hence methods of order 5 or more are inappropriate because of poor error behaviour. These methods are described in more detail in section 5.7.

In another approach, van der Houwen and Sommeijer (1988)) consider the use of an s-stage implicit Runge-Kutta corrector of order p in conjunction with the trivial predictor based on the updated value y_n at the end of a step. If the Runge-Kutta method given by (2.7.2) is used as a corrector and this method is corrected $l-1$ times and the update is computed at the end of the step then it can be written as an ls-stage explicit Runge-Kutta method given by

$$\begin{array}{c|ccccc} 0 & 0 & & & & \\ c & A & 0 & & & \\ c & 0 & A & 0 & & \\ \vdots & \vdots & & \ddots & \ddots & \\ c & 0 & \dots & 0 & A & 0 \\ \hline & 0 & \dots & 0 & 0 & b^T \end{array} \tag{5.5.6}$$

Because a general predictor-corrector method can be written in block explicit multivalue format, standard order and stability theories described in Chapter 3 can be used. In particular, for the class of methods described by (5.5.6) Jackson and Nørsett (1991) and Burrage (1991, 1993a) have shown that each application of a corrector increases the order of the overall method by one until the order of the corrector is reached. In particular if the starting procedure and predictor in (5.5.2) are of order p_1 and l corrections are performed with a Runge-Kutta corrector of order p then the order of (5.5.3) is $\min\{p, p_1 + l\}$. A consequence of this is that if a trivial predictor is used and a high-order method of order 8, say, is required then eight corrections are needed. Burrage (1993a) has shown that by choosing the coefficients of the corrector appropriately it is possible to obtain an increase in the order of more than 1 after each correction if a non-trivial predictor is used.

However, as noted in Chapter 3, the efficacy of any explicit method is strongly dependent on the size of the local error coefficients. Consequently, Burrage (1991, 1993a) has developed a theory which enables an error analysis of any standard prediction-correction scheme. Because of the complex mixing of predictor and corrector errors, Burrage has shown that efficient parallel methods are unlikely to be constructed without a rigorous analysis of the local error terms. This is described more fully in section 5.8.

Numerical results

In order to judge the efficacy of some of these approaches, a method of order 8 based on the trivial predictor and a five-stage Lobatto corrector of order 8 which is corrected eight times (called LOB8) is compared with the thirteen stage DORP7(8) Runge-Kutta method (see Prince and Dormand (1981)). This latter method is generally recognized as being a very efficient sequential method for tolerances within the range 10^{-6} to 10^{-13}.

These two methods have been implemented with the same variable step strategy in Occam2 on a single T800 transputer running at 20MHz on the two body problem, on the interval $[0,2\pi]$, with eccentricity =0.6, given by

$$
\begin{aligned}
y_1' &= y_3, \quad y_1(0) = 1-\epsilon \\
y_2' &= y_4, \quad y_2(0) = 0 \\
y_3' &= \frac{-y_4}{(y_1^2+y_2^2)^{3/2}}, \quad y_3(0) = 0 \\
y_4' &= \frac{-y_2}{(y_1^2+y_2^2)^{3/2}}, \quad y_4(0) = \sqrt{\frac{1+\epsilon}{1-\epsilon}}.
\end{aligned}
$$

The results obtained for a variety of different tolerances $10^{-2},\ldots,10^{-12}$ are given in Table 5.2 (with the time being given in terms of the number of ticks of the computer clock).

Comparing LOB8 and DORP7(8) it can be seen that LOB8 takes approximately three times as many function evaluations, and a factor varying between 1.5 and 3.3 times as many ticks of the computer clock. Furthermore, as the error tolerance is made more stringent, the global error in LOB8 becomes approximately four times as large as that for DORP7(8). This problem is a very smooth one, and taking these results as typical the following observations can be made.

If function evaluations are expensive, so that they dominate computation, then LOB8 is approximately three times as expensive as DORP7(8). But LOB8 is only a four-processor method (since the five-stage Lobatto method has as its first intermediate approximation the trivial calculation $Y_1 = y_n$), and so the maximum speed-up is (ignoring communication time) at most $\frac{4}{3}$. On the other hand, if function evaluations are cheap then the more reliable figure is computer time and then at lax tolerances LOB8 is roughly only twice as expensive, and so the maximum speed-up is 2 in this case. It should be noted that van der Houwen and Sommeijer (1988) obtained similar results using a four-stage Gauss method of order 8 (called Gauss8) as a corrector.

Table 5.2 *DORP8 and LOB8*

DORP8	Tol	Steps	Functions	Time	Global error
	(-2)	8	128	2131	1.79(-2)
	(-4)	12	192	2127	4.09(-4)
	(-6)	18	318	2713	4.46(-6)
	(-8)	28	496	3529	6.15(-8)
	(-10)	48	816	5012	9.97(-10)
	(-12)	82	1102	6321	9.10(-12)
LOB8	Tol	Steps	Functions	Time	Global error
	(-2)	9	372	3283	2.07(-2)
	(-4)	13	564	4304	7.07(-4)
	(-6)	20	976	6500	1.12(-5)
	(-8)	33	1572	9654	2.52(-7)
	(-10)	53	2276	13347	3.63(-9)
	(-12)	90	3784	21295	3.78(-11)

Convergence and stability

Van der Houwen et al. (1994) have extended the work on parallel block methods based on the fixed-point iteration of (5.5.3) and a prediction of the form (5.5.2). It is clear from (5.5.3) that the matrix L has a crucial role in determining the convergence rate of the correction process. For example, for the standard linear test problem it is easily seen that

$$\varepsilon^{(k)} = z^k L^k \varepsilon^{(0)}, \quad \varepsilon^{(k)} := Y^{(k)} - Y.$$

Given l iterations, then in the maximum norm the maximum stepsize that can be taken to ensure convergence is given by

$$h_l = \theta_l / \rho(\frac{\partial f}{\partial y}), \tag{5.5.7}$$

where

$$\theta_l^l := 1/ \parallel L^l \parallel_\infty . \tag{5.5.8}$$

Thus the **region of convergence** in the complex plane is given by

$$C_l = \{z : |z| < \theta_l\}. \tag{5.5.9}$$

On the other hand a stability condition must also be imposed on (5.5.2) and (5.5.3) so that

$$\sigma(h\frac{\partial f}{\partial y}) \subset S \cap C_l \tag{5.5.10}$$

and so there is no point in constructing methods for which S (the stability region) is much larger than C_l. However, large convergence regions can improve convergence speed.

As a consequence, van der Houwen et al. (1994) attempt to construct effective correctors with large values for θ_l and reasonably large stability intervals on the negative real and imaginary axes. The abscissae points are chosen to be the Radau II quadrature points and the above three quantities are calculated for various Adams-type correctors (those methods with $A_1 = (0, e)$) with or without output formula, as well as more general correctors. They note that by introducing an external stage (output formula) the stability can be improved, but at the cost of an additional function evaluation. Van der Houwen et al. (1994) develop a five-processor, order 8 method based on an Adams-type corrector and show a speed-up of about 2.7 over the 8(7) Runge-Kutta method of Prince and Dormand (1981) based on the number of sequential right-hand side function evaluations needed to obtain a certain accuracy on two simple test problems.

Burrage and Jackiewicz (1995) construct block methods of the form (5.5.5) where the underlying corrector is an s-stage, r-step multistep Runge-Kutta method. If such a method is of Radau type (with $c_s = 1$) then the maximum attainable order is $2s + r - 2$. The advantage of this approach is that since the stage order of these methods is $s + r - 1$ the predictor can also be of order $s + r - 1$ and so only $s - 1$ corrections are needed to obtain maximal order. Because of their high order, these methods are effective when stringent tolerance is imposed. Indeed these methods turn out to be very effective in a sequential environment, as high order methods are easily achieved with quite low values of s and r.

It is clear from (5.5.7)-(5.5.10), that the predictor-corrector approach described by (5.5.2) and (5.5.3) is essentially a nonstiff approach. This can be further seen by analysing, for example, the linear stability properties of (5.5.6). Thus if a Runge-Kutta method is corrected l times from the trivial predictor then the stability polynomial is given by

$$p(z) = 1 + zb^\top e + \dots z^{l+1} b^\top A^l e.$$

If the order of the corrector is w and $l = w - 1$ then

$$p(z) = 1 + z + \dots z^w/w! \tag{5.5.11}$$

The stability regions associated with (5.5.11) can easily be evaluated. If R denotes the stability interval $[-R, 0]$ along the negative real axis and I denotes the stability interval $[-I, I]$ along the imaginary axis then values

Table 5.3 *stability intervals*

w	1	2	3	4	5	6	7
R	2	2	2.52	2.78	3.22	3.55	3.95
I	0	0	1.73	2.82	0	0	1.76

of R and I for various values of w are given in Table 5.3. It can be seen that R only increases slowly as w increases and for a low-order method $w = 4$ appears to be a suitable value. However, this shows that methods such as (5.5.6) are totally unsuited for stiff problems.

Diagonally implicit iterations

Predictor-corrector methods for stiff problems can easily be constructed and placed in block multivalue format as block lower triangular rather than strictly block lower triangular. Thus, for example, van der Houwen et al. (1992b) have constructed a family of methods, called PDIRKs, based on a Runge-Kutta corrector and a diagonally implicit iteration scheme. Hence, if the Runge-Kutta corrector is stiffly accurate, then this iteration scheme with l iterations can be written as

$$\begin{aligned} Y^{(k+1)} - h(D \otimes I)F(Y^{(k+1)}) &= e \otimes y_n + h((A - D) \otimes I)F(Y^{(k)}) \\ y_{n+1} &= e_s^T Y^{(l)}. \end{aligned} \tag{5.5.12}$$

Here D is a diagonal matrix which can be chosen to obtain good stability behaviour and the method can be interpreted as a diagonally implicit prediction-correction scheme in which the corrector is split by a positive diagonal matrix. In a more general setting by writing the underlying corrector in (5.5.3) as

$$Y = (A \otimes I)Y_n + h(L_1 \otimes I)F(Y_n) + h((L - D) \otimes I)F(Y) + h(D \otimes I)F(Y)$$

a fixed-point iteration scheme similar to (5.5.12) can be used.

Because D is diagonal these iterations in (5.5.12) can be performed in parallel with similar implementational costs to sequential DIRKs. Although in this case the stage order is at most 2 an analysis by van der Houwen and Sommeijer (1993) has shown that it is possible to choose D by minimizing the spectral radius of the amplification matrix over the spectrum of the eigenvalues associated with the problem, which results in a much improved performance than this limit on the stage order might otherwise have suggested. Diagonally implicit iterations are described in greater detail in section 5.10.

5.6 Parallel one-block methods

Many extant methods (Miranker and Liniger (1967), Shampine and Watts (1969), Worland (1976), Chu and Hamilton (1987)) can be simply described in multivalue form. However, van der Houwen and Sommeijer (1992) have used the general format described by (5.3.5) to derive more general formulas. They restrict the class of methods studied to methods of the form

$$\begin{array}{c|cc} I & 0 & 0 \\ A & B & L \\ \hline \end{array} \tag{5.6.1}$$

where, if $L = 0$, this method can be interpreted as a predictor method.

By requiring that all components in (5.6.1) have order w, this implies that (5.6.1), as a multivalue method, has stage order w. Hence, from (3.2.15), with $q(\tau^j) = (c-e)^j$, the order conditions for (5.6.1) are given by

$$A(c-e)^j + j[B(c-e)^{j-1} + Lc^{j-1}] - c^j = 0, \quad j = 0, \ldots, w. \tag{5.6.2}$$

It is necessary for these methods to be zero-stable and a very simple way of doing this is to assume that A is the zero matrix apart from the last column which is a column of unity values. This matrix will henceforth be denoted by $\mathbf{E}$ and such methods will, as in van der Houwen and Sommeijer (1992), be called **Adams-type methods**. An example of the general family of explicit Adams-type methods with two values per block and order 2 is given by

$$\begin{array}{cc|cc} 1 & 0 & & \\ 0 & 1 & & \\ \hline 0 & 1 & \frac{-c^2}{2(1-c)} & \frac{c(2-c)}{1-c} \\ 0 & 1 & \frac{-1}{2(1-c)} & \frac{3-2c}{2(1-c)} \end{array} \tag{5.6.3}$$

Here $c \neq 1$. Van der Houwen and Sommeijer (1992) show that this method has order 3 at the step points if $c = \frac{5}{3}$. The choice $c = 0$ reduces to the Adams-Bashforth method, while the choice $c = 2$ reduces to a method of Miranker and Liniger (1967).

The order of (5.6.3) can be increased by at least one if the requirement that the method be of Adams-type is weakened. In this case van der Houwen and Sommeijer (1992) construct a method of stage order 3 given by

$$
\begin{array}{cc|cc}
1 & 0 & & \\
0 & 1 & & \\
\hline
\frac{c^2(c-3)}{(c-1)^3} & \frac{3c-1}{(c-1)^3} & \frac{c^2}{(c-1)^2} & \frac{c}{(c-1)^2} \\
\frac{3c-5}{(c-1)^3} & \frac{(c+1)(c-2)}{(c-1)^3} & \frac{2-c}{(c-1)^2} & \frac{(c-2)^2}{(c-1)^2}
\end{array}
\tag{5.6.4}
$$

It is easily shown that this method is zero-stable if $c \leq 1-\sqrt{3}$ or $c \geq 1+\sqrt{3}$ and the choice $c = \frac{1}{2}$ reduces to the method of Chu and Hamilton (1987).

Van der Houwen and Sommeijer (1992) also construct a family of implicit Adams-type methods of order 4 with two values per block given by

$$
\begin{array}{cc|cccc}
1 & 0 & & & & \\
0 & 1 & & & & \\
0 & 1 & \frac{c^3}{12(c-1)} & \frac{c(c^2-6c+6)}{12(1-c)} & \frac{c(c^2-6c+6)}{12(1-c)} & \frac{c^3}{12(c-1)} \\
0 & 1 & \frac{1-2c}{12(1-c)(1-2c)} & \frac{6c^2-10c+3}{12c(c-1)} & \frac{2c-3}{12c(c-1)} & \frac{6c^2-14c+7}{12(c-1)(c-2)} \\
\hline
\end{array}
$$

The choice of $c = 1 - \frac{1}{\sqrt{5}}$ guarantees order 5 at the step points. Van der Houwen and Sommeijer (1992) suggest an implementation of this method in $PE(CE)^l$ mode where the predictor is as in (5.6.3). They also generalize this construction procedure to methods with three and four values per block.

Sommeijer et al. (1992) consider a different approach to constructing implicit block methods of the form (5.6.1) by allowing L to be a diagonal matrix. Such methods are said to be **parallel block methods**, and represent a decoupling of the derivative coefficient matrix. Thus given s processors each processor has to evaluate a component of $F(Y_n)$ and solve an implicit system of equations whose dimension is that of the original system. If this is done by some form of the modified Newton-Raphson method then each processor needs the matrix $M_i = I - hl_{ii}J_i$ and its LU factorization, where J_i is some approximation to the Jacobian of the problem evaluated at a suitable point. If, for example, $J_1, J_2, \ldots, J_s$ are all approximated by the Jacobian, J, evaluated at one point say y_n then

$$
\begin{aligned}
M_i &= (1 - \frac{l_{ii}}{l_{11}})I + \frac{l_{ii}}{l_{11}} M_1, \quad i = 2, \ldots, s, \\
M_1 &= I - hl_{11}J.
\end{aligned}
\tag{5.6.5}
$$

In this case if there are more than s processors available it may be possible for additional processors to compute M_1 and then to communicate this to the M_i. This will require considerable communication and may only be feasible if other computation can be overlapped with the communication of M_1. However, if it is assumed that

$$l_{ii} = \theta, \quad i = 1, \ldots, s, \tag{5.6.6}$$

so that $L = \theta I$, then $M_i = M_1$, $i = 2, \ldots, s$. Hence only one LU factor need be computed per step, rather than s as in the case of (5.6.4).

The order properties of parallel block methods can be studied via (5.6.2) with L diagonal. The stability matrix for such methods is given by

$$M(z) = (I - zL)^{-1}(A + zB)$$

and it is clear that a necessary condition for A-stability (assuming some irreducibility conditions) is that the $l_{ii} > 0$, $i = 1, \ldots, s$. In addition, from Theorem 3.3.1 it is easily seen that a necessary condition for L-stability is $\rho(L^{-1}B) = 0$.

Sommeijer et al. (1992) construct a two-point, A-stable method of order 3 given by

$$A = \begin{pmatrix} 0 & 1 \\ 0 & 1 \end{pmatrix}, \ B = \begin{pmatrix} \frac{147}{220} & \frac{161}{220} \\ \frac{-50}{33} & \frac{23}{66} \end{pmatrix}, \ L = \begin{pmatrix} \frac{7}{10} & 0 \\ 0 & \frac{13}{8} \end{pmatrix}, \ c = \begin{pmatrix} 2.1 \\ 1 \end{pmatrix}$$

and also construct a three-point, A-stable method of order 4. Sommeijer et al. (1992) attempt to construct higher-order methods that are A-stable but because of the plethora of parameters are only able to construct order 5 methods that are almost A-stable.

Chartier (1993) has adopted a different viewpoint to Sommeijer et al. (1992) by dramatically restricting the free parameter space by setting

$$B = 0,$$

which also ensures stability at infinity. In addition, by setting

$$c = (1, \ldots, s)^\top$$

and L (diagonal) such that

$$l_{ii} = (1 + c_i)/\theta,$$

Chartier (1993) attempts to choose θ so that families of L-stable methods of stage order $s - 1$ are constructed.

In doing this, Chartier modifies the order conditions of (5.6.1) to get more general order conditions. By writing (5.6.1) as

$$Y_{n+1} = (A \otimes I)Y_n + h(B \otimes I)F(Y_n) + h(L \otimes I)F(Y_{n+1}) \tag{5.6.7}$$

then (5.6.7) can of course be written as a multivalue method in which the vector $(Y_n^\top, hF(Y_n)^\top)^\top$ is carried from step to step. In this case the correct value function is given by $z(x,h)$ where $z(x,h) = (u^\top, v^\top)^\top$ has components with

$$\begin{aligned} u^\top &= (y^\top(x+(c_1-1)h), \ldots, y^\top(x+(c_s-1)h)) \\ v^\top &= (hy'^\top(x+(c_1-1)h) \ldots, hy'^\top(x+(c_s-1)h)). \end{aligned}$$

One of the abscissae is chosen to equal 1, in order to get an update value at the end of a step.

Recalling the definitions of the A-type, B-type and C-type simplifying assumptions given in (3.2.10), (3.2.12) and (3.2.13) then if $A(w-1)$ holds so that

$$q(t) = q(\tau^p), \quad \forall \rho(t) \leq p, \quad p \leq w-1, \tag{5.6.8}$$

where

$$q(\tau^p) = \left((c_1-1)^p, \ldots, (c_s-1)^p, p(c_1-1)^{p-1}, \ldots, p(c_s-1)^{p-1}\right)^\top,$$

then from Theorem 3.2.5, (5.6.7) will be of order w if

$$A(c-e)^j + j(B(c-e)^{j-1} + Lc^{j-1}) = c^j, \quad j = 0, \ldots, w-1 \tag{5.6.9}$$

and there exist vectors q_1 and q_2 such that

$$\begin{pmatrix} A-I & B \\ 0 & -I \end{pmatrix} \begin{pmatrix} q_1 \\ q_2 \end{pmatrix} = \begin{pmatrix} c^p - (c-e)^p - pLc^{p-1} \\ -p(c-e)^{p-1} \end{pmatrix}.$$

This final condition is of course equivalent to

$$(A-I)q_1 = c^p - (c-e)^p - pB_2(c-e)^{p-1} - pLc^{p-1}. \tag{5.6.10}$$

Sommeijer et al. (1992) assume that q_1 in (5.6.10) is given by

$$q_1 = (c-e)^p,$$

while Chartier (1993) does not make this assumption and this gives additional flexibility. By noting that A must always have at least one unity eigenvalue then there always exists a projection matrix E_1 such that

$$E_1(A-I) = 0. \tag{5.6.11}$$

For the methods under construction, E_1 is of rank 1 so that (5.6.10) and (5.6.11) impose only a single constraint on the method parameters, in this case θ. Thus methods of order s are obtained. In fact the values $s = 2, (\theta = 4)$ and $s = 4, (\theta = 5)$ give L-stable methods of order s with smallest error constants.

Chartier has proven that there exist L-stable methods of order s for $s \leq 6$ and conjectured (based on numerical evidence) that there exist L-stable methods of order s for $s \leq 11$. The advantage of these schemes is that s processors can be exploited to implement them with the same effective computational cost as a BDF method of order s, but the parallel block methods have vastly superior stability properties. This represents an attempt to use parallelism to improve the robustness of algorithms. Similar approaches will be further discussed in section 6.5.

5.7 Limits on parallelism

Tam (1989) noted that many proposed explicit parallel methods have substantially smaller stability regions than sequential methods of the same order, so that if a problem is stability-bound, then no matter how many processors are available these parallel methods will not be efficient. Tam (1989) also noted that in terms of comparing methods, it is the scaled stability region and not the stability region that is significant. Thus if S is the stability region associated with some s-stage method, then the **scaled stability region** S_c is defined to be

$$S_c = \{\frac{1}{s}z : z \in S\}.$$

The reason for this of course is that it is the scaled stepsize $\frac{h}{s}$ that measures the efficiency of a method in terms of the time stepped per unit work.

In comparing scaled stability regions, Tam (1989) uses the approach of Jeltsch and Nevanlinna (1981, 1982) by measuring the largest disk passing through the origin that can be inscribed in the scaled stability region. This is a sensible approach because the disk is significant in understanding the error propagation of methods when applied to nonlinear problems. Tam (1989) has extended the work of Jeltsch and Nevanlinna (1982), who have studied sufficient conditions for a sequential method to have an **optimal scaled stability region** (a region that cannot be properly contained in that of another method), to parallel methods. The essential result here is that for sequential or parallel methods, the largest disk that can be inscribed in a scaled stability region is the unit circle centred on $(-1, 0)$. In this sense the Euler method is optimal. However, this does not take into account problems which may have localized spectra such as those arising from parabolic or hyperbolic problems.

Given this result, Tam (1989, 1992a, 1992b) suggests constructing parallel methods which have scaled stability regions as good as, or almost as good as, existing sequential methods but with possibly superior order or smaller error constants. Taking this to its logical conclusion, Tam (1989, 1992a) constructs a family of one-stage block methods whose stability polynomial is equal to perfect powers of that of the Euler method and a family of two-stage block methods whose stability polynomial is equal to perfect powers of the second-order Taylor series method (Tam 1992b). Thus a p-processor implementation would have stability polynomials, respectively,

$$(w-(1+z))^p, \quad (w-(1+z+z^2/2))^p.$$

Because these stability polynomials are perfect powers, Tam notes that there is a subtle difference between the stability regions on the boundary between, for example, the one-stage block method and the Euler method. This shows that it is not the stability polynomial but the minimal polynomial of the stability matrix that determines a stability region.

An example of a family of one-stage, two-processor methods of order two is given in Tam (1989) as

$$\begin{aligned} y_{n+3/2} &= (1-b)y_{n+1/2} + by_n + \tfrac{h}{4}((8+b)f_{n+1/2} + (b-4)f_n) \\ y_{n+1} &= (1-a)y_{n+1/2} + ay_n + \tfrac{h}{4}((3+a)f_{n+1/2} + (a-1)f_n). \end{aligned}$$

When $a=1$, $b=0$ this method has stability polynomial $(w-(1+z))^2$.

The principal local truncation coefficient of this method can be shown to have the form

$$(-7,-1)^\top \frac{h^3}{24} y^{(3)}(\theta),$$

but Tam notes that there is a subtle interplay between the two components which results in an accelerated error growth for certain ranges of eigenvalues associated with the Jacobian of the problem. By extending the theory given in Hairer et al. (1987), which describes the relationship between local and global errors, to block methods, Tam shows that it is possible to write down a differential equation which describes how the global errors in the two components $(e_1(x), e_2(x))$ vary as a function of x. In fact for the model test problem (2.4.6) it can be shown, up to $O(h^3)$, that

$$\begin{pmatrix} e_1'(x) \\ e_2'(x) \end{pmatrix} = qB \begin{pmatrix} e_1(x) \\ e_2(x) \end{pmatrix} - \frac{1}{24} \begin{pmatrix} 7 \\ 1 \end{pmatrix} y^{(3)}(x), \tag{5.7.1}$$

where B is the matrix in the second-order method of Tam that multiplies the derivative term. For the choice $a=1$, $b=0$ this matrix is defective and it can be shown from (5.7.1) that $e_2(x)$ behaves as $O(x^2)e^{\lambda x}$. For eigenvalues along the imaginary axis this gives $O(x^2)$ growth rather than

$O(x)$ growth. In fact, for a p-processor implementation there would be an $O(x^p)$ growth in this case.

Tam (1989) also gives an example of a two-stage method given by

$$
\begin{aligned}
y^p_{n+3/2} &= \frac{1}{7}(3y_{n+1/2} + 4y_n) + \frac{h}{7}(15f_{n-1/2} - 6f_n) \\
y^p_{n+1} &= -4y_{n+1/2} + 5y_n + h(2f_{n-1/2} + f_n) \\
y_{n+3/2} &= y_{n+1/2} + \frac{7h}{18}f^p_{n+3/2} + \frac{h}{18}(15f_{n+1/2} - 4f_n) \\
y_{n+1} &= y_n + \frac{1}{6}f^p_{n+1} + \frac{h}{6}(4f_{n+1/2} + f_n),
\end{aligned}
$$

which has stability polynomial $(w - (1 + z + z^2/2))^2$. Thus the stability region is the interior of the stability region of the second-order Taylor series method. In this case a global error analysis implies that the error propagation matrix is the identity matrix and the accelerated error growth is much less severe than in the one-stage case.

Tam (1989) has compared the p-processor, two-stage method with the order $p+1$ Adams PECE method on DETEST for $p = 2$, 3 and 4 and for $p = 4$ obtains a possible speed-up of about 2.5 based on comparison of function evaluations. An additional advantage of the two-stage block family of methods is that each processor only uses the function of the predicted value computed by itself, since the appropriate scaling matrices are diagonal. This is in contrast to the block methods of Shampine and Watts (1969) or the predictor-corrector multiblock methods of Chu and Hamilton (1987) which require two broadcasts per step for the corrector stage. Naturally, the approach of Tam (1989) can also be extended to multiblock methods.

Skeel and Tam (1989) also use the general framework of multivalue methods to obtain results about the likely performance of parallel methods when applied to nonstiff problems. As has already been noted, if at each step of a method, r values are carried from step to step and s stages are computed with each stage involving up to p independent function evaluations which are performed in parallel on p processors, then this method can be written as a multivalue method in which the derivative coefficient matrix (B_1) is block strictly lower triangular and is of dimension ps, while the update matrix B_2 is $r \times ps$. If this method is now applied to the standard test problem (2.4.6) then the $r \times r$ stability matrix given by (3.3.1) is a matrix whose entries are polynomial of at most degree s. Hence Skeel and Tam (1989) are able to show that given any preconsistent, p processor method with $q := \text{rank}(B_2) < p$ then there exists a preconsistent q processor method with the same stability matrix.

An important consequence of this is the following result:

Theorem 5.7.1. *There is no advantage in having more processors than the number of saved values.*

On the basis of this result there appears to be little future for parallelism based on one-step methods. However, this result is only valid for the standard linear test problem. For more general nonlinear problems the local discretization error is a complicated mixture of elementary differentials whose error coefficients can be manipulated to obtain some advantages, as will be seen in sections 5.8 and 5.9.

5.8 An error analysis

In an attempt to explain the numerical results given in Table 5.2 in section 5.5, and with the view to designing effective numerical methods for the numerical solution of nonstiff problems by controlling the magnitude and the nature of the truncation coefficients, Burrage (1991, 1993a) has developed a comprehensive theory based on the use of Butcher series which allows the analysis of the local error of any method of the form (5.5.6). Burrage (1991) has applied this technique to an analysis of the local behaviour of predictor-corrector methods based on the trival predictor and an implicit Runge-Kutta corrector and has compared the local error behaviour of the Gauss8 method with the local error behaviour of DORP7(8). He has shown that the principal error coefficients of both these methods have comparable magnitude (although this is a function of the large number of corrections that are required in the predictor-corrector implementation), and that the exact difference depends on the type of norm used to measure the error coefficients and the nature of the elementary differentials that appear in the local error expansion.

In order to show how this can be done, this general error analysis technique is now described and, as a particular example, will be applied to the method given in (5.5.1).

Recalling now the definitions of order and error as applied to multivalue methods and in order to avoid tensor product notation it will be assumed, without loss of generality, that the differential equation is scalar. Thus suppose that the correct value function has an expansion as a Butcher series given by

$$z(x,h) = B(q, y(x))$$

and that the internal vectors at each iteration, Y, and the local error, l_{n+1}, have Butcher series expansions

$$Y = B(k, y(x_n)), \quad l_{n+1} = B(e, y(x_n)).$$

Consider now the general predictor-corrector method described by (5.5.2), (5.5.3) and (5.5.4), with $B_1 = 0$, which can be written in block explicit multivalue form as

$$\begin{array}{c|ccccc} I & 0 & & & & \\ A_0 & L_0 & 0 & & & \\ A_1 & 0 & L & 0 & & \\ \vdots & \vdots & & \ddots & \ddots & \\ A_1 & 0 & \dots & 0 & L & 0 \\ \hline A & 0 & \dots & 0 & 0 & B \end{array} \tag{5.8.1}$$

with sl stages. Here it is assumed that $l-2$ corrections are performed as well as an updating stage. The following result is a direct consequence of the order analysis presented in Chapter 3.

Theorem 5.8.1. *Let $z(x,h)$ be the correct value function associated with (5.8.1) with an error expansion $B(q, y(x))$ as a Butcher series and let Y_n have the Butcher series expansion $B(k_0, y(x_n))$ and if l_{n+1}, the local error vector, has an expansion as $B(e, y(x_n))$ then*

$$l_{n+1} = \sum_{t\in T^*} e(t)[F(t)]y(x_n)\frac{h^{\rho(t)}}{\rho(t)!}, \tag{5.8.2}$$

where for $t = [t_1, \dots, t_v]$

$$e(t) = \sum_{u\subseteq t} \binom{\rho(t)}{\rho(u)} \frac{\alpha(u,t)}{\alpha(t)} q(u) - Aq(t) - \rho(t)B\prod_{i=1}^{v} k_{l-1}(t_i), \tag{5.8.3}$$

with

$$\begin{aligned} k_1(t) &= A_0 q(t) + \rho(t)L_0\prod_{i=1}^{v} k_0(t_i) \\ k_{j+1}(t) &= A_1 q(t) + \rho(t)L_1\prod_{i=1}^{v} k_j(t_i), \quad j = 1,\dots,l-2. \end{aligned} \tag{5.8.4}$$

It should be noted that if Y_n has order p in all its components then

$$k_0(t) = (c-e)^{\rho(t)}, \quad \forall \rho(t) \le p.$$

In the case of Runge-Kutta methods of the form given in (5.5.6), Theorem 5.8.1 is simplified. This simplification is given in the following result.

Corollary 5.8.2. *If a Runge-Kutta corrector is applied to the trivial predictor with $l-1$ corrections then the local error is given by*

$$l_{n+1} = \sum_{t \in T^*} e(t)[F(t)]y(x_n)\frac{h^{\rho(t)}}{\rho(t)!},$$

where for $t = [t_1, \ldots, t_v]$

$$e(t) = 1 - \rho(t)b^T \prod_{i=1}^{v} k_{l-1}(t_i), \qquad (5.8.5)$$

with

$$\begin{aligned} k_0(\phi) &= e, \quad k_0(t) = 0, \quad \rho(t) > 0 \\ k_{j+1}(t) &= \rho(t)A\prod_{i=1}^{v} k_j(t_i), \quad j = 0, \ldots, l-2. \end{aligned} \qquad (5.8.6)$$

Proof: By writing a Runge-Kutta method in multivalue format with

$$A_0 = A_1 = A = (\,0 \quad e\,), \quad B = \begin{pmatrix} 0^\mathsf{T} \\ b^\mathsf{T} \end{pmatrix},$$

where 0 is the $s \times (s-1)$ matrix of zeros, and with

$$q(\phi) = 1, \quad q(t) = 0, \quad \rho(t) > 0,$$

the result is immediate from Theorem 5.8.1. □

Corollary 5.8.2. can now be applied to study the behaviour of the local error of (5.5.6) in which an arbitrary Runge-Kutta method of order p is corrected $p-1$ times to give a predictor-corrector method of order p.

In order to accomplish this, the height, $h(t)$, of a tree t must be defined.

Definition 5.8.3. *A tree $t = [t_1, \ldots, t_v]$ is said to be of height $h(t)$ if $h(t) = 1 + \max_{j=1,\ldots,v} \{h(t_j)\}, \quad h(\tau) = 1$.*

Let t_w denote an arbitrary tree of order w; then using Corollary 5.8.2 it is easy to show by an induction argument that

$$k_{p-1}(t) = k(t), \quad \forall t \text{ with } \rho(t) \le p-1$$

$$k_{p-1}(t_p) = \begin{cases} k(t), & h(t) \neq p-1 \\ 0, & \text{otherwise} \end{cases}$$

$$k_{p-1}(t_{p+1}) = \begin{cases} k(t), & h(t) < p-1 \\ 0, & \text{otherwise,} \end{cases}$$

where the $k(t)$ are the terms associated with the underlying Runge-Kutta method. Thus $k(\phi) = e$ and for $t = [t_1, \ldots, t_v]$

$$k(t) = \rho(t) A \prod_{i=1}^{v} k(t_i).$$

Consequently, if

$$\hat{l}_{n+1} = \sum_{\rho(t) \geq p+1} \hat{e}(t)[F(t)]y(x_n) \frac{h^{\rho(t)}}{\rho(t)!}$$

is the local truncation error for the underlying Runge-Kutta corrector of order p then the local truncation error for (5.5.6) can be written as

$$\begin{aligned} l_{n+1} &= \frac{h^{p+1}}{(p+1)!}\Bigg(\sum_{h(t_{p+1})<p} \hat{e}(t_{p+1})[F(t_{p+1})]y(x_n) + (f')^p f(y(x_n)) \Bigg) \\ &+ \frac{h^{p+2}}{(p+2)!}\Bigg(\sum_{h(t_{p+2})<p} \hat{e}(t_{p+2})[F(t_{p+2})]y(x_n) \\ &+ \sum_{h(t_{p+2}) \geq p} [\bar{F}(t_{p+2})]y(x_n) \Bigg) + O(h^{p+3}), \end{aligned} \quad (5.8.7)$$

where the summation denotes summation over all possible trees of the appropriate order.

In studying (5.8.7) it is seen that the influence of the predictor occurs in the trees of largest height. The general trend is that as more corrections are performed the order of the predictor-corrector method approaches the order of the corrector. However, when the number of corrections is such that the order cannot increase then the effect of more corrections is to shift the errors due to the predictor further away from the principal error term. Thus, for example, if an additional correction is performed then it is easy to show that the local error is, in this case,

$$\begin{aligned} l_{n+1} &= \frac{h^{p+1}}{(p+1)!}\Bigg(\sum_{\rho(t)=p+1} \hat{e}(t_{p+1})[F(t_{p+1})]y(x_n) \Bigg) \\ &+ \frac{h^{p+2}}{(p+2)!}\Bigg(\sum_{h(t_{p+2})<p+1} \hat{e}(t_{p+2})[F(t_{p+2})]y(x_n) \end{aligned} \quad (5.8.8)$$

$$+ \quad (f')^{p+1} f(y(x_n)) \Big) + O(h^{p+3}).$$

Here the principal error term is due purely to the errors in the corrector while the first contribution of the predictor occurs only as one term in the h^{p+2} term. It is important to analyse the error terms in the error expansion at least as far as the first contribution from the predictor. It may happen (as it does in (5.8.7) and also does, as will be seen below, for the Abbas and Delves method given in (5.5.1)) that the principal error term depends only on the corrector and is quite small while the second term, which can have a mix of predictor and corrector errors, may be large and dominate the first term.

The error analysis presented above has been extended by a number of authors to give a complete characterization of the local errors when the implicit equations defining the internal components of a Runge-Kutta method are solved by a variety of techniques. These include the modified Newton method, in which the Jacobian is updated only at the beginning of each step, the exact Newton method and the modified Newton method when the Jacobian is perturbed by some $O(h)$ term. This latter variant corresponds to a regular updating but not at every step. This extension has some relevance in a parallel environment since the structure of the approximation to the Jacobian can often be exploited in parallel computation. Jackson et al. (1992) have considered what happens in the case of nonstiff problems, while van Dorsselaer and Spijker (1992) have given an analysis for stiff dissipative systems. Kvaernø(1992) has extended this work to DAEs of index one. The analysis especially in the DAE case is quite technical and will not be given here but the basic result in the nonstiff case is that for a Runge-Kutta method of order w and stage order r with initial predictor of order q to the internal stages then the order in the update after k iterations takes on the following values:

$$\begin{aligned} \min\{w, q+k+1\} &\quad - \quad \text{modified Newton} \\ \min\{w, t+k+1\} &\quad - \quad O(h) \\ \min\{w, 2^k(t+1)+q-t-1\} &\quad - \quad \text{exact Newton,} \end{aligned}$$

where $t = \min\{r, q\}$.

Massive parallelism via prediction-correction?

(5.5.1) has been suggested as a method which can offer the possibilities of massive parallelism in that the local error has the form

$$\frac{-sh^3}{12} y^{(3)}(x_n) + O(h^4).$$

Apparently the local error does not grow cubically in s but only linearly and so it seems natural to take s (which represents the number of processors) as large as possible. This claim is now investigated in some detail.

The correcting method of (5.5.1) has the update included as part of the correction process, and so the corrector can be written in the form

$$\begin{array}{c|cccccc} 0 & 0 & & & & \\ 1 & 1/2 & 1/2 & & & \\ 2 & 1/2 & 1 & 1/2 & & \\ \vdots & \vdots & \vdots & \ddots & \ddots & \\ s & 1/2 & 1 & \cdots & 1 & 1/2 \\ \hline \end{array}$$

Denoting the abscissae of this method by $\hat{c}$ and the Runge-Kutta matrix by $\hat{A}$ then from (5.8.5) and (5.8.6) of Corollary 5.8.2 it is easy to show that

$$\begin{aligned} k_2(\tau^2) &= 2\hat{A}\hat{c}, & k_2(\tau^3) &= 3\hat{A}\hat{c}^2, & k_2([\tau^2]) &= 0, \\ k_3(\tau^3) &= 3\hat{A}\hat{c}^2, & k_3([\tau^2]) &= 6\hat{A}^2\hat{c}, & k_3(\tau^4) &= 4\hat{A}\hat{c}^3, \\ k_3([\tau,\tau^2]) &= 8\hat{A}\hat{C}\hat{A}\hat{c}, & k_3([\tau^3]) &= 12\hat{A}^2\hat{c}^2, & k_3([[\tau^2]]) &= 0, \end{aligned}$$

where $\hat{C} = \text{diag}(0, 1, \ldots, s)$.

Defining $b^\top = e_{s+1}^\top \hat{A}$ then the local error of (5.5.1) is from Corollary 5.8.2 given by

$$l_{n+1} = \frac{h^3}{3!}(s^3 - 3b^\top \hat{c}^2) y^{(3)}(x_n) + \frac{h^4}{4!} \sum_{\rho(t)=4} e(t)[F(t)]y(x_n) + O(h^5),$$

where

$$\begin{aligned} e([\tau^4]) &= s^4 - 4b^\top \hat{c}^3 &= -s^2 \\ e([\tau,\tau^2]) &= s^4 - 4b^\top \hat{c}^3 &= -s^2 \\ e([\tau^3]) &= s^4 - 12b^\top \hat{A}\hat{c}^2 &= -2s^2 \\ e([[\tau^2]]) &= s^4 & \\ s^3 - 3b^\top \hat{c}^2 &= -s/2. & \end{aligned}$$

Thus although the principal error is only linear in the blocksize there is a contribution from the tall tree in the h^4 term of $\frac{(sh)^4}{24}(f')^3 f$. In most cases this term will dominate the principal error term so that the prospects of massive parallelism with s large is clearly an illusion in this particular example. In fact this example is symptomatic of all prediction-correction processes and limits the potential parallelism of all such approaches to at best a modest one.

Error behaviour

It is instructive to compare the error behaviour of some of the high-order predictor-corrector Runge-Kutta methods with existing methods for solving non-stiff problems on a sequential machine. A very popular and effective class of methods are the Runge-Kutta pairs of Dormand and Prince, and in particular, DP45 (7 stages) given in Dormand and Prince (1980) and PD56 (8 stages) and PD78 (13 stages), given in Prince and Dormand (1981). The principal error coefficients for each of these three methods will be compared with those for the predictor-corrector methods of orders 5, 6 and 8 based on the trivial predictor and a corrector which is either the three-stage Radau IIA method of order 5 or the three and four-stage Gauss methods of orders 6 and 8, respectively.

From (5.8.7) the principal error coefficients for the predictor-corrector methods are the same as the principal error coefficients for the corrector method apart from the term corresponding to the tallest tree. In each of the three cases being considered here the C-type simplifying assumptions associated with the corrector methods will collapse the error terms of the corrector down to a small number of non-collapsible trees. Thus denoting the three predictor-corrector methods by Radau5, Gauss6 and Gauss8 it is easy to show that the number of distinct error terms in the principal error coefficients are 5, 10 and 22, respectively, as compared with the 21, 48 and 286 trees of orders 6, 7 and 9, respectively.

For any collocation method of order $p \geq s$ satisfying $B(p)$ and $C(s)$ it can be shown that the error terms for the h^{p+1} term have the form

$$\hat{e}(t) = \gamma_t(1 - (p+1)b^\top c^p), \quad \forall \rho(t) = p + 1, \tag{5.8.9}$$

where γ_t is a rational number depending on tree t. Thus for Radau5, Gauss6 and Gauss8, tables are presented for the frequency (f) and the γ_t of each of the non-collapsible trees.

In the case of the Radau5 method, the distinct trees are

$$\tau^6, \quad [\tau, \tau^4], \quad [\tau^5], \quad [[\tau^4]], \quad [[[[\tau^2]]]],$$

which will be denoted by $\hat{t}_1, \ldots, \hat{t}_5$.

In the case of the Gauss6 method, the distinct trees are

$$\tau^7, \quad [\tau^2, \tau^4], \quad [\tau, \tau, \tau^4], \quad [\tau, \tau^5], \quad [\tau, [\tau^4]], \quad [\tau^6]$$

$$[[\tau, \tau^4]], \quad [[\tau^5]], \quad [[[\tau^4]]], \quad [[[[[\tau^2]]]]],$$

which will be denoted by $T_1, \ldots, T_{10}$.

Table 5.4 *Radau5 method*

Tree	$\hat{t}_1$	$\hat{t}_2$	$\hat{t}_3$	$\hat{t}_4$	$\hat{t}_5$
f	6	5	5	3	1
γ	1	2	5	6	

Table 5.5 *Gauss6 method*

Tree	T_1	T_2	T_3	T_4	T_5	T_6	T_7	T_8	T_9	T_{10}
f	11	4	4	5	4	7	4	5	4	1
γ	1	8/3	4/3	5/2	10/3	6	8	15	20	

Finally, in the case of the Gauss8 method, the distinct trees are

$$\tau^9,\quad [\tau^3,\tau^5],\quad [\tau,\tau^2,\tau^5],\quad [\tau,\tau,\tau,\tau^5],\quad [\tau,\tau,\tau^6],\quad [\tau^2,\tau^6],$$

$$[\tau,\tau,[\tau^5]],\quad [\tau^2,[\tau^5]],\quad [\tau,\tau^7],\quad [\tau,[\tau,\tau^5]],\quad [\tau,[\tau^6]],\quad [\tau,[[\tau^5]]],$$

$$[\tau^8],\quad [[\tau^2,\tau^5]],\quad [[\tau,\tau,\tau^5]],\quad [[\tau,\tau^6]],\quad [[\tau,[\tau^5]]],$$

$$[[\tau^7]],\quad [[[\tau,\tau^5]]],\quad [[[\tau^6]]],\quad [[[[\tau^5]]]],\quad [[[[[[[\tau^2]]]]]]],$$

which will be labelled by $t_1,\ldots,t_{22}$.

Table 5.6 *Gauss8 method*

Tree	t_1	t_2	t_3	t_4	t_5	t_6	t_7	t_8	t_9	t_{10}	t_{11}
f	49	18	9	9	11	11	9	9	19	9	11
γ	1	15/4	5/4	5/4	2	4	5/2	5/2	7/2	35/8	7
Tree	t_{12}	t_{13}	t_{14}	t_{15}	t_{16}	t_{17}	t_{18}	t_{19}	t_{20}	t_{21}	t_{22}
f	9	27	9	9	11	9	19	9	11	9	1
γ	35	8	10	10	8	20	28	35	56	70	

Note that when computing the norm of the principal error coefficients, f is to be multiplied by γ as well as by the quadrature error term $1-(p+1)b^{\top}c^p$ for all trees except the tallest tree. Finally, note that for the Radau5, Gauss6 and Gauss8 methods the absolute value of the quadrature error term can be shown to be, respectively, $\frac{1}{100}$, $\frac{1}{400}$, $\frac{1}{4900}$.

The norms of the principal error coefficients are computed from the above tables and are presented in Table 5.7 along with the norms of the principal error coefficients of the higher-order Prince and Dormand pairs which were given to the author by Sharp in a private communication. The errors have been scaled by the appropriate factorial factor.

Table 5.7 *norms of error coefficients*

Method	$\|e(t_{p+1})\|_2$	$\|e(t_{p+1})\|_1$	$\|e(t_{p+1})\|_\infty$
DP45	3.9910^{-4}	7.3510^{-4}	2.7810^{-4}
PD56	2.3310^{-4}	1.1810^{-3}	8.8210^{-5}
PD78	4.5110^{-6}	4.2510^{-5}	1.0410^{-6}
Radau5	1.4210^{-3}	2.3810^{-3}	1.3810^{-3}
Gauss6	2.0610^{-4}	3.3110^{-4}	1.9810^{-4}
Gauss8	2.8310^{-6}	4.6810^{-6}	2.7610^{-6}

Conclusions

The errors due to the predictor-corrector methods seem to compare well with the Dormand and Prince pairs especially as the order increases. On the other hand, for some mildly stiff problems for which $(f')^p f$ can grow rapidly, the predictor error can dominate the corrector error. This is seen from Table 5.7 in which the infinity norm of the Radau5 and Gauss8 methods are much larger than the corresponding infinity norm for DP45 and PD78.

Thus it seems that the high-order methods considered here, such as those based on Gauss or Radau correctors, do have suitable error properties, but at the cost of requiring many iterations if the trivial predictor is used. But it is important to note from (5.8.7) that the error coefficients associated with the next power of h have a strong contribution from the predictor which the Dormand and Prince Runge-Kutta pairs do not have. In fact it is easy to show that the number of trees of order $p+2$ with height p or more is $p+1$. Thus the h^{p+2} error contribution from the predictor in the 1-norm is $\frac{p+1}{(p+2)!}$ as compared with an h^{p+1} contribution of $\frac{1}{(p+1)!}$. Thus the error contribution from the predictor takes a long time to die away.

The usefulness of the theoretical tools developed in this section is seen by presenting some results from van der Houwen and Sommeijer (1988) who have implemented a variable-step, predictor-corrector Runge-Kutta method based on the s-stage (with $s = 4$ and 5), order $2s$ Gauss-Legendre corrector corrected $2s-1$ times. They compare this with the same variable-step implementation of the 8(7) Prince and Dormand method on three relatively simple nonlinear test problems. In the case that the corrector is an order 8 method, a very rough conclusion that can be drawn from their results is that between 0.63 and 0.96 of the number of function evaluations of the PD pair is needed to get the same accuracy with their predictor-corrector pair, although, of course, this figure depends on both the problem and the accuracy required. This factor can be explained by arguing that since the errors of both methods are comparable, approximately the same stepsizes would be taken with both implementations and so efficiencies can be compared by comparing function evaluations per integration step.

In the case of the PD78 method there are 13 function evaluations per step while for the Gauss8 method there are eight function evaluations per step for each of the four processors. The ratio of $\frac{8}{13}$ is in reasonable agreement. Van der Houwen and Sommeijer also compare the 10-stage Gauss method with the PD78 implementation and in this case approximately half the number of function evaluations per processor are needed. This improvement is, of course, due to the smaller error constants which would allow the Gauss10 method to take a slightly larger stepsize than in the PD78 case. However, one must be wary of situations for which the predictor error, because of its size relative to the individual error terms due to the corrector, dominates.

Thus, in conclusion, both practical and theoretical results suggest that a parallel implementation based on a trivial predictor and a Runge-Kutta corrector with a large number of correctors would yield, at best, a meagre speed-up. If the Runge-Kutta corrector is replaced by a multistep Runge-Kutta corrector (which can have a higher order for the same amount of work), some improvement is possible. However, it appears that a high-order multistep-type predictor is needed to reduce the number of corrections. As a consequence of this and similar results van der Houwen and Sommeijer (1992) construct predictor-corrector methods of up to order 8 and with at most one correction. These methods have a number of free parameters and these are chosen either by requiring reasonably large intervals of stability or by minimising the error in the corrector. But they do not consider stability regions, so that these methods may not be suitable for problems with eigenvalues near the imaginary axis, and, in addition, they do not analyse the contribution that the predictor makes to the overall error.

5.9 High order predictors

The previous analysis demonstrates the importance of examining more than just the principal error term in the local error expansion of any predictor-corrector method. For methods based on the trivial predictor and a high order corrector an error analysis indicates the satisfactory behaviour of the error, but this is a function of the large number of corrections that have to be performed per step.

It is the influence of the predictor error for the tall trees which prevents any type of massive parallelism or even moderate parallelism for methods based on a Runge-Kutta corrector and the trivial predictor. After l corrections there is a principal local error term arising from the predictor of the form $\frac{h^{l+1}}{(l+1)!}(f')^l f(y(x_n))$, so that, necessarily, l must be large to obtain any sort of satisfactory error behaviour.

Perhaps a more realistic requirement in the design of accurate and efficient methods is to restrict the number of corrections to at most two. In this case the predictor should be a multistep predictor of sufficiently high order. For these high-order predictions, the influence of the predictor is capable of being partially modified by the choice of the method parameters (see Burrage (1990b, 1993a), for example).

With the view to constructing efficient predictor-corrector methods with only one or two corrections, Burrage (1993a) has extended the local error analysis used in Burrage (1991) to obtain formulas for the local error expansion of the class of methods based on using an s-stage collocation Runge-Kutta method of order $p > s$ as a corrector, while the predictor is obtained by interpolating the derivative values. Since the stage order of the corrector is s the predictor will also have order s. This is an example of using an Adams-type predictor where the prediction is based only on the value of the approximation at the end of the step and the s intermediate derivative approximations. This class of methods can be written as

$$\begin{aligned} Y_{n+1}^{(0)} &= e \otimes y_n + h(A_p \otimes I)F(Y_n^{(l)}) \\ Y_{n+1}^{(k+1)} &= e \otimes y_n + h(A \otimes I)F(Y_{n+1}^{(k)}), \quad k = 0, \ldots, l-1 \\ y_{n+1} &= e_s^\top Y_{n+1}^{(l)}, \end{aligned} \tag{5.9.1}$$

where A is the Runge-Kutta matrix, and A_p is the prediction matrix. It should be noted that in this formulation $Y_0^{(0)}$ is to be generated from y_0 by some appropriate starting procedure.

Since the underlying s-stage Runge-Kutta method is assumed to satisfy $C(s)$ and if V, W and $\hat{V}$ are the $s \times s$ matrices whose jth columns are, respectively, c^{j-1}, c^j/j and $(c-e)^{j-1}$, then a predictor based on the s past derivative values $hf(Y_1), \ldots, hf(Y_s)$ gives

$$A = WV^{-1}, \quad A_p = W\hat{V}^{-1}. \tag{5.9.2}$$

If (5.9.1) is applied with l corrections to the standard linear test problem then the stability matrix is given by

$$M(z) = E + \sum_{j=1}^{l} A^j E z^j + z^{l+1} A^l A_p, \quad E = [0\, e].$$

Furthermore, if the corrector is a collocation method and $l \leq s$ then

$$M(z) = E + \sum_{j=1}^{l} E_j z^j / j! + z^{l+1} A^l A_p, \tag{5.9.3}$$

where E_j is the matrix with zeros in the first $s-1$ columns and c^j in the last column.

In order to see how the abscissae and the number of corrections can affect the stability of a method with the form (5.9.1), some stability information is presented in Table 5.8, where R and I have the same meaning as in Table 5.3.

Table 5.8 *stability results*

Method	I	R
LOB5(1)	0.1	0.8
LOB5(10)	4.3	4.3
LOB8(1)	0.1	0.7
LOB8(2)	0.3	1.1
NC8(1)	0.7	0.7

Here, LOB5 and LOB8 represent predictor-corrector methods of the form (5.9.1) based on the Lobatto abscissae with five and eight stages, respectively, while NC8 is a similar method but based on equidistant abscissae with eight stages. The number in brackets after the method represents the number of corrections.

These results and others elsewhere (Burrage (1990b)) suggest that for a fixed number of corrections the choice of the free parameters (the abscissae) do not unduly affect the size of the stability region (although they can have some effect along the imaginary axis, as in NC8(1) and LOB8(1)). The most important factor in the size of the stability region is the number of corrections. Thus a family of methods based on only one correction are unlikely to be efficient for problems which are mildly stiff because the stepsize will be constrained by stability considerations rather than accuracy.

The analysis of section 5.8 and, in particular, Corollary 5.8.2 can be easily modified to allow an analysis of the local error behaviour of (5.9.1). In this case the local error vector (for all the components) is given by

$$l_{n+1} = \sum_{t \in T^*} \bar{e}(t)[F(t)]y(x_n)\frac{h^{\rho(t)}}{\rho(t)!} \tag{5.9.4}$$

where, for $t = [t_1, \ldots, t_v]$,

$$\bar{e}(t) = c^{\rho(t)} - k_l(t), \tag{5.9.5}$$

and

$$\begin{aligned} k_0(\phi) &= e, \quad k_0(t) = \rho(t)A_p\prod_{i=1}^{v}\tilde{k}(t_i) \\ k_{j+1}(\phi) &= e, \quad k_{j+1}(t) = \rho(t)A\prod_{i=1}^{v}k_j(t_i), \; j = 0,\ldots,l-1. \end{aligned} \tag{5.9.6}$$

Here it has been assumed that $Y_0^{(0)}$ has a Butcher series expansion as $B(\tilde{k}, y(x_0))$, and that the measure of the local truncation error is given by $e_s^\top l_{n+1}$, where $e_s^\top = (0,\ldots,0,1)$.

The most obvious way to generate $Y_0^{(0)}$ from the initial value is to use the trivial predictor in conjunction with a Runge-Kutta method corrected a certain number of times. Using the results in Burrage (1990b, 1993a), or modifying (5.9.6) with $A_p = 0$, then it is easy to show that if the corrector has stage order p and p corrections are performed, as part of the starting procedure, then

$$\begin{aligned} \tilde{k}(t) &= (c-e)^{\rho(t)}, \quad \forall\rho(t) \le p, \\ \tilde{k}(t_{p+1}) &= (p+1)A(c-e)^p, \quad h(t_{p+1}) \ne p, \\ \tilde{k}(t_{p+1}) &= 0, \quad h(t_{p+1}) = p, \end{aligned} \tag{5.9.7}$$

while if $p+1$ corrections are performed then

$$\tilde{k}(t) = (c-e)^{\rho(t)}, \quad \forall\rho(t) \le p, \quad \tilde{k}(t_{p+1}) = (p+1)A(c-e)^p, \tag{5.9.8}$$

where t_{p+1} denotes an arbitrary tree of order $p+1$.

Explicit expressions for the local error can now be derived in terms of the local error of the Runge-Kutta collocating corrector method. Let the Runge-Kutta corrector have a local error vector for all the stages given by

$$\hat{l}_{n+1} = \sum_{t\in T^*} \hat{e}(t)[F(t)]y(x_n)\frac{h^{\rho(t)}}{\rho(t)!} \tag{5.9.9}$$

where, for $t = [t_1,\ldots,t_v]$,

$$\hat{e}(t) = c^{\rho(t)} - k(t), \tag{5.9.10}$$

with

$$k(\phi) = e, \quad k(t) = \rho(t)A\prod_{i=1}^{v}k(t_i). \tag{5.9.11}$$

Then it is easy to show by a simple induction argument that

$$\begin{aligned} k_l(t) &= c^{\rho(t)}, \quad \forall \rho(t) \le p, \quad l = 0, 1, \ldots, \\ k_l(t) &= k(t), \quad \forall \rho(t) \le p + l. \end{aligned} \tag{5.9.12}$$

Burrage (1990b, 1993a) has obtained formulas for the k_l for trees of order greater than $p+l$, but they become increasingly more complicated as the order of the tree increases. Rather than do this in general, only the case $l = 1$ and $p \ge 2$ will be considered here. In order to simplify the notation $[t_p]_l$ will be used to denote the tree $[\ldots[t_p]\ldots]$, where the square brackets are repeated l times and $b(t)$ will be called the **branch number** of t and represents the number of branches emanating from the root of t. From (5.9.6) and (5.9.7)

$$k_0(t_{p+1}) = (p+1)A_p(c-e)^p, \quad k_0([t_{p+1}]) = (p+2)A_p\tilde{k}(t_{p+1}),$$

$$k_0(t_{p+2}) = (p+2)A_p(c-e)^{p+1}, \quad b(t_{p+2}) \ge 2$$

while

$$\begin{aligned} k_1(t_{p+1}) &= k(t_{p+1}) \\ k_1([t_{p+1}]) &= (p+2)Ak_0(t_{p+1}) \\ k_1(t_{p+2}) &= (p+2)Ac^{p+1}, \quad b(t_{p+2}) \ge 2 \\ k_1([[t_{p+1}]]) &= (p+3)Ak_0([t_{p+1}]) \\ k_1([t_{p+2}]) &= (p+3)Ak_0(t_{p+2}), \quad b(t_{p+2}) \ge 2 \\ k_1([\tau, t_{p+1}]) &= (p+3)ACk_0(t_{p+1}) \\ k_1(t_{p+3}) &= (p+3)Ac^{p+2}, \quad t_{p+3} \text{ otherwise.} \end{aligned}$$

Taking $p = s$ in the above, it is now possible to write down the local truncation error for (5.9.1) after one correction. Here, for example, $\sum[F(t_{s+1})]y(x_n)$ represents the sum over all elementary differentials corresponding to all trees of order $s+1$.

Prediction:

$$\begin{aligned} l(x_{n+1}) &= \tfrac{h^{s+1}}{s!}G_{p1}\sum[F(t_{s+1})]y(x_n) \\ &\quad + \tfrac{h^{s+2}}{(s+1)!}\Big(G_{p2}\textstyle\sum_{b(t_{s+2})\ge 2}[F(t_{s+2})]y(x_n) \\ &\quad + G_{p3}\textstyle\sum[F([t_{s+1}])]y(x_n)\Big) + O(h^{s+3}), \end{aligned}$$

where

$$\begin{aligned} G_{p1} &= 1/(s+1) - e_s^\top A_p(c-e)^s \\ G_{p2} &= 1/(s+2) - e_s^\top A_p(c-e)^{s+1} \\ G_{p3} &= 1/(s+2) - e_s^\top A_p\tilde{k}(t_{s+1}). \end{aligned}$$

Correction:

$$
\begin{aligned}
l(x_{n+1}) \quad = \quad & \tfrac{h^{s+1}}{s!} G_{c1} \textstyle\sum [F(t_{s+1})] y(x_n) \\
& + \tfrac{h^{s+2}}{(s+1)!} G_{c2} \textstyle\sum_{b(t_{s+2})>1} [F(t_{s+2})] y(x_n) \\
& + \tfrac{h^{s+2}}{(s+1)!} G_{c3} \textstyle\sum [F([t_{s+1}])] y(x_n) \\
& + \tfrac{h^{s+3}}{(s+2)!} G_{c4} \textstyle\sum_{b(t_{s+3})>1,\, t_{s+3} \neq [\tau, t_{s+1}]} [F(t_{s+3})] y(x_n) \\
& + \tfrac{h^{s+3}}{(s+2)!} G_{c5} \textstyle\sum [F([\tau, t_{s+1}])] y(x_n) \\
& + \tfrac{h^{s+3}}{(s+2)!} G_{c6} \textstyle\sum_{b(t_{s+2}) \geq 2} [F([t_{s+2}])] y(x_n) \\
& + \tfrac{h^{s+3}}{(s+2)!} G_{c7} \textstyle\sum [F([[t_{s+1}]])] y(x_n) + O(h^{s+4}),
\end{aligned}
$$

where

$$
\begin{aligned}
G_{c1} &= 1/(s+1) - e_s^{\top} A c^s \\
G_{c2} &= 1/(s+2) - e_s^{\top} A c^{s+1} \\
G_{c3} &= 1/(s+2) - (s+1) e_s^{\top} A A_p (c-e)^s \\
G_{c4} &= 1/(s+3) - e_s^{\top} A c^{s+2} \\
G_{c5} &= 1/(s+3) - (s+1) e_s^{\top} A C A_p (c-e)^s \\
G_{c6} &= 1/(s+3) - (s+2) e_s^{\top} A A_p (c-e)^{s+1} \\
G_{c7} &= 1/(s+3) - (s+2) e_s^{\top} A A_p \tilde{k}(t_{s+1}).
\end{aligned}
$$

The values for $\tilde{k}(t_{s+1})$ are given in (5.9.7) and (5.9.8) (with p=s), depending on whether s or $s+1$ initial corrections are performed.

Numerical results

Table 5.9 *results for the two body problem*

LOB8(1)	Tol	Steps	Functions	Time	Global error
	(-2)	18	329	4808	6.18(-3)
	(-4)	26	476	6512	1.48(-4)
	(-6)	44	812	9825	2.31(-6)
	(-8)	73	1232	13212	5.09(-9)
	(-10)	120	1904	18050	4.04(-10)
	(-12)	1034	14651	101513	3.31(-11)

In an attempt to improve upon the efficiency of LOB8 (see Table 5.2) the local truncation coefficients of the one-corrector method is minimized in some norm, under the assumption that the corrector has order $s+1$. The contribution of error terms higher than order $s+3$ are ignored with the hope that they contribute significantly less to the local error than the error terms of orders $s+2$ and $s+3$.

The 1-norm of the G_{cj} is minimized over a particular subclass of the parameter space, and the ensuing one-corrector method (New8) is compared with LOB8(1) and NC8(1) on the 2 body problem in a variable-stepsize program using local extrapolation and a starting procedure which consists of s corrections by the appropriate s-stage Runge-Kutta corrector. The results are presented in Table 5.9.

New8	Tol	Steps	Functions	Time	Global error
	(-2)	17	329	4804	6.39(-2)
	(-4)	22	399	5428	6.26(-4)
	(-6)	34	644	7858	2.72(-7)
	(-8)	52	903	10280	4.16(-8)
	(-10)	79	1302	13390	3.25(-10)
	(-12)	127	2072	19684	2.57(-12)

NC8(1)	Tol	Steps	Functions	Time	Global error
	(-2)	17	315	4664	5.30(-3)
	(-4)	27	490	6657	1.25(-4)
	(-6)	44	805	9654	2.24(-6)
	(-8)	74	1239	13136	1.57(-8)
	(-10)	121	1918	18243	3.31(-10)
	(-12)	2810	39557	270656	9.72(-11)

Figure 5.6 gives potential speed-ups over DORP8 based on function evaluations from results in Tables 5.2 and 5.9 for LOB8, LOB8(1) and New8, denoted, respectively, by the dash-dot, dash and solid lines. Here it is assumed that LOB8 requires only four processors, while New8 requires eight processors. The potential speed-ups ignore load balancing factors and communication times as well as, in the New8 case, time to recalculate method coefficients. Since this problem is typical of many small nonstiff problems some conclusions can be drawn from these results.

Conclusions

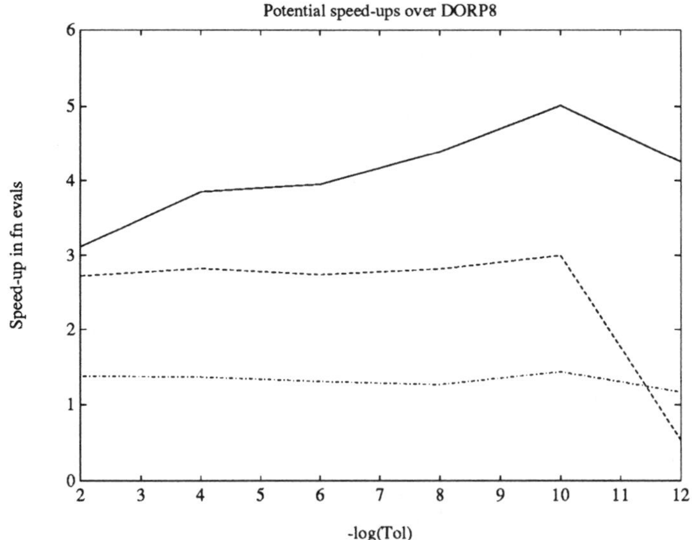

FIG. 5.6. potential speed-ups

• Methods based on one correction by either a Newton-Cotes or a Lobatto collocation method give approximately the same performance. This suggests that there is little to choose in terms of efficiency between collocation methods which have their abscissae in [0, 1]. However, if two corrections are performed the Lobatto methods are slightly more efficient at stringent tolerances (see Burrage (1990b)) because of their higher order.

• LOB8 needs approximately 1.2 times as many function evaluations as LOB8(1) but LOB8(1) takes approximately 1.5 times as long as LOB8 in ticks of the computer clock. This is because LOB8 is a Runge-Kutta method while LOB8(1) is a multivalue method and a stepsize change means recalculation of the method coefficients. In a parallel implementation this recalculation of coefficients can be spread over a number of processors.

• LOB8(1) is between 1.2 and 1.5 times as expensive (except at very lax tolerances) as New8, with similar results for computer time, while LOB8 takes between 1.2 and 2 times as many function evaluations as New8. In addition, New8 is much more efficient at stringent tolerances.

• New8 takes between 2.5 and 1.6 times as many function evaluations as DORP8, so that an eight-processor environment there is a possible maximum speed-up of $8/1.6 \approx 5$ depending on the error tolerance.

• A two-correction implementation can gain even better results than those presented above. But in this case, a completely different set of error coefficients have to be minimized.

The results of this section have shown that it is possible to carefully design predictor-corrector methods with suitable stability and local error properties but at the cost of considerable theoretical analysis. However, there is a natural limit to the types of speed-ups possible with all of these approaches, as illustrated by the Abbas and Delves method. At best, this predictor-corrector process will result in moderate speed-up over the best extant explicit sequential methods. These conclusions, and others, are discussed more fully in a review article by Burrage (1993b).

Parallel implementation

In fact, actual parallel performance can be considerably less than these theoretical bounds. This has been demonstrated by Kalvenes (1989) who has implemented predictor-corrector methods, based on the trivial predictor and Gauss corrector, of orders four-twelve on DETEST on a shared-memory computer with eight processors (the Alliant FX/8).

Denoting these methods by Gauss$w$, where w denotes the order of the predictor-corrector method, Kalvenes (1989) shows numerically that Gauss12, Gauss10 and Gauss8 have at least as good error properties as DORP8. This confirms the previous theoretical analysis. However, over all tolerances, DORP8 was the most efficient code even though it ran on only one processor, but at stringent tolerances (10^{-9}) Gauss12 was superior to DORP8. A difficulty with these numerical comparisons is that the problems in DETEST are not large enough and the function evaluations not computationally intensive enough to make meaningful comparisons in a parallel setting.

Rauber and Rünger (1994c) have considered various SPMD implementations of iterated Runge-Kutta methods based on the trivial predictor and a Runge-Kutta corrector in a distributed memory MIMD environment. A SPMD programming model is constructed which allows the prediction of runtimes and speed-up values before implementation, thus avoiding unnecessary implementations if the predicted performance is unsatisfactory. The model, and the implementation on an Intel iPSC/860, specifies the subtasks, explicit synchronization points and data exchanges – which are expressed with standard communication primitives. These primitives include

• node-to-node,

• single-node broadcast of the same message to all processors,

• single-node scatter of a separate message to all processors,

- multinode broadcast of a simultaneous single-node broadcast for all processors,
- total exchange of an individual message from every processor to every other processor.

For the iPSC/860, machine-specific timing models have been developed for these communication primitives and a timing model is then applied to a number of different parallel implementations. These implementations mainly differ in terms of workload and data distribution. One implementation attempts to reduce communication times by delaying function evaluations to the next iteration of the corrector method. The authors claim that in many cases predicted and measured execution times are quite accurate. Their results show that these predictor-corrector methods are only suited for a parallel implementation if the function is sufficiently complex compared with the time to execute a multibroadcast, and even then parallel performance is poor.

5.10 Diagonally implicit iteration

It has already been seen that on parallel computers, predictor-corrector methods based on implicit multistage correctors such as Runge-Kutta methods or multivalue methods can be effective. However, because of limited stability regions such an approach is only suitable for nonstiff systems. Van der Houwen et al. (1992b) have, however, constructed families of methods which are highly stable, and hence suitable for stiff systems, based on diagonally implicit iteration of fully implicit Runge-Kutta methods. This approach can be extended to more general classes of methods but in this section only the Runge-Kutta case will be considered.

The basic idea is to split the coefficient matrix by a splitting

$$A = D + (A - D)$$

and this leads to an implicit iteration of the form $(k = 0, \cdots, l-1)$

$$Y^{(k+1)} - h(D \otimes I)F(Y^{(k+1)}) = e \otimes y_n + h((A-D) \otimes I)F(Y^{(k)}), \quad (5.10.1)$$

with a possible update by either

$$y_{n+1} = y_n + h(b^\top \otimes I)F(Y^{(l)}) \quad (5.10.2)$$

or

$$y_{n+1} = (e_s^\top \otimes I)Y^{(l)}, \text{ with } \quad b^\top = e_s^\top A, \quad (5.10.3)$$

depending on whether the underlying method is stiffly accurate.

A predictor for (5.10.1) can be computed from

$$Y^{(0)} - h(B \otimes I)F(Y^{(0)}) = e \otimes y_n + h(L \otimes I)F(e \otimes y_n). \qquad (5.10.4)$$

The case $D = 0$, corresponding to the explicit case, has already been discussed but by letting D be an arbitrary diagonal matrix with nonnegative diagonal elements an efficient implicit formulation is obtained. Van der Houwen et al. (1992b) identify three types of methods based on the form of the predictor. These are called type A, B and C methods and are characterized by $B = L = 0$, $B = D$ and $L = 0$, $B = D$, respectively. The general form can be written in block form as

$$\begin{array}{c|ccccc}
 & 0 & & & & \\
 & L & B & & & \\
 & 0 & A-D & D & & \\
 & \vdots & \ddots & \ddots & \ddots & \\
 & 0 & & 0 & A-D & D \\
\hline
 & 0 & \dots & 0 & 0 & b^{\top}
\end{array} \qquad (5.10.5)$$

It is easily seen that after a fixed number of iterations these methods belong to the class of DIRKs and on parallel computers are effectively SDIRKs. (Van der Houwen et al. (1992b) call these methods **parallel diagonal implicitly iterated Runge-Kutta methods** (PDIRKs). As in the explicit case in which $D = 0$, an iteration increases the order of the overall method by one until the order of the corrector is attained. For this reason, PDIRKs possess embedded formulas of lower order and can be used simply in a variable-stepsize, variable-order setting.

One of the advantages of these methods over DIRKs in general is that the order of most extant DIRKs is limited to 6 (restricted only by the difficulty in construction). However, van der Houwen et al. (1992b) have constructed highly stable PDIRK methods up to order 10. One disadvantage of this approach is the large number of sequential stages (corrections), which is, for example, eight for an L-stable PDIRK of order 8. On the other hand, later stages are cheap compared with the first stage, because the s LU factorizations associated with the linearization of $Y_i - hd_{ii}f(Y_i), i = 1, \dots, s$ are retained and there are accurate initial iterates, for solving the associated implicit stages, available from the previous iteration.

The number of corrections needed to attain a prescribed order depends on the order of the predictor given in (5.10.4) and whether the method is stiffly accurate or not. The following result given in van der Houwen et al. (1992b) describes this situation and is easily proved from the standard order conditions of Runge-Kutta methods.

Theorem 5.10.1. *If the order of the underlying Runge-Kutta corrector is w then the order of (5.10.1) after l corrections will be*
$\min\{w, l+r\}$
$\min\{w, l+r+1\}$ *if* $(L+B)e = Ae$
$\min\{w, l+r+2\}$ *if* $(L+B)e = Ae,\ BAe = A^2e,$
where $r = 0$ *if the method is stiffly accurate and* $r = 1$, *otherwise.*

It should be noted that the choice $B = D$ has some implementational advantages in that the LU factorizations used in solving for the predictor can then be used at all subsequent iterations in the correction stage, while the choice $B = 0$ has further advantages in that the predictor is purely an explicit stage. These represent the two natural choices for efficient implementation.

Stability

Thus given a high-order correction, the question that remains is one of choosing the matrix D. This can be done on the basis of stability considerations. The simplest choice is $D = dI$, $d > 0$. This choice has further implications in that the diagonal elements of (5.10.5) are constant for all processors and iterations. Thus only one LU factorization is needed per integration step and, as in the parallel block method case, this LU factorization can be done on a separate set of processors and communicated to the processors involved in computing the internal stages for (5.10.1).

An application of (5.10.1) to the standard linear test equation gives two different stability functions depending on whether the method is stiffly accurate or not. Denoting the associated stability functions after l iterations by $R_l^s(z)$ and $R_l(z)$ for the stiffly accurate and nonstiffly accurate case, respectively, then it is easy to show that

$$\begin{aligned} R_l^s(z) &= e_s^\top (I - zD)^{-1} Q_l(z) e \\ R_l(z) &= 1 + zb^\top (I - zD)^{-1} Q_l(z) e, \end{aligned}$$

where

$$\begin{aligned} Q_l(z) &= \sum_{j=0}^{l-1} P^j + P^l (I - zD)(I - zB)^{-1}(I + zL) \\ P(z) &= z(A - D)(I - zD)^{-1}. \end{aligned} \qquad (5.10.6)$$

In the case that $B = D = dI$, and if $l = w$, where w is the order of the underlying corrector, then R_l^s and R_l are functions only of d and are independent of the corrector. Using a Gauss corrector, van der Houwen et al. (1992b) have constructed L-stable PDIRKs of order w with w effective stages for $w \leq 6$ and $w = 8$, and have similarly constructed L-stable PDIRKs of order w with $w+1$ effective stages with $w \leq 8$ and $w = 10$ based on a Radau IIA corrector. In both cases the stiffly accurate formulation of (5.10.3) was used. Van der Houwen et al. (1992b) have also constructed A-stable methods of orders $w \leq 4$ and $w = 6$ with $w - 1$ effective stages based on a nonstiffly accurate formulation.

In general, the L-stability properties of PDIRKs are superior to those of SDIRKs, which were investigated by Wolfbrandt (1977). Wolfbrandt showed that there exist L-stable SDIRKs of orders w with $w \leq 6$ and $w = 8$. Although the stage order of PDIRKs and DIRKs are the same, in general the stiffly accurate formulation of (5.10.3) is to be preferred. This is because of the fact that although for general stiff problems there is an order reduction phenomenon to the stage order of a method, in the case of singularly perturbed problems the classical order of a stiffly accurate method may still dominate the global error as the perturbation parameter decreases. This behaviour was described more fully in Chapter 3 and is the reason for choosing the stiffly accurate formulation of (5.10.3).

Numerical results

Van der Houwen et al. (1992b) have made some comparisons with LSODE (Hindmarsh (1980)) and SIMPLE (Nørsett and Thomsen (1986)) and a stiffly accurate, L-stable PDIRK method of order 7 based on a four-stage Radau IIA corrector with D chosen as $D = dI$, $d = 0.1690246379$. Error control within the PDIRK implementation is based on taking the difference of the embedded formulas, while the Newton process to solve for $Y^{(j)}$ is started with an initial guess of $Y^{(j-1)}$. The implementation is on a four-processor Alliant FX/4, which is a shared-memory machine.

On a three-dimensional chemical kinetics problem due to Robertson the PDIRK implementation is, for comparable accuracies, about three times as fast as SIMPLE and 1.6 times as fast as LSODE on one processor and on four processors there is a further speed-up of a factor between two and three. However, on a van der Pohl equation, LSODE on one processor and PDIRK on four processors have approximately the same performance while SIMPLE experiences difficulties if moderate or stringent accuracies are required.

Sommeijer (1992) has extended these tests to a 15-dimensional problem in circuit analysis (Hairer et al. (1989)). In this case the PDIRK implementation is modified so that prediction is performed by collocating the Radau IIA quadrature points, while D is chosen so that

$$\rho(P(\infty)) = \rho(I - D^{-1}A) = 0. \qquad (5.10.7)$$

The implementation on the four-processor Alliant FX/4 was approximately 1.5 to 2 times as fast as LSODE and RADAU5 (Hairer and Wanner (1991)) running on one processor.

Although this testing is far from comprehensive or conclusive some general comments can be made.

- The comparisons with SIMPLE at stringent tolerances are in many ways unfair because this code is only effective at lax tolerances whereas the PDIRK methods are very accurate with small local truncation errors.
- The relatively poor performance of PDIRK on the van der Pohl problem is due to the fact that it is a multistage method and the Jacobian of the van der Pohl problem is rapidly varying. Most multistage methods perform poorly on such problems.
- The PDIRK approach is more robust than the LSODE approach because of superior stability properties.
- It would be interesting to compare the PDIRK approach with STRIDE rather than RADAU5. RADAU5 is restricted to order 5 and for large problems the LU factorizations require a factor of 5 times more computational effort over both PDIRK and STRIDE. Thus the performances of PDIRK on four processors and STRIDE on one processor would be expected to be similar for large problems.

Choosing the splitting matrix

Van der Houwen and Sommeijer (1991) have explored the effect of choosing D such that (5.10.7) holds. By noting that if $Y(x_{n+1}) = (y(x_n + c_1h), \ldots, y(x_n + c_sh))^\top$ then

$$Y(x_{n+1}) - Y^{(k)} = (Y(x_{n+1}) - Y_{n+1}) + (Y_{n+1} - Y^{(k)}),$$

which to a first-order approximation can be written as

$$Y(x_{n+1}) - Y^{(k)} = Y(x_{n+1}) - Y_{n+1} + P^k(Y_{n+1} - Y^{(0)}), \qquad (5.10.8)$$

where

$$P = P(hDJ) := (I - hDJ)^{-1}(A - D)hJ$$

and J denotes the Jacobian of the problem.

Clearly, if the underlying corrector has order w and $Y_{n+1} - Y^{(0)} = O(h^p)$ then (5.10.8) implies

$$Y(x_{n+1}) - Y^{(l)} = O(h^{w+1}) + O(h^{l+p})$$

and this result shows that the order increases with each iteration until the order of the corrector is obtained. Although the order cannot increase beyond the order of the corrector it may be efficacious to continue the iteration process because there is no guarantee that the stability properties of (5.10.5) will be comparable to the corrector.

Van der Houwen and Sommeijer (1991) use (5.10.8) to accelerate the rate of convergence of the iteration process. By considering the standard linear test equation where $z = hq$ runs through the spectrum of hJ, the matrix P becomes

$$P = z(I - zD)^{-1}(A - D).$$

It is clear that rapid convergence of the nonstiff components will take place if $\rho(A - D) = 0$. On the other hand, for stiff components P behaves like $I - D^{-1}A$ and van der Houwen and Sommeijer (1991) suggest a stiff iteration by minimizing $\rho(I - D^{-1}A)$ with $D > 0$. Some numerical testing by van der Houwen and Sommeijer (1991) suggests that this is a more robust approach than choosing $D = dI$ and that the phenomenon of order reduction is in fact suppressed within a few iterations for this choice. In fact an even better approach may be to minimize $\rho(P)$ over all $z \in \mathbb{C}^-$, as this will lead to rapid convergence for problems with a wide spread of eigenvalues with negative real part.

Van der Houwen and Sommeijer (1991) also consider a slight generalization to (5.10.1) of the form

$$Y^{(k+1)} - h(D \otimes I)F(Y^{(k+1)}) = e \otimes y_n + ha \otimes f(y_n) + h((A - D) \otimes I)F(Y^{(k)}),$$

where

$$y_{n+1} = y_n + hb_0 f(y_n) + h(b^\top \otimes I)F(Y^{(l)})$$

or

$$y_{n+1} = e_s^\top Y^{(l)},$$

again depending on whether a stiffly accurate formulation is required. This represents a correction process based on an $(s + 1)$-stage Runge-Kutta method of the form

$$\begin{array}{c|cc} 0 & 0 & 0 \\ c & a & A \\ \hline & b_0 & b^\top \end{array}$$

These methods can have stage order $s+1$ and are called Lagrange methods by van der Houwen and Sommeijer (1991).

By analysing the stability function associated with (5.10.1) and (5.10.2) it can be shown that diagonal iteration is not suitable for iterating nonstiffly accurate correctors, although if the $Y^{(0)}$ are defined implicitly then this situation can change. Nevertheless, van der Houwen and Sommeijer (1991) construct two, three and four-processor methods based on either a

Lagrangian corrector or a Radau IIA corrector in which $\rho(I - D^{-1}A)$ is minimized.

In the case of the two-stage Radau IIA method of order three it can be shown that

$$D = \text{diag}\left(\frac{4-\sqrt{6}}{6}, \frac{4+\sqrt{6}}{10}\right)$$

gives $\rho(I - D^{-1}A) = 0$ and that the method characterized by (5.10.1) and (5.10.2) is A-stable after one correction. For the three-stage and four-stage Radau IIA correctors, (5.6.4) is A-stable after, respectively 5 and 7 corrections. In these cases

$$D = \text{diag}\left(\frac{4365}{13624}, \frac{1032}{7373}, \frac{1887}{5077}\right), \quad D = \text{diag}\left(\frac{3055}{9532}, \frac{531}{5956}, \frac{1471}{8094}, \frac{1848}{7919}\right).$$

A similar analysis can be performed for the Lagrangian corrector. These methods have the advantage of a one higher stage order than the usual Runge-Kutta formulation but lack, in general, the superlinear convergence of the Radau IIA correctors for the nonstiff components.

In solving the l implicit equations associated with (5.10.1) there are in fact a number of possible strategies. The approaches discussed so far in this section assume that for each $k = 0, 1, \ldots, l-1$ a modified Newton iteration process takes place until convergence occurs. This may be expensive and does not take into account the special structure of (5.10.1) in which $Y^{(k)}$ is a very good approximation to $Y^{(k+1)}$. Thus van der Houwen and Sommeijer (1993) suggest performing one Newton iteration per correction stage. This approach can be viewed as a modified Newton method for solving the internal stages of the corrector based on a block diagonal approximation to the Jacobian.

5.11 Prediction-correction based on general splittings

As has been seen, the standard approach to prediction-correction, results in methods having a relatively small stability region (unless a large number of corrections are performed) and, in general, large error coefficients (unless great care is taken) because of the complex mixing of predictor and corrector errors. A possible way to improve upon this is by the use of splitting techniques, which is a popular approach in waveform relaxation methods.

Returning to the general form of a predictor-corrector method given in (5.5.3), it can be noted that the corrector is based on the implicit method

$$Y_{n+1} = Z_n + h(L \otimes I)F(Y_{n+1}). \tag{5.11.1}$$

The technique for solving this set of nonlinear equations depends on the size of the Lipschitz constant associated with the problem. However, the standard technique, if the problem is at all stiff, is to use some variant of the modified Newton method. Let the predicted value to Y be given by (5.5.2), namely

$$Y_{n+1}^{(0)} = (A_0 \otimes I)Y_n + h(L_0 \otimes I)F(Y_n). \tag{5.11.2}$$

Then the modified Newton method as applied to (5.11.1) gives

$$\begin{aligned} M_{k,n}\Delta_k &= -Y_{n+1}^{(k)} + Z_n + h(L \otimes I)F(Y_{n+1}^{(k)}) \\ Y_{n+1}^{(k+1)} &= Y_{n+1}^{(k)} + \Delta^{(k)}, \quad k = 0, 1, \cdots, \end{aligned} \tag{5.11.3}$$

where $I = I_s \otimes I_m$, $M_{k,n} = I - h(L \otimes I)J_{k,n}$ and $J_{k,n}$ is some approximation to the Jacobian at $Y_{n+1}^{(k)}$. In fact the $J_{k,n}$ are block diagonal matrices with $J_{k,n} = \text{diag}(f'(Y_1^{(k)}), \ldots, f'(Y_s^{(k)}))$ in the full Newton case and with $J_n = \text{diag}(f'(y_n), \ldots, f'(y_n))$ being an example of a modified Newton scheme.

Thus (5.11.3) gives rise to a general iteration scheme of the form

$$M_{k,n}Y_{n+1}^{(k+1)} = (M_{k,n} - I)Y_{n+1}^{(k)} + Z_n + h(L \otimes I)F(Y_{n+1}^{(k)}). \tag{5.11.4}$$

In the case that

$$M_{k,n} = I, \quad \forall k, n, \tag{5.11.5}$$

(5.11.4) gives the standard prediction-correction approach which is just fixed-point iteration, while if

$$M_{k,n} = I - h(L \otimes I)J_n, \quad \forall k, \tag{5.11.6}$$

where $J_n = \text{diag}(f'(y_n), \ldots, f'(y_n))$ then (5.11.4) represents a modified Newton approach.

The $M_{k,n}$ can be chosen intermediate to the choices in (5.11.5) and (5.11.6) in an attempt to obtain both good convergence properties and cheap implementation in a parallel environment.

Defining

$$\varepsilon_{n+1}^{(k+1)} = Y_{n+1}^{(k+1)} - Y_{n+1} \tag{5.11.7}$$

then a linearization of the problem gives

$$\varepsilon_{n+1}^{(k+1)} = R_{k,n}\varepsilon_{n+1}^{(k)}, \quad R_{k,n} = I - M_{k,n}^{-1}(I - h(L \otimes I)J_n). \tag{5.11.8}$$

Another way of viewing this is to apply the underlying corrector in (5.11.1) to the linear problem

$$y'(x) = J(x)y$$

which gives

$$P_n Y_{n+1}^{(k)} = Z_n, \quad k = 1, \ldots, l, \quad P_n = I - h(L \otimes I)J_n.$$

Thus the choice of $M_{k,n}$ in (5.11.5) and (5.11.6) represents a preconditioning of the matrix P_n which will enable an acceleration of (5.11.4). If the eigenvalue structure of the underlying problem is known (and this is often the case, for example, for problems arising from parabolic partial differential equations by the method of lines) then polynomial preconditioning is a well-known procedure for accelerating the convergence.

For example, two possibilities are

$$\begin{array}{ll} \text{TYPE I}: & M_k^{-1} = \alpha_k I \Rightarrow R_{k,n} = I - \alpha_k P_n \\ \text{TYPE II}: & M_k^{-1} = \alpha_k I - \beta_k P_n \Rightarrow R_{k,n} = I - \alpha_k P_n + \beta_k P_n^2. \end{array} \tag{5.11.9}$$

Suppose now that the eigenvalues of the Jacobian of f are real and lie in the interval $[-q, 0]$, $q > 0$. Then the rate of convergence over l iterations can be maximized by minimizing $\rho\left(\left(\prod_{k=1}^{l} R_{l-k,n}\right)^{\frac{1}{l}}\right)$.

Some analysis using Chebyshev polynomials (Burrage (1993d)) gives that the spectral radii of the amplification matrices, as functions of $z = hq$, are minimized with

$$\begin{array}{l} \alpha = \frac{2}{1+v}, \quad \rho(R) = 1 - \alpha \\ \alpha = \frac{8(1+v)}{1+6v+v^2}, \quad \beta = \frac{\alpha}{1+v}, \quad \rho(R) = 1 - \alpha + \beta \\ \alpha_k = \frac{4}{4v+s_k^2(1-v)^2}, \quad \beta_k = \frac{\alpha_k}{1+v}, \quad s_k^2 = \sin^2\left(\frac{(2k-1)\pi}{4p}\right). \end{array} \tag{5.11.10}$$

Here $v = \det(I + zL_2)$, and it is also assumed that in the first two cases the same α and β are used over all the iterations, while in the third case these parameters are allowed to vary.

For the trapezoidal rule, for example, $v = 1 + z/2$ and so in the first two cases of (5.11.10), the spectral radii of R are given by

$$\begin{aligned} \rho_1(R) &= \frac{z}{z+4} \\ \rho_2(R) &= \frac{z^2}{z^2 + 16z + 32} = \frac{(\rho_1(R))^2}{2 - (\rho_1(R))^2}. \end{aligned} \tag{5.11.11}$$

Since $z > 0$, the scheme based on (5.11.4), (5.11.9) and (5.11.10) and the trapezoidal rule is a convergent one, and the smaller the value for z the quicker the convergence. On the other hand, for very stiff problems, z can become large and so convergence will be slow in this case. Furthermore, the effect of introducing another matrix-vector operation, as in the type II method, effectively halves the number of iterations as can be seen from (5.11.11). Table 5.10 gives values for $\rho_1(R)$ and $\rho_2(R)$ for different values of z, as well as the expected number of iterations (its) to reduce an initial error by a factor of 10^3. The advantage of using a type II method is clearly seen for the case of a moderately stiff problem.

Table 5.10 *spectral radii*

z	$\rho_1(R)$	its_1	$\rho_2(R)$	its_2
6	0.6	14	0.257	5
96	0.96	170	0.8546	44
996	0.996	1723	0.984	434

The advantage of this approach is that the implementation properties are similar to explicit methods but the stability properties are similar to A-stable implicit methods. In particular, Burrage (1993c) has shown that by choosing the M_k appropriately considerable improvements in efficiency can be gained over the standard prediction-correction approach, especially for problems that are moderately stiff. This has been demonstrated by using a method based on the Euler predictor and a trapezoidal corrector given by

$$\begin{aligned} y_{n+1}^{(0)} &= y_n + hf(y_n) \\ y_{n+1}^{(k+1)} &= (I - M_k^{-1})y_{n+1}^{(k)} + M_k^{-1}(y_n + hf(y_n)/2) + M_k^{-1}f(y_{n+1}^{(k)})/2, \end{aligned} \tag{5.11.12}$$

where M_k^{-1} is either of type I or type II, and $P_n = I - \frac{h}{2}f'(y_n)$.

The computational savings with these methods arise from the fact that the corrections can be calculated from simple matrix-vector operations. No solutions of linear systems are required which can be difficult to program efficiently in a parallel environment. This approach is particularly appropriate for SIMD computers such as the MasPar.

An implementation on a MasPar

Method (5.11.12) has been implemented on two problems on a 4K processor MasPar MP1 sited at the University of Queensland using a fixed-stepsize scheme with MPFortran. The first problem is the linear problem

$$u' = (N+1)^2 Qu, \quad u(0) = 1, \tag{5.11.13}$$

where Q is a block-tridiagonal matrix of the form (I_N, T, I_N) and T is the tridiagonal matrix $(1, -4, 1)$. An appropriate value for q for this problem is $8(N+1)^2$. The second problem is the one-dimensional form of the Brusselator (which is given in (9.4.1) and (9.4.2)), and which is converted to a system of ordinary differential equations by the method of lines using central differencing. It is of dimension $2N$ and is given by

$$\begin{aligned} u_i' &= B + u_i^2 v_i - (A+1)u_i + \hat{\alpha}(u_{i+1} - 2u_i + u_{i-1}) \\ v_i' &= Au_i - {u_i}^2 v_i + \hat{\alpha}(v_{i+1} - 2v_i + v_{i-1}). \end{aligned} \tag{5.11.14}$$

Rather than present a number of tables recording the computational results, these results will be summarized in the following remarks:

1. In the case of (5.11.13) some care must be taken in choosing N. For example, if $N = 65$ and the number of available processors is $64 \times 64 = 4096$ then the computational time will be approximately twice as long as in the case $N = 64$. This is because the MasPar automatically layers the computational grid into memory so that the $N = 65$ case requires two layers. This is done automatically and does not require programmer intervention. On the other hand the time for solving any N^2-dimensional problem (where $N \leq 64$) should be approximately the same, but can depend on the machine load.

2. Although the implementation of (5.11.12) on (5.11.13) requires only Level 2 BLAS operations of the form Qv, where Q is as in (5.11.13) and v is a vector defined on all the elements of the computational grid, it is important to structure the problem so this is done efficiently. This is achieved by representing v as an $N \times N$ matrix and forming Qv as a sequence of EOSHIFTs given by

```
EOSHIFT(v,SHIFT=-1,BOUNDARY=f1,DIM=1) +
EOSHIFT(v,SHIFT=-1,BOUNDARY=f2,DIM=2) - 4.0*v +
EOSHIFT(v,SHIFT=+1,BOUNDARY=f3,DIM=2) +
EOSHIFT(v,SHIFT=+1,BOUNDARY=f4,DIM=1)
```

Here SHIFT represents a shift up or down the computational grid of the vector v, DIM represents column or row shifts and $f1$, $f2$, $f3$, $f4$ the boundary conditions. This segment of MPFortran represents a direct mapping of the five point central differencing of the computation nodes to the processor grid.

3. For the one-dimensional version of the Brusselator given in (5.11.14), there are two coupled vectors each of dimension N and these are automatically layered as two row vectors onto the MasPar topology. For a type II

implementation, a Jacobian matrix has to be evaluated at each time step. Since the Jacobian matrix has a simple block-tridiagonal structure with the identity matrix as the off-diagonal blocks, the forming of the vector product of the Jacobian times each of the two vectors representing the components of the problem is easily done as a sequence of two `EOSHIFT`s columnwise for each vector.

4. (5.11.13) and (5.11.14) has also been solved by a block method of size two based on a two-stage Radau corrector of order 3. In this case two approximations (one a third of the way along the integration step and one at the end of the integration step) are computed per processing element. The computational time is, as expected, approximately twice that for the trapezoidal corrector, since the convergence properties of this method are similar to those for the trapezoidal approach. This can be seen from (5.11.10) by noting that in the Radau case, $v = 1 + 2z/3 + z^2/24$.

5. Since both these problems arise from the semi-discretization of a partial differential equation by a five-point or three-point differential operator, the formation of the associated matrix-vector products can be done very easily in MPFortran using `EOSHIFT`s, because the operations can be performed locally. For problems with less structure, an implementation on SIMD machine would be more inefficient because global communication would be needed.

6. It is for this reason that no general robust implementation of a parallel ordinary differential algorithm has been attempted for a SIMD machine. For such machines good performance is likely only on problems with a very regular, possibly constant, structure.

6

DIVERSE APPROACHES TO PARALLELISM

In this chapter a number of different approaches to developing parallel algorithms for differential systems will be discussed. Some of these approaches will offer a modest performance while some offer the possibility of massive parallelism. In the case of linear systems of either initial value or boundary value type there is a possibility of massive parallelism by exploiting the structure of the ensuing linear systems through such techniques as cyclic reduction and block linear algebra and various approaches are discussed in sections 6.1, 6.2 and 6.3. A second approach to parallelism involves exploiting parallelism to improve the robustness and efficiency of existing algorithms and this approach is described in sections 6.4 and 6.5, with particular reference to extrapolation and deferred correction techniques. Finally the concept of parallelism across the steps will be introduced and a number of different approaches will be analysed from parallel shooting to predictor-corrector techniques. The chapter will conclude with some applications to other types of differential systems including second-kind Volterra equations and Volterra integro-differential systems. Thus this chapter will cover the following material:

- section 6.1: exploiting massive parallelism in the solution of linear systems of initial value problems;
- section 6.2: exploiting massive parallelism in the solution of linear systems of boundary value problems;
- section 6.3: the use of Krylov subspace techniques to reduce the dimension of the original problem and then decouple this into independent processes by certain multistage methods;
- section 6.4: the use of extrapolation and frontal defect correction techniques to obtain a modest parallelism;
- section 6.5: exploiting parallelism to improve the robustness and efficiency of existing algorithms;
- section 6.6: a brief introduction to a number of other techniques which are not easily classifiable including applications to atmospheric problems and the use of deflation techniques as described in Chapter 4;

- section 6.7: an introduction to the concept of parallelism across the steps through such techniques as multiple shooting, the concurrent computation of blocks of approximations and operator splitting;
- section 6.8: an introduction to both sequential and parallel techniques for solving delay differential equations;
- section 6.9: an extension of parallel techniques to such problems as second-kind Volterra equations and Volterra integro-differential systems.

6.1 Linear IVPs

Consider the general class of LIVPs given by

$$y'(x) = Q(x)y(x) + g(x), \quad y(x_0) = y_0 \in \mathbb{R}^m. \tag{6.1.1}$$

If any s-stage Runge-Kutta method given by (2.7.2) is applied to (6.1.1) with $Q(x) = Q$ over one step then it is easy to show that the following linear recurrence relation results:

$$\begin{aligned} y_{n+1} &= (I_m + b^\top \otimes Z(I - A \otimes Z)^{-1}e)y_n \\ &+ h(b^\top \otimes g_n + b^\top \otimes Z(I - A \otimes Z)^{-1}A \otimes g_n), \end{aligned} \tag{6.1.2}$$

where

$$\begin{aligned} g_n &= (g(x_n + c_1 h)^\top, \ldots, g(x_n + c_s h)^\top)^\top, \\ Z &= hQ. \end{aligned}$$

If (6.1.2) is to be solved over N steps with constant stepsize then it can be written in the form

$$\begin{aligned} y_0 &= b_0 \\ y_{n+1} &= R(Z)y_n + b_{n+1}, \quad n = 0, \ldots, N-1, \end{aligned} \tag{6.1.3}$$

where

$$b_{n+1} = h(b^\top \otimes g_n + b^\top \otimes Z(I - A \otimes Z)^{-1}A \otimes g_n). \tag{6.1.4}$$

Here $R(z) = \frac{p(z)}{q(z)}$ is the stability function associated with the method, and $R(Z)$ is $q^{-1}(Z)p(Z)$. Hence (6.1.4) can be written in the form of (6.1.3) as

$$\begin{aligned} y_0 &= b_0 \\ Q(Z)y_{n+1} &= P(Z)y_n + b_{n+1}, \quad n = 0, \ldots, N-1, \end{aligned}$$

which avoids explicitly calculating $Q(Z)^{-1}$.

Thus if any Runge-Kutta method is applied to an arbitrary linear problem a linear recurrence relation of the form

$$\begin{aligned} y_0 &= b_0 \\ Q_{n+1}y_{n+1} &= -P_{n+1}y_n + b_{n+1}, \quad n = 0, \ldots, N-1 \end{aligned} \tag{6.1.5}$$

is always obtained.

Written in the form (6.1.5) there seems to be no obvious source of parallelism. However, (6.1.5) can be written as the block-bidiagonal system ($Cy = b$) of dimension $m(N+1)$ given by

$$\begin{pmatrix} I & & & \\ P_1 & Q_1 & & \\ & \ddots & \ddots & \\ & & P_N & Q_N \end{pmatrix} \begin{pmatrix} y_0 \\ y_1 \\ \vdots \\ y_N \end{pmatrix} = \begin{pmatrix} b_0 \\ b_1 \\ \vdots \\ b_N \end{pmatrix}. \tag{6.1.6}$$

Such bidiagonal systems arise frequently in many other areas of scientific computation and special techniques for solving these problems were described in Chapter 4. The more general problem given in (6.1.1) can also be considered in a similar way and this is described in section 6.7.

6.2 Linear BVPs

While the main thrust of this monograph is directed at parallel methods for IVPs there has been some recent work on parallel algorithms for boundary value problems which result in a study of parallel linear solvers for systems akin to (6.1.6). Hence, this section will address recent developments in this area.

The general form of a BVP is given by

$$y' = f(x, y), \quad x \in (x_0, x_f), \quad g(y(x_0), y(x_f)) = 0. \tag{6.2.1}$$

Traditional numerical methods for solving such problems tend to be more inherently parallelizable than those for IVPs, because these methods are usually global in that approximations to the solution are computed simultaneously at a number of points. These approximations are then refined by some iterative process. In contrast, methods for IVPs tend to be local and sequential with the solution being updated from step to step.

Thus the traditional approach to solving (6.2.1) is to generate a mesh

$$x_0 < x_1 < \ldots < x_N = x_f$$

and to apply a numerical method such as a Runge-Kutta method to (6.2.1) on this mesh. In the case of an s-stage implicit Runge-Kutta method where

$s > 1$, a secondary mesh arises on each subdivision $[x_i, x_{i+1}]$ corresponding to the internal abscissae associated with the method. The choice of the method here is very important because unlike initial value problems a well-conditioned and stable boundary value problem can have both rapidly increasing and decreasing modes. In order for a method to cope appropriately with this, it should be A-stable with the stability boundary lying on the imaginary axis. Examples of such methods are symmetric methods and consequently the Gaussian methods are often the basis of successful BVP solvers (see, for example, the code COLSYS (Ascher et al. (1981)). However, as remarked by Ascher and Chan (1991), there are difficulties with symmetric methods for some stiff problems because there is no damping of fast decreasing or increasing modes.

If some variant of Newton's method is used in conjunction with a numerical method then a series of linear BVPs (LBVPs) is obtained on each subinterval $[x_i, x_{i+1}]$ of the form

$$y'(x) = Q(x)y(x) + g(x) \tag{6.2.2}$$

subject to the linear boundary conditions

$$\beta_a y(x_i) + \beta_b y(x_{i+1}) = \beta, \tag{6.2.3}$$

(see, for example, Ascher and Chan (1991) or Paprzycki and Gladwell (1990a)).

Of course the $y(x_i)$ and $y(x_{i+1})$ in (6.2.3) are unknown and so Ascher and Chan (1991) have proposed a multiple shooting technique which is inherently parallel. (Note that this approach was first proposed by Nievergelt (1964) in conjunction with IVPs.) Thus, as Ascher and Chan (1991) note, on each of N subintervals an IVP has to be solved to approximate a fundamental solution $y_i(x)$ and a particular solution $v_i(x)$ of (6.2.2) on each interval of the mesh of the form

$$\begin{aligned} Y_i' &= Q(x)Y_i, \quad Y_i(x) \in \mathbb{R}^{m\times m}, \quad x \in [x_i, x_{i+1}], \quad Y_i(x_i) = I \\ v_i' &= Q(x)v_i + g(x), \quad x \in [x_i, x_{i+1}], \quad v_i(x_i) = 0. \end{aligned}$$

Hence by applying the boundary conditions and continuity at the grid points the following set of recurrence relations must be solved for $y_0, \ldots, y_N$

$$\begin{aligned} Y_i(x_{i+1})y_i + v_i(x_{i+1}) &= y_{i+1}, \quad i = 0, \ldots, N-1 \\ \beta_a y_0 + \beta_b y_N &= \beta. \end{aligned} \tag{6.2.4}$$

Similarly, if a one-step finite difference method is used to solve (6.2.2) and (6.2.3) then replacing the derivative term on the interval $[x_i, x_{i+1}]$ leads to methods of the form

$$\frac{y_{i+1} - y_i}{x_{i+1} - x_i} = \theta_i y_i + \Psi_i y_{i+1} + f_i.$$

Here $\theta_i = \Psi_i = \frac{1}{2}Q(x_i + \frac{1}{2})$, $f_i = g(x_i + \frac{1}{2})$ in the case of the box scheme, and $\theta_i = \frac{1}{2}Q(x_i)$, $\Psi_i = \frac{1}{2}Q(x_{i+1})$, $f_i = \frac{1}{2}(g(x_i) + g(x_{i+1}))$ in the case of the trapezoidal rule (see Wright (1990), for example).

Thus for both multiple shooting and one-step finite differences, a set of recurrence relations of the form

$$\begin{aligned} Q_{i+1}y_{i+1} &= -P_{i+1}y_i + b_i, \quad i = 0, \dots, N-1 \\ \beta_a y_0 + \beta_b y_N &= \beta \end{aligned} \tag{6.2.5}$$

arises and this can be written in block form ($Cy = b$)

$$\begin{pmatrix} P_1 & Q_1 & & & \\ & P_2 & Q_2 & & \\ & & \ddots & \ddots & \\ & & & P_N & Q_N \\ \beta_a & & & & \beta_b \end{pmatrix} \begin{pmatrix} y_0 \\ y_1 \\ \vdots \\ y_N \end{pmatrix} = \begin{pmatrix} b_0 \\ b_1 \\ \vdots \\ b_{N-1} \\ \beta \end{pmatrix}. \tag{6.2.6}$$

Such systems will be called **almost block-bidiagonal (ABBD)** systems.

In the case of multiple shooting

$$Q_i = -I, \quad P_{i+1} = Y_i(x_{i+1}), \quad i = 0, \dots, N-1$$

and if the boundary conditions are separated with

$$\beta_a = \begin{pmatrix} 0 \\ \beta_0 \end{pmatrix}, \quad \beta_b = \begin{pmatrix} \beta_1 \\ 0 \end{pmatrix}$$

then by a suitable arrangement of C in (6.2.6) in which the last row becomes the first, C can be written as

$$\begin{pmatrix} \beta_0 & & & & & \\ P_1 & -I & & & & \\ & P_2 & -I & & & \\ & & \ddots & \ddots & & \\ & & & P_N & -I \\ & & & & \beta_1 \end{pmatrix}. \tag{6.2.7}$$

Such systems are called by Paprzycki and Gladwell (1989), **almost block-diagonal (ABD)** systems.

In the sequential case, various pivoting algorithms based on Gaussian elimination have been studied and, in particular, it is known for systems of the form given in (6.2.6) that the bound on element growth is exponential only in the bandwidth (see Wright (1990) and Bohte (1975), for example). Wright (1990) has obtained bounds on the relative error of the approximate solution of (6.2.6) obtained either by Gaussian elimination with row partial pivoting or by Gaussian elimination with alternate row and column elimination.

It should also be noted that in the case of the linear systems arising from LIVPs, cyclic reduction is a stable process under very mild restrictions. However, this is not the case for LBVPs, as Ascher and Chan (1991) have pointed out.

Thus consider performing cyclic reduction on an almost block-bidiagonal (ABBD) system, and as a particular example take the blocked system

$$\begin{pmatrix} X & X & & & & & \\ & X & X & & & & \\ & & X & X & & & \\ & & & X & X & & \\ & & & & X & X & \\ & & & & & X & X \\ X & & & & & & X \end{pmatrix}$$

where X indicates an $m \times m$ block and $+$ will indicate a fill-in block. A three-processor implementation would give the following sequence.

$$\begin{pmatrix} X & X & & & & \\ + & 0 & X & & & \\ & & X & X & & \\ & & + & 0 & X & \\ & & & & X & X \\ X & & & & + & X \end{pmatrix} \to \begin{pmatrix} X & X & & & & \\ X & & X & & & \\ + & & 0 & X & & \\ + & & 0 & X & X & \\ & & & X & 0 & X \\ X & & & + & 0 & X \end{pmatrix} \to$$

$$\begin{pmatrix} X & X & & & & \\ X & & X & & & \\ X & & & X & & \\ X & & & 0 & X & \\ + & & & 0 & & X \\ X & & & 0 & & X \end{pmatrix} \to \begin{pmatrix} X & X & & & & \\ X & & X & & & \\ X & & & X & & \\ X & & & & X & \\ X & & & & & X \\ X & & & & & 0 \end{pmatrix}.$$

Unfortunately, as Ascher et al. (1988) pointed out, this algorithm is in general not stable even if the underlying BVP is well conditioned. Even if the above approach is applied to the separated boundary condition model in (6.2.7) the process of finding a stable, efficient parallel algorithm is not a simple one.

Ascher and Chan (1991) offer one very simple approach in that, rather than solve $Cy = b$ where C is given by (6.2.7), the normal equation

$$C^{\top}Cy = C^{\top}b$$

is solved. $C^{\top}C$ is now block-tridiagonal and symmetric positive definite and there are stable, parallel cyclic reduction algorithms for such problems (see Dongarra and Johnsson (1987), for example). A drawback of this approach is that the condition number of the original system is squared, and the amount of storage is approximately doubled, and of course the squaring of the condition number becomes quite significant if the BVP is very stiff.

In the case of LBVPs with separated boundary conditions, rather than apply the tearing algorithms of Dongarra and Johnsson (1987) and Dongarra and Sameh (1984) which hold for banded systems and which introduces additional overheads, Paprzycki and Gladwell (1989) have modified this algorithm to ABD systems. Paprzycki and Gladwell (1990b) have also considered solving ABD systems based on the use of Level 3 BLAS with a block decomposition and an appropriate interchange strategy.

In an attempt to avoid the poor stability properties that can plague parallel algorithms based on Gaussian elimination, Wright (1990) has developed a structured QR factorization algorithm which applies equally well to separable or non-separable boundary conditions. The first step consists of partitioning the system (6.2.6) into p sections of approximately equal size, which are then processed independently.

The first partition, augmented with the right-hand side has the form

$$C = \begin{pmatrix} P_1 & Q_1 & & & & b_1 \\ & P_2 & Q_2 & & & b_2 \\ & & \ddots & \ddots & & \vdots \\ & & & P_k & Q_k & b_k \end{pmatrix}.$$

An orthogonal matrix $\Theta_1 \in \mathbb{R}^{2m\times 2m}$ is found such that

$$\Theta_1^{\top}\begin{pmatrix} Q_1 \\ P_2 \end{pmatrix} = \begin{pmatrix} R_1 \\ 0 \end{pmatrix}$$

where $R_1 \in \mathbb{R}^{m\times m}$ is upper triangular and a new matrix C_1 given by

$$C_1 = \begin{pmatrix} \Theta_1^\top & 0 \\ 0 & I \end{pmatrix} C = \begin{pmatrix} S_1 & R_1 & T_1 & & & & \bar{b}_1 \\ S_2 & 0 & \bar{Q}_2 & & & & \bar{b}_2 \\ & & P_3 & Q_3 & & & b_3 \\ & & & \ddots & \ddots & & \vdots \\ & & & & P_k & Q_k & b_k \end{pmatrix}$$

is then formed. This step is repeated $k-1$ times by finding a series of orthogonal matrices $\Theta_i \in \mathbb{R}^{2m \times 2m}$ such that

$$\Theta_i^\top \begin{pmatrix} \bar{Q}_i \\ P_{i+1} \end{pmatrix} = \begin{pmatrix} R_i \\ 0 \end{pmatrix}$$

until a system of the form

$$\begin{pmatrix} S_1 & R_1 & T_1 & & & & \bar{b}_1 \\ S_2 & 0 & R_2 & T_2 & & & \vdots \\ \vdots & & \ddots & \ddots & \ddots & & \vdots \\ S_{k-1} & & & \ddots & R_{k-1} & T_{k-1} & \vdots \\ \hat{P}_1 & & & & 0 & \hat{Q}_1 & \bar{b}_k \end{pmatrix} \tag{6.2.8}$$

is obtained. This process can be performed on each partition. From each of these partitions a reduced system can be formed which has the structure

$$\begin{pmatrix} \hat{P}_1 & \hat{Q}_1 & & \\ & \ddots & \ddots & \\ & & \hat{P}_p & \hat{Q}_p \\ B_a & & & B_b \end{pmatrix}. \tag{6.2.9}$$

Once this system has been solved the remaining variables can be found by back-substitution. The form of (6.2.9) is identical to the original form and so this process can be applied recursively to obtain smaller and smaller systems. The number of recursions and the number of processors used at each level is difficult to resolve and depends on the number available, and the nature of the architecture.

Wright (1990) notes that this scheme is a standard Householder QR factorization and so traditional stability analyses apply, and that the operation count for this algorithm is the same as the normal equations approach of Ascher and Chan (1991) although their approach is less stable.

6.3 The Krylov subspace techniques

It is known that there is an exact solution to (2.2.4) namely

$$y(x) = e^{xQ}y_0 + \int_0^x e^{(x-s)Q}g(s)ds \tag{6.3.1}$$

or alternatively

$$y(x+h) = y(x) + \int_0^h e^{(h-s)Q}(y(x+s) + Qy(x))ds. \tag{6.3.2}$$

The computation of this solution is, of course, an extension of the matrix exponential problem and it is well known that there are at least 19 dubious ways to compute this quantity (Moler and van Loan (1978)). However, in the cases in which Q is large and even sparse, it is inappropriate to compute e^{hQ} and so the solution should be formed without computing the matrix exponential. This can be done by noting from (6.3.1) that the solution only needs the computation of $e^{hQ}v$. Thus the solution can be evaluated if it is possible to compute the exponential of a matrix times a vector v, say, efficiently.

This problem has been considered by Gallopoulos and Saad (1989) in which they project the problem onto a Krylov subspace K_k of dimension k generated by

$$K_k(Q, v) = \text{span}\{v, Qv, \ldots, Q^{k-1}v\}.$$

The Arnoldi algorithm (the Lanczos algorithm for nonsymmetric matrices) can be implemented in parallel to generate an orthonormal basis V_k and an upper Hessenberg matrix H, which represents the projection of Q onto K_k with respect to this basis such that

$$e^Q v \approx \| v \|_2 V_k e^H e_1, \tag{6.3.3}$$

where $e_1 = (1, 0, \ldots, 0)^T$. Of course one of the problems here is the appropriate and automatic determination of the dimension of the subspace.

A Runge-Kutta method can now be used to approximate $e^H v$ by either a polynomial or rational Padé approximation. Since the problem is likely to be stiff, a real pole Padé approximation can be used (so that the underlying method is multiply implicit). In this case the stability function associated with the method can be written as

$$R(z) = 1 + \sum_{j=1}^{s} \frac{a_j}{z - \lambda_j}$$

so that when approximating $e^H v$ by $R(H)v$ a series of linear systems of the form

$$(H - \lambda_j I_k)x_j = v, \quad j = 1, \ldots, s$$

is solved in parallel with the solution computed from

$$v + \sum_{j=1}^{s} a_j x_j.$$

Gallopoulos and Saad (1989) then use (6.3.2) with a fixed stepsize h and fixed k. The difficulty here is in knowing how to choose h and k appropriately and whether there should be an adaptive procedure for both h and k. Gallopoulos and Saad (1989) also note that Padé approximations are accurate near the origin but may be inaccurate far away. Thus because differential systems that arise from the semi-discretization of parabolic equations have a wide spread of eigenvalues on the real negative half plane a Chebyshev approximation approach may be suitable.

Gallopoulos and Saad (1990) note that (6.3.1) can also be written as

$$y(x + h) = e^{hQ}y(x) + \int_0^h e^{(h-s)Q}g(x + s)ds \tag{6.3.4}$$

and that a quadrature formula of the form

$$\sum_{j=1}^{r} w_j e^{(h-s_j)Q}g(x + s_j)$$

can be used to approximate the second term. Thus at each step $r + 1$ Krylov subspace approximations are needed and there is a trade-off here between accuracy and efficiency. The smaller the value for r, the smaller the value of h.

In fact if the representation (6.3.2) is used then only r Krylov subspace approximations are needed per step. However, this representation has different stability properties to the first approach. The difference scheme for (6.3.2) based on a quadrature approximation can be written in the form

$$y_{n+1} = (I + h\sum_{j=1}^{k} w_j e^{(h-s_j)Q})y_n + r_n$$

and it is easily seen that in the scalar case with $Q = q$ the trapezoidal rule is only stable if $h|q| \leq 2$ while the midpoint rule is stable for all

h. More generally, Gallopoulos and Saad (1990) show that the Gauss and open Newton Cotes quadrature formulas are stable for all h.

Of course a real pole Padé approximation can be replaced by a diagonal Padé approximation when computing $e^H e_1$. This gives greater accuracy at each step but there are larger overheads in dealing with the complex poles (see Leyk and Stewart (1993), for example).

One of the drawbacks of the Padé approximation approach discussed above is that they are only accurate near the origin. In addition, if hQ has widely spread eigenvalues then the computation of the rational approximants involves solving ill-conditioned linear systems. Some of these difficulties can be avoided by scaling and squaring. Thus by noting that

$$e^{hQ} = (e^{hQ/2^l})^{2^l},$$

then

$$e^{hQ} \approx (R(hQ/2^l))^{2^l}$$

gives acceptable accuracy even for small Padé orders. Here l should be not too large since rounding errors can dominate the computed squares.

Sidje (1994) has considered the linear IVP (2.2.4) which describes the evolution of the state probability vector of a Markov process having a very large and sparse infinitesimal generator, and attempts to construct efficient parallel methods for such problems. The crucial point in this construction is to note that if $e^{hQ}v$ is approximated by $p_{k-1}(Q)v$, where p_{k-1} is some polynomial of degree $k-1$, then this will result in an element of the Krylov subspace $K_k(Q, v)$. Thus the approximation problem can be viewed as finding an element of $K_k(Q, v)$ that approaches $e^Q v$.

The crucial element of all Krylov subspace techniques involves the computation of an orthonormal basis $V_k = (v_1, \ldots, v_k)$ of $K_k(Q, v)$. This is usually performed by the Arnoldi algorithm which yields

$$QV_k = V_{k+1}H,$$

where H is an upper Hessenberg matrix. However, parallelism here is restricted to using matrix-vector products, inner products and **saxpys**. An alternative approach which gives more parallelism is through the classical Gram-Schmidt process which unfortunately can become unstable. However, Björck (1994) notes that the stability can be restored through one iteration of a reorthogonalization process – but this is at twice the cost.

Another approach is to compute the QR decompositions of the columns appearing in the representation of K_k, but this is inappropriate since the condition number of K_k increases dramatically with increasing k. A more appropriate technique is to consider a different basis representation, such as a Newton basis, of the form

$$\text{span}\{v, (Q - \lambda_1 I)v, \ldots, \prod_{j=1}^{k-1}(Q - \lambda_j I)v\}.$$

Here the λ_j are chosen to improve the conditioning of the process. It was shown in Bai et al. (1992) that a suitable choice for the λ_j is based on the Leja ordering of the Leja points (see Sidje (1994), for example). In particular, the λ_j can be chosen as the eigenvalues of the Hessenberg matrix obtained from the first step of the Arnoldi process.

Thus by letting X_k be a matrix with k columns given by

$$X_k = (v, (Q - \lambda_1 I)v, \ldots, \prod_{j=1}^{k-1}(Q - \lambda_j I)v)$$

the desired orthonormal basis for K_k can be computed by finding the QR decomposition of X_k. Thus an efficient parallel implementation involves:

- forming $(Q - \lambda I)v$ using sparse procedures;
- a parallel QR factorization of X_k to give $Q_k R_k = X_k$;
- a sequential computation of the associated Hessenberg matrix and $w = e^{hH_k} e_1$
- a parallel computation of $Q_k w$.

Sidje notes that although there are efficient parallel implementations for dense BLAS, less attention has been paid to sparse BLAS with little exploitable structure in a distributed-memory environment. Thus Sidje develops a sparse matrix-vector product routine taking into account various load balancing and storage issues. Because the third stage of the above process can be viewed as the implementation of a simple one-step method for a linear IVP, Sidje suggests strategies for varying h based on the same heuristics used in standard ODE solvers. There is also the possibility of adapting the dimension of the Krylov subspace. Performance figures (on a Paragon) of this approach are presented in Sidje (1994) and these suggest that the above algorithm is a promising one.

6.4 Extrapolation

Extrapolation a is very natural technique for exploiting modest forms of parallelism. For example, Burrage and Plowman (1990) have considered a parallel implementation of the smoothed midpoint rule. The basic idea here is that any differential method can be applied over a time interval $[t_n, t_n+H]$ with a series of differing constant stepsizes $h_i = \frac{H}{n_i}, i = 1, \ldots, p$. Each n_i in the stepsize sequence is an integer and there are a number of possible ways of generating the n_i such as through doubling or by a harmonic Romberg sequence. If the underlying method has an even power series expansion of the stepsize in the global error, then Richardson extrapolation can be performed to generate high-order methods of any desired accuracy. An efficient parallel implementation requires the grouping of the p methods to ensure that each processor has approximately the same amount of work. This approach is well suited for a parallel implementation when the problem size is large or function evaluations are very costly but again the parallelism is strictly limited.

Lustman et al. (1992) consider the solution of general problems by extrapolation, in which the system is solved independently on each processor by Gragg's modified midpoint rule using different stepsizes. The results are then combined by either polynomial or rational extrapolation to obtain higher accuracies. The advantage of the Gragg approach is the even power expansion of the global error as a function of h.

Thus given p processors, the jth processor solves (2.1.2) using stepsize $h_j = \frac{1}{j}H, j = 1, \ldots, p$. Since the first processor uses the largest stepsize it will finish its task first, the second processor second, and so on. Thus as soon as the second processor is finished, the first processor can start computing the elements in the first column of the extrapolation table, with the second processor working on the second column as soon as two entries in the previous column are ready. For eight processors, Lustman et al. (1992) give the following processor loading described in Table 6.1.

Table 6.1 *processor loading*

processor	1	2	3	4	5	6	7	8
	T_{00}	T_{01}	T_{02}	T_{03}	T_{04}	T_{05}	T_{06}	T_{07}
	T_{10}							
	T_{11}	T_{20}						
	T_{12}	T_{21}	T_{30}					
	T_{13}	T_{22}	T_{31}	T_{40}				
	T_{14}	T_{23}	T_{32}	T_{41}	T_{50}			
	T_{15}	T_{24}	T_{33}	T_{42}	T_{51}	T_{60}		
	T_{16}	T_{25}	T_{34}	T_{43}	T_{52}	T_{61}	T_{70}	

In some rudimentary testing, Lustman et al. (1992) show that as the stepsize is decreased the speed-up of an eight-processor hypercube implementation versus the time required by one processor to execute the same task approaches 8. However, the authors do not compare their parallel implementation with other efficient solvers. It should also be noted that this parallel extrapolation approach only improves the order of the process and limited benefits would be expected in a massively parallel setting. This was borne out in a transputer implementation of an extrapolation code by Burrage and Plowman (1990) which again obtained only modest speed-ups.

In a different approach, Evans and Megson (1987) note that the construction of extrapolation tables is similar to operations on certain matrix equations which are lower triangular, and this technique can be exploited using systolic arrays.

Rauber and Rünger (1994a, 1994b) have considered the implementation of extrapolation methods on distributed MIMD machines and have given various load balancing options as well as a performance analysis for the hypercube.

Simonsen (1990) has constructed interpolants for various extrapolation methods including both the Euler methods and the class of symmetric two step methods, and also reports on the numerical behaviour of a parallel implementation of an extrapolation method using the smoothed midpoint rule. Two versions of this code (one using macrotasking and the other using microtasking) were implemented on a two-processor Cray X-MP on DETEST. Vectorization was exploited and the parallel implementation used the fact that all elements in the first column of the extrapolation table are independent of one another and each element in a column is independent of all other elements in the same column. Close to optimal speed-ups were obtained on the two-processor machine.

A similar approach occurs in a parallel implementation of iterated defect correction algorithms (Augustyn and Ueberhuber (1992)). Iterated defect correction is a technique for iteratively improving the accuracy of numerical procedures and has as its theoretical basis the work of Frank and Ueberhuber (1978).

The basic idea of iterated defect correction is as follows.

I. First solve an IVP with a cheap method, then estimate the defect in this answer and use the residual to construct a neighbouring problem whose solution is known explicitly.

II. The cheap method is then used to solve the neighbouring problem and the difference between the exact solution and this numerical approximation serves as an estimate of the error in the original problem.

III. This estimate can be used to correct the base solution and this procedure is iterated. After a small number of iterations, accuracy can no longer be improved.

Parallelism here is obtained by subdividing the region of integration into k subintervals $[x_0, x_1], \ldots, [x_{k-1}, x_k]$ and applying a frontal-type approach as described for frontal prediction-correction techniques in Chapter 5. Thus processor i $(i = 1, \ldots, p)$ will compute the solution on $[x_{i-1}, x_i]$, the first defect on $[x_{i-2}, x_{i-1}]$, the second defect on $[x_{i-3}, x_{i-2}]$, etc. But since only a small number of corrections are needed to attain convergence, this gives a very limited form of parallelism. In fact this approach has exactly the same limitations as the frontal predictor-corrector approach where only a small number of corrections can effectively be utilized. On the other hand this approach is a viable one if parallelism is being used to improve the robustness and efficiency of existing sequential algorithms.

6.5 Using parallelism for robustness

There are a number of ways in which parallelism can be exploited without necessarily designing new algorithms. In particular, the approaches through extrapolation and deferred correction described in the previous section can be interpreted as an attempt to use parallelism to improve the robustness and accuracy of extant algorithms. A different approach could involve taking linear combinations of different sequential methods running on different processors to improve stability and accuracy, while Enright and Higham (1991) have used small-scale parallelism in defect evaluation and error control to improve reliability.

Since it is generally accepted that only modest gains in efficiency are possible in terms of exploiting small-scale parallelism in the solution of nonstiff IVPs, Enright and Higham (1991) show how to exploit parallelism by improving reliability and functionality rather than efficiency when using any explicit Runge-Kutta method of order w.

This is done by taking several smaller substeps in conjunction with the main step which allows free interpolation. This means that methods can be used which do not have a built-in interpolant or an embedded formula pair and so there are significantly fewer function evaluations. The error control is based on defect sampling and it can be shown that if $k-1$ parallel substeps are taken with at least $k-1$ processors and if

$$w < 2k - 1$$

then both the interpolant and the error control mechanism satisfy very strong reliability conditions. Of course, if extra processors are available a parallel Runge-Kutta method could be used over these additional processors.

Defect control works by assuming that on the interval $[x_n, x_n + h]$ there is a continuous approximation p_n such that

$$p_n(x_n + \theta h) \approx y(x_n + \theta h), \quad \theta \in (0, 1].$$

By defining the defect $\delta_n(x)$ by

$$\delta_n(x) = p_n'(x) - f(x, p_n(x)),$$

Enright (1989) suggested an error control mechanism in which the defect is sampled at a single point θ_1 on every step and $\| \delta_n(x_n + \theta_1 h) \|$ is used to approximate $\max_{[0,1]} \| \delta_n(x + \theta h) \|$. In order to obtain robust schemes, however, special interpolants are needed (see Higham (1989)) which are considerably more expensive in a sequential environment. This is not the case in the parallel implementation.

Thus by assuming that the k parallel steps are of length $\sigma_i h$, $i = 1, \ldots k$, $(\sigma_k = 1)$, and defining

$$g(\theta) = \sum_{i=1}^{k} \sigma_i^{w+1} d_i(\theta),$$

where $d_i(\sigma_j) = 0, j \neq i$, $d_i(\sigma_i) = 1$, $d_i'(\sigma_j) = 0$ then Enright and Higham (1991) show

$$\| \delta_n(x_n + \theta h) \| = \frac{\| l_{n+1} \|}{h} |g'(\theta)| + O(h^{w+1}),$$

where l_{n+1} is the local truncation error of the underlying method. The idea now is to sample the defect at $\theta = \theta_1$ where $|g'(\theta)|$ is maximized. By assuming that the parallel steps are well spaced out, Enright and Higham (1991) determine optimum values for θ_1 for $k = 4$ with orders 5 and 6, and for $k = 5$ with orders 7 and 8.

Clearly this approach is well load-balanced and there is little communication between processors so that a near optimal level of parallelism is expected if function evaluations dominate the overall integration cost, and this is borne out by some numerical tests.

6.6 Other techniques

There are of course many approaches to developing parallel techniques for IVPs which do not necessarily fit into the classifications presented here. For example, another way in which parallelism can be exploited is in the automatic generation of Taylor series methods. Such work would be an extension of that given by Barton, Willers and Zahar (1971).

Automatic differentiation

Recently, there has been work on techniques for automatic differentiation (see Griewank (1991), for example). Possible approaches include a **forward** and a **backward** mode approach. In the former, the forward mode can be propagated simultaneously with function evaluation, whereas in the backward approach there are greater memory overheads. This is because reverse elimination can begin only after the function has been fully evaluated which means all intermediate calculations must be kept in memory. On the other hand, the forward mode generates much greater fill-in in terms of vertex elimination (it is closely related to the sparse Gaussian elimination problem) than the reverse mode approach.

Automatic differentiation techniques can have significant impact in the solution of differential-algebraic equations which requires efficient and accurate techniques for evaluating derivatives of implicit functions. However, there appears to be little work in this area and the application of automatic differentiation techniques to a parallel environment is also still in its infancy (see Bischoff (1991), for example).

Some attention has also been paid to the development of parallel software tools for aiding in the implementation of parallel algorithms for the solution of differential equations. Thus, Dimitriadis and Karplus (1988a, 1988b) have developed a general set of parallel tools for solving large systems of differential equations including model generation, algorithm selection, automatic decomposition of the model, partitioning and scheduling.

Convolution algorithms

Boglaev (1993) exploits a convolution approach in which the general nonlinear problem is linearized by a modified Newton technique and the Jacobian is approximated by different constant matrices on various subintervals. These constant matrices are then diagonalized so that the nonlinear system is reduced to a sequence of scalar linear problems on the subinterval decomposition of the form (2.2.4). The solution to these problems can then be written in convolution form which can then be evaluated efficiently in parallel by, for example, FFT algorithms. Because the diagonalization process is costly, this technique seems appropriate for problems of modest size.

Symmetrical forms

Abou-Rabia and Boglaev (1994) suggest exploiting parallelism by solving ODE systems in symmetrical form, with one form placed on each processor. Given one independent variable and m solution components there are $m+1$ symmetrical forms, obtained from the original system by replacing the independent variable by, in turn, each of the m component variables. The equations can be written as

$$\frac{dy_j}{dy_i} = \frac{f_j(y_0, y_1, \ldots, y_m)}{f_i(y_0, y_1, \ldots, y_m)}, \quad i = 0, 1, \ldots, m, \quad j = 0, 1, \ldots, m,$$

where y_0 denotes the independent variable. The advantage of this approach is that it can happen that some of the forms can be much less stiff than the original formulation.

Abou-Rabia and Boglaev (1994) then suggest a parallel implementation in which, periodically, the master processor compares the current integral curve length of all the processors and reinitializes all processors to the right bound value of the processor with maximal length of the integral curve. Thus processors can skip sections of their solution trajectory and the algorithm effectively behaves as if it is integrating the form with the least amount of stiffness. Some numerical tests are given by the authors on trivial problems of at most dimension 3. It remains to be seen how this approach behaves near equilibrium points and what happens if m is large as this implies an equivalently large number of processors.

Atmospheric problems

Finally in this section two new applications for solving differential systems are considered and while there is no particular use of parallelism here they are both capable of exploiting parallelism and at the same time they represent new efficient sequential approaches. The first application involves an application to the modelling of air pollution.

Air pollution modelling can involve the numerical integration of the associated atmospheric chemical kinetics equation at a vast number of grid points. Such a system is very stiff and can be written in the form

$$y'(x) = P(x, y) - L(x, y)y \tag{6.6.1}$$

where L is a diagonal matrix and P is a nonnegative vector (see Verwer and Simpson (1994), for example). A traditional approach for solving such problems is based on quasi-steady-state approximation (QSSA) methods of the form

$$y(x + h) = e^{-hL}y(x) + (I - e^{-hL})L^{-1}P.$$

This is exact if P and L are constant and the efficacy of this approach relies on the fact that P and L are usually slowly varying.

QSSA methods are cheap to implement and have good stability properties but are not particularly accurate unless the lumping of related chemical reactions take place. As Verwer and Simpson (1994) note, if this lumping is done carefully then QSSA methods can be more efficient than general-purpose stiff solvers. Nevertheless, the lumping process can be very time-consuming. For this reason, Verwer (1993) and Verwer and Simpson (1994) have designed special-purpose implicit methods which take advantage of the structure associated with (6.6.1). These methods are based on the variable-step, second-order BDF method and the implicit equations are solved by the nonlinear Gauss Seidel method. Since the nonlinear system can be written in fixed-point form

$$y = F(y) := (I + \beta hL(x, y))^{-1}(Y + \beta hP(x, y)),$$

where β is a method parameter and Y some predictor, the iterations take the form

$$y_i^{(k+1)} = F_i(y_1^{(k+1)}, \ldots, y_i^{(k+1)}, y_{i+1}^{(k)}, \ldots, y_m^{(k)}), \quad i = 1, \ldots, m. \tag{6.6.2}$$

This is just the nonlinear Gauss-Seidel method and of course can be replaced by the nonlinear Jacobi method which would then offer considerable scope for exploiting parallelism. Verwer and Simpson (1994) have shown numerically that for chemical kinetic problems of the form (6.6.1) only a very few number of iterations are needed to obtain good convergence behaviour. Of course, correcting to convergence (as in Verwer (1993)) means a larger stepsize can be taken, but as Verwer and Simpson (1994) note, there are restarts at sunset and sunrise as certain concentrations go to zero and it is therefore more advantageous to take smaller stepsizes. As with waveform relaxation methods the convergence of (6.6.2) is strongly affected by the ordering of the components, but for problems of the form (6.6.1) an ordering in terms of decreasing stiffness seems to be appropriate. Because this approach is Jacobian-free with low memory requirements it is very efficient at relative error tolerances of 1%, which is an appropriate level due to the uncertainties in the modelling process, compared with general-purpose stiff solvers.

Deflation techniques for ODEs

The second approach described in this section represents a generalization of the deflation techniques, described in Chapter 4 for linear systems of equations, to differential systems. Large differential equations are often stiff and can be characterized by rapidly changing and slowly changing modes. The techniques of Shroff and Keller (1993), Jarausch (1993) and Burrage et al. (1994a, 1994b) can thus be applied directly to the differential equation system to produce an automatic partitioning technique into stiff and nonstiff methods which can then be solved, as appropriate, by implicit and explicit numerical methods. Some attempts at partitioning in this way have been done previously but never in a truly adaptive manner which these new techniques may now allow.

For example, consider the application of the implicit Euler method to (2.1.1). Then this can be written as a nonlinear difference equation

$$F(y_{n+1}) = 0, \quad F(y) = y - y_n - h_n f(y). \tag{6.6.3}$$

The technique of Shroff and Keller (1993) can be applied directly to (6.6.3), but if the modified Newton method is used, then at each step point a sequence of linear systems of the form

$$\begin{aligned} Q_n \Delta_n &= y_{n+1}^{(k)} - y_n - h_n f(y_{n+1}^{(k)}) \\ y_{n+1}^{(k+1)} &= y_{n+1}^{(k)} - \Delta, \quad k = 0, 1, \ldots, l-1 \\ Q_n &= I - h_n f'(y_n) \end{aligned} \tag{6.6.4}$$

has to be solved.

At a given step it is customary to perform an LU factorization of Q_n at the first iteration, so that for the remaining $l-1$ iterations only backwards and forward substitutions are needed in (6.6.4). However, if at the next step h_n is changed but the Jacobian is not changed (as is often the case in many stiff codes) these LU factors cannot be reused.

On the other hand, if the deflation process based on Richardson iteration (rather than Jacobi iteration) is used then not only can the deflation process continue across the iterations in one integration step but also across many integration steps (as long as the Jacobian is kept constant throughout this region of integration). Hence continuous deflation has an advantage over LU factorization in this respect.

In order to see why this is the case note that $J_{n+1} = J_n$ implies

$$Q_{n+1} = r_n Q_n + (1 - r_n) I, \quad r_n = \frac{h_{n+1}}{h_n}. \tag{6.6.5}$$

Now in the notation of Chapter 4 the iteration matrices for Q_n and Q_{n+1} using (M, N) splittings are given by

$$\begin{aligned} H_n &= M_n^{-1} N_n = I - M_n^{-1} Q_n \\ H_{n+1} &= M_{n+1}^{-1} N_{n+1} = I - M_{n+1}^{-1} Q_{n+1}. \end{aligned} \tag{6.6.6}$$

Hence (6.6.5) and (6.6.6) imply

$$H_{n+1} = I - M_{n+1}^{-1}(r_n M_n (I - H_n) + (1 - r_n) I).$$

Thus if λ and x are an eigenvalue and associated eigenvector of H_n then

$$H_{n+1}x = x - r_n(1-\lambda)M_{n+1}^{-1}M_n x - (1-r_n)M_{n+1}^{-1}x.$$

Thus x will be an eigenvector of H_{n+1} if x is an eigenvector of both M_{n+1} and M_n, which in turn implies that x is an eigenvector of Q_n as well. This can be achieved, for example, if

$$\begin{aligned} H_n &= p(Q_n) \\ H_{n+1} &= q(Q_n), \end{aligned}$$

where p and q are arbitrary polynomials of degree 1 or more satisfying $p(0) = 1, q(0) = 1$. The simplest such polynomials occur if

$$M_n = \omega_n I, \quad M_{n+1} = \omega_{n+1} I, \tag{6.6.7}$$

where ω_n, $\omega_{n+1} \in \mathbb{R}$ in which case

$$H_n = I - \frac{1}{\omega_n}Q_n, \quad H_{n+1} = I - \frac{1}{\omega_{n+1}}Q_{n+1}. \tag{6.6.8}$$

These iteration schemes are of course just examples of Richardson iteration, and the remaining implementation question concerns the choice of ω_n and ω_{n+1}.

Given that the Jacobian associated with (2.1.1) has all eigenvalues with nonpositive real part, then all the eigenvalues of Q_n will have positive real part.

Thus let λ denote any eigenvalue of Q_n. From (6.6.8) it is easily seen that $\rho(H_n) < 1$ if and only if

$$\omega_n = \theta \max_{\lambda \in \sigma(Q_n)} \left\{ \frac{|\lambda|^2}{2Re(\lambda)} \right\}, \quad \theta > 1, \quad Re(\lambda) > 0. \tag{6.6.9}$$

Thus given that a suitable ω_n has been found satisfying (6.6.9), a value for ω_{n+1} can be chosen by using (6.6.5) and (6.6.9) namely

$$\omega_{n+1} > \max_{k \in \sigma(Q_{n+1})} \left\{ \frac{|k|^2}{2Re(k)} \right\} \tag{6.6.10}$$

or

$$\omega_{n+1} > \frac{r}{2}\left(2t + \frac{|\lambda|^2 - t}{Re(\lambda) + t}\right), \quad t = \frac{1-r_n}{r_n}, \quad \lambda \in \sigma(Q_n). \tag{6.6.11}$$

The choice of ω_n and ω_{n+1} can have significant impact on the behaviour of the deflation technique. If ω_n and ω_{n+1} are large enough, convergence is guaranteed, but on the other hand this implies many eigenvalues are clustered around one and convergence will be slow even if many eigenvalues are deflated. In fact, numerical testing has shown that if ω_n and ω_{n+1} are chosen so that the conditions (6.6.10) and (6.6.11) are violated then convergence can be much faster. This is because the eigenvalues are much less clustered and although divergence will initially take place if those eigenvalues causing divergence are deflated sufficiently quickly the convergence of the deflation scheme can rapidly accelerate.

6.7 Parallelism across the steps

It has already been seen how parallelism across the method can be achieved by computing the stages of some multistage method on different processors, while parallelism across the system can be introduced by performing function evaluations on different processors. In parallelism across the steps the aim is to carry out calculations of blocks of steps concurrently. A prime example of this is the concept of multiple shooting in which parallelism is reduced at the cost of redundant computations. This was first explored by Nievergelt (1964) in relationship to IVPs.

Shooting methods

For linear problems, Nievergelt (1964) proposed a stabbing approach in which the interval of integration is subdivided and then a number of different problems are solved on each interval concurrently. Interpolation is then performed in parallel. Nievergelt demonstrated a bounded analysis in the linear case but even in this case as m becomes large the number of solutions grows exponentially. Consequently, this approach is now considered not to be practical, especially for nonlinear problems. However, it is perhaps the first instance of a specific parallel implementation for a restricted class of problems.

Both Kiehl (1993) and Khalaf and Hutchinson (1992) have considered extensions of this approach through the use of parallel multiple shooting techniques. In the former, Kiehl (1993) formulates an initial value problem as a boundary value problem in which all the boundary conditions are at the left and then uses BVP multiple shooting techniques. Given that the dimension of the problem is small this process works well for linear problems or for nonlinear problems in which good initial approximations are available.

The concept of parallelism across the step has already been introduced in section 6.1 when a Runge-Kutta method was applied to the linear test problem (2.2.4) to obtain a block-bidiagonal system of linear equations (6.1.6), which can then be solved by techniques such as cyclic reduction. For the more general problem (6.1.1), the interval of equation can be split into N subintervals with processor j solving an associated problem on $[x_{j-1}, x_j]$ which involves computing the fundamental solution on that interval.

Now the solution to (6.1.1) on $[x_0, T]$ is given by

$$y(x) = R(x, x_0)y_0 + \int_{x_0}^{x} R(x, s)g(s)ds, \qquad (6.7.1)$$

where $R(x, x_0)$ is the resolvent matrix satisfying the differential system

$$R'(x, x_0) = Q(x)R(x, x_0), \quad R(x, x_0) = I.$$

The solution given in (6.7.1) can also be written as

$$y(T) = R(T, x)(R(x, x_0)y_0 + \int_{x_0}^{x} R(x, s)g(s)ds) + \int_{x}^{T} R(T, s)g(s)ds.$$

Thus if the interval of integration is split into N intervals then N differential systems of the form

$$z'(x) = Q(x)z(x) + g(x), \quad z(x_{j-1}) = 0, \quad x \in [x_{j-1}, x_j],$$

together with the resolvent equation

$$R'(x, x_{j-1}) = Q(x)R(x, x_{j-1}), \quad R(x, x_{j-1}) = I,$$

have to be solved. The solution can then be computed recursively from

$$y(x_j) = R(x_j, x_{j-1})y(x_{j-1}) + z(x_j), \quad j = 1, \ldots, N.$$

Now $y(T)$ can also be computed by

$$y(T) = z(x_n) + \sum_{j=0}^{N-1}(\prod_{l=0}^{N-1-j} R(x_{j+l+1}, x_{j+l}))z(x_j).$$

Given $N = 2^m$, for some m, this last step can be efficiently computed on a hypercube using a modification of recursive doubling (see Stone (1973)). Lustman et al. (1991) have considered an implementation of this approach on an Intel hypercube using the box scheme.

Interprocessor communication occurs at the recursive doubling stage and Lustman et al. (1991) consider various communication protocols including "send on request", "broadcast data as soon as it is ready" and "message typing". In this last protocol, messages are stored in processor memory with an identifier and are only accessed as appropriate. Lustman et al. (1991) claim from numerical tests that "message typing" appears to be the most efficient.

Grouping solutions across steps

Fei (1986) has developed parallel algorithms based on linear multistep methods, and, in particular, Adams and BDF methods. These iterative methods, which exploit parallelism across the steps, are shown to converge under certain restrictions on the stepsize, length of window, and the initial iteration. This approach is further developed by He and Wang (1988) who derive schemes that are second-order convergent.

Bellen and Zennaro (1989) attempt to exploit parallelism across the steps for arbitrary nonlinear problems and note that one step of any one-step method applied to (2.1.2) can be viewed as a nonlinear difference equation of the form

$$y_{n+1} = F_{n+1}(y_n), \quad n = 0, \dots, N-1. \tag{6.7.2}$$

In order to exploit parallelism across, say, p steps, Bellen and Zennaro (1989) view (6.7.2) as a fixed-point problem in the space of $N+1$ vector sequences. Thus defining $u = (y_0, y_1, \dots, y_N)$, (6.7.2) is equivalent to the fixed-point problem

$$u = \phi(u), \tag{6.7.3}$$

where $\phi(u) = (u_0, F_1(u_0), \dots, F_N(u_{N-1}))^\top$. (6.7.3) can be solved either by standard fixed-point iteration, or a faster iterative scheme such as modified Newton, or Steffensen's algorithm can be used, depending upon the complexity of the Jacobian evaluations. Bellen and Zennaro (1989) consider only Steffensen's algorithm which at the $(k+1)$th iteration takes the form

$$\begin{aligned} u_0^{(k+1)} &= u_0^{(k)} \qquad (6.7.4)\\ u_{n+1}^{(k+1)} &= F_{n+1}(u_n^{(k)}) + \Delta_{n+1}(u^{(k)})(u_n^{(k+1)} - u_n^{(k)}), \ n = 0, \dots, N-1, \end{aligned}$$

where $\Delta(u^{(k)})$ is some appropriate approximation to the Jacobian. This is now a linear recurrence relation with parallel complexity $\log_2(N)$. In addition, F_{n+1} and Δ_{n+1} can be computed in parallel.

After each iteration, a new exact value is computed so that the true solution y is computed after N iterations for any $u^{(0)}$. However, in order to gain enhanced performance a locally quadratic convergence is needed, which the Steffensen algorithm gives. Bellen and Zennaro (1989) note that the smaller the first and second order partial derivatives of F_n, the faster the convergence and the larger the convergence neighbourhood of the Steffensen approach. They also suggest that if these conditions hold and the F_n are expensive to calculate then there is some potential for computing a large number of steps in parallel more quickly than by using (6.7.2) directly.

However, Bellen et al. (1990a) note that this approach has some disadvantages in that the mesh is fixed and there is no information about the error behaviour of the method. Consequently, Bellen et al. (1990a) perform a moving mesh approach in which the mesh is changed after each iteration. There are a number of difficulties here in constructing an initial mesh which fits a given step tolerance as well as deriving a set of corresponding initial guesses, and then deciding at each iteration which of the unaccepted values are suitable to be interpolated and reiterated and which must be replaced. This and other matters are discussed in some detail in Bellen et al. (1990a). Although some simulations on a scalar machine are given, no parallel implementation is attempted, although it is noted that this approach is a suitable one for a SIMD environment.

However, there seem to be a number of drawbacks with this variable mesh approach:

- there are considerable communication overheads,
- equitable load balancing is complicated,
- the neighbourhood of the exact solution where convergence takes place can be small. In order to increase this, a smaller block of values must be used which affects the parallel performance. This effect has an analogue with the use of windowing in waveform relaxation techniques, discussed in Chapter 7.

Chartier (1993) has modified the approach of Bellen and Zennaro (1989) in the case of dissipative problems and he proves some global convergence results when Newton's method rather than Steffensen's method is used. An implementation on a hypercube is made.

A different approach that could be used in the solution of (6.7.3) is based on the deflation process described in Chapter 4. Since the deflation process automatically accelerates the convergence of the fixed-point process associated with (6.7.4) by a coupled iteration with a modified Newton solve some of the difficulties mentioned above can possibly be avoided. In particular, the restriction on having to have a small block of values in order to get satisfactory convergence may no longer hold because of the accelerated convergence of the deflation process.

Prediction-correction across the steps

It is clear now from the comments of Chapter 5 that parallelism across the method will achieve only modest speed-ups in performance (about 2 – 8), irrespective of whether a large number of processors are used or not. Furthermore, speed-ups at the upper end of this range are only likely to be attained if some form of preconditioning can be used to reduce the number of corrections (see van der Houwen (1991) and Burrage (1993b, 1993d), for example). Van der Houwen et al. (1993a, 1993b) therefore suggest to improve parallel performance by applying parallelism across the steps in both a nonstiff (PIRK methods) and stiff (PDIRK) setting.

This approach is similar to that described in Bellen et al. (1990a). However, the approach of Bellen et. al (1990a) is based on a Steffensen iteration process while that of van der Houwen et al. (1993a) is based on a Gauss-Seidel-type iteration. This Gauss-Seidel iteration process appears to be a more robust one, although it is at the cost of less massive parallelism. Some simple numerical testing suggests PIRK methods exploiting a parallelism across the steps could achieve speed-up factors of up to 15 compared with the best sequential nonstiff codes, while PDIRK methods exploiting parallelism across the steps could achieve speed-up factors of between 7 and 10 over, for example, LSODE (see Sommeijer (1992), for example). However, these speed-ups are based only on the number of sequential function evaluations to get a desired accuracy. A parallel implementation may reduce these speed-up factors significantly.

In what follows, the nonstiff case will be considered first. Thus suppose that N steps of some method

$$Y_{n+1} = (A \otimes I)Y_n + h(L \otimes I)F(Y_{n+1}), \quad n = 0, \dots N-1 \qquad (6.7.5)$$

with a fixed stepsize h are needed to integrate a problem over some integration interval. Given that l fixed-point iterations of (6.7.5) are needed per integration step the conventional predictor-corrector approach over N steps can be described as

ALGORITHM : CONVENTIONAL(l,N)

do $n = 1$ TO N

do $k = 1$ TO l

$$Y_n^{(k)} = (A \otimes I)Y_{n-1}^{(l)} + h(L \otimes I)F(Y_n^{(k-1)})$$

enddo

Here $Y_0^{(l)}$ and $Y_n^{(0)}$ are provided by, respectively, a starting procedure and a predictor formula.

Perhaps the simplest way of exploiting parallelism across the steps is to interchange the loop structure in the above thus giving

ALGORITHM : INTERCHANGE(l,N)

do $k = 1$ TO l

do $n = 1$ TO N

$$Y_n^{(k)} = (A \otimes I)Y_{n-1}^{(k-1)} + h(L \otimes I)F(Y_n^{(k-1)})$$

enddo

If the underlying corrector is a Runge-Kutta method then $A = E$ and the starting procedure and predictor can be estimated by

$$\begin{aligned} Y_0^{(j)} &= e \otimes y_0, \quad j = 0, \ldots, l-1 \\ Y_n^{(0)} &= (E \otimes I) Y_{n-1}^{(l)}, \quad n = 1, \ldots, N. \end{aligned}$$

Thus given the initial guesses $Y_n^{(0)}, n = 1, \ldots, N$, all stage vectors $Y_n^{(1)}$ are computed concurrently, then all $Y_n^{(2)}$, etc. Given sN processors the number of sequential stages is l. Clearly this process represents a Jacobi-type iteration, but as already described for the method of Abbas and Delves (1989), this process will suffer from very poor convergence because of the poor prediction process. The convergence can be improved by windowing, but the window size may have to reduce to the stepsize in order to obtain appropriate convergence behaviour. Consequently, van der Houwen et al. (1993a) suggest a Gauss-Seidel-type iteration process which takes the form

Algorithm : GAUSS-SEIDEL(l,N)

do $k = 1$ TO l
do $n = 1$ TO N
$\quad Y_n^{(k)} = (A \otimes I) Y_{n-1}^{(k)} + h(L \otimes I) F(Y_n^{(k-1)})$
enddo

This modification introduces a dependency in the steps direction and the iterates can then be computed according to various orderings. Van der Houwen et al. (1993a) suggest representing the $Y_n^{(k)}$ in the (n, k) plane and this can lead to row ordering (k constant), column ordering (n constant) or diagonal ordering ($i = n + k$ constant) as illustrated by Figure 6.1. Here all iterates identified with the same shading are computed concurrently.

In the diagonal ordering all iterates labelled, for example, by $i + 2$ are computed concurrently after the $i+1$ iterates are available. Although there is much less parallelism compared with the Jacobi process, the convergence properties are much improved. This process has some similarities to the frontal methods described previously and its effectiveness will be limited by the number of corrections needed per step to get suitable accuracy. Clearly this approach requires sl processors and $l + N$ sequential steps.

The predicted values $Y_n^{(0)}$ can be computed in many different ways, but for ease of analysis and implementation, van der Houwen et al. (1993a) suggest a prediction of the form

$$Y_n^{(0)} = (A_0 \otimes I) Y_{n-1}^{(r_n)}.$$

Here r_n can either take the value l or is such that $Y_{n-1}^{(r_n)}$ represents a safe starting point for increasing the step index. In this case, r_n is determined

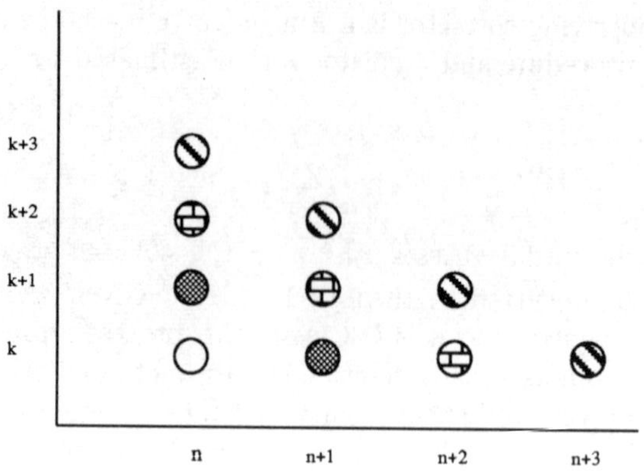

FIG. 6.1. parallelism across the step

adaptively and will depend on the step index. A suitable value for A_0 is given by $A_0 = E$, so that $Y_n^{(0)}$ represents the trivial predictor. Alternatively, the $Y_n^{(0)}$ can represent a Lagrange predictor based on extrapolating the collocating polynomial associated with the underlying corrector.

Stability can be analysed in the usual way by relating $Y_n^{(1)}$ to $Y_{n-1}^{(1)}$ given a one-step predictor. However, a study of the convergence regions associated with this process requires some modification from the usual approach. By defining

$$\begin{aligned} \varepsilon_n^{(k)} &:= Y_n^{(k)} - Y_n, \quad n = 1, \ldots, N \\ \varepsilon^{(k)} &:= (\varepsilon_1^{(k)}, \ldots, \varepsilon_N^{(k)})^\top \end{aligned} \tag{6.7.6}$$

it is easily seen that for the standard linear test equation, the Gauss-Seidel algorithm gives

$$\varepsilon_n^{(k)} - A\varepsilon_{n-1}^{(k)} = zL\varepsilon_n^{(k-1)}.$$

Hence

$$\varepsilon^{(k+1)} = zM\varepsilon^{(k)}, \tag{6.7.7}$$

where M is an $N \times N$ block matrix, with each block being of order s, given by

$$M = P^{-1}Q,$$

with

$$Q = L \otimes I_N, \quad P = (-A, I_s, 0).$$

In the case of an Adams-type corrector, $A = E$, and since $E^k = E$ for any k, then it is easily shown that

$$M = \begin{pmatrix} L & & & \\ L_1 & L & & \\ \vdots & \ddots & \ddots & \\ L_1 & \dots & L_1 & L \end{pmatrix}, \quad L_1 = EL.$$

The region of convergence, C, is given by

$$C = \{z : \rho(zM) < 1\}$$

and because the nonzero spectrum of M is the same as L for the Adams-type correctors

$$C = \{z : \rho(zL) < 1\}.$$

It should be noted that for $s \leq 5$ the Gauss methods have correspondingly smaller values of $\rho(L)$ compared with the Radau IIA methods. Because these values are relatively small it is in fact the stability condition imposed by the predictor that limits the stepsize, rather than the convergence region.

In the case of dynamic iteration, the above error analysis has to be modified and so does the dependency diagram. For example, in the case that $r_n = 3$, Figure 6.1 is modified to give Figure 6.2 in which all iterates identified with the same shading are computed concurrently.

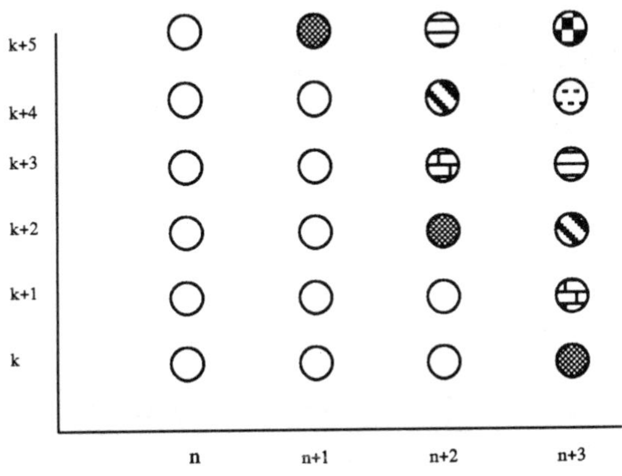

FIG. 6.2. dynamic iteration across the steps

Van der Houwen et al. (1993a) implement the dynamic process in a sequential environment with a rudimentary stepsize control strategy and investigate the accuracy behaviour as a function of the averaged number of iterations and the number of concurrently computed iterates p. For an order 8 implementation with $p = 8$, maximum speed-up factors over DOPRI8 of at most 10 appear possible, obtained at very stringent tolerances.

However, these estimates are only based on a function evaluation count and ignore communication costs and load balancing difficulties.

When exploiting parallelism across the steps for stiff problems, van der Houwen et al. (1993b) replace (6.7.5) by a PDIRK formula and this leads to iteration schemes of the form

$$\begin{aligned} Y_n^{(1)} - h(D^* \otimes I)F(Y_n^{(1)}) &= (E \otimes I)Y_{n-1}^{(1)} \\ Y_n^{(j)} - h(D \otimes I)F(Y_n^{(j)}) &= (E \otimes I)Y_{n-1}^{(r_n)} + h((A-D) \otimes I)F(Y_n^{(j-1)}). \end{aligned} \tag{6.7.8}$$

As for the nonstiff case, $r_n = j-1$ leads to a Jacobi-type iteration, while $r_n = j$ gives a Gauss-type iteration. Here D^* and D are diagonal matrices. A natural choice is $D^* = D$ so that the prediction and correction formulas share the same LU decomposition. In this case a predictor stage order of $s-1$ is attained for a Radau IIA corrector. However, unlike the standard predictor-corrector implementation, step-parallel methods require the predictor to be stable over the integration interval. Given that the underlying corrector is A-stable, the stability region of any step-parallel method is the intersection of the convergence region of the corrector and the stability region of the predictor. However, van der Houwen et al. (1993b) note that the choice for the predictor in (6.7.8) does not lead to predictors with good stability properties and suggest modifying the predictor to give a two-step formula of the form

$$Y_n^{(1)} - h_n(D^* \otimes I)F(Y_n^{(1)}) = (E_1 \otimes I)Y_{n-1}^{(1)} + (E_2 \otimes I)Y_{n-2}^{(1)},$$

which has improved stability properties.

The convergence region of these implicit schemes can be studied in exactly the same manner as that which leads to (6.7.6) and (6.7.7). In this case it can be shown that

$$\varepsilon_n^{(k)} = K_n \varepsilon_{n-1}^{(r_n)} + P_n \epsilon_n^{(k-1)},$$

where

$$K_n = (I - z_n D)^{-1} E, \quad P_n = z_n D (I - z_n D)^{-1} (D^{-1}A - I), \quad z_n = h_n q.$$

Hence

$$\varepsilon^{(k)} = Q(z)\varepsilon^{(k-1)}, \quad z = (z_1, \ldots, z_N)^{\mathsf{T}},$$

where Q is an $N \times N$ block iteration matrix.

It is easily shown that for the Jacobi iteration, $Q = Q_J$ is block-bidiagonal of the form

$$\begin{pmatrix} P_1 & & & \\ K_2 & P_2 & & \\ & K_3 & P_3 & \\ & & \ddots & \ddots \end{pmatrix}, \tag{6.7.9}$$

while for Gauss-Seidel iteration, $Q = Q_G$ is block lower triangular of the form

$$\begin{pmatrix} P_1 & & & & \\ K_2P_1 & P_2 & & & \\ K_3K_2P_1 & K_3P_2 & P_3 & & \\ K_4K_3K_2P_1 & K_4K_3P_2 & K_4P_3 & P_4 & \\ \vdots & \vdots & \vdots & & \ddots \end{pmatrix}. \tag{6.7.10}$$

The condition for convergence is thus the same for the underlying PDIRK and so the stability of the step-parallel methods considered here is determined only by the predictor.

However, the rate of convergence of these step-parallel methods may be much worse than the underlying PDIRK because of ill conditioning of Q due to, with fixed stepsizes, Q having s eigenvalues of multiplicity N. If all stepsizes are different then the conditioning improves but convergence may still be slow. However, it can be shown from (6.7.9) and (6.7.10) (van der Houwen et al. (1993b)) that at infinity

$$Q_J(z) = Q_G(z) = (I - D^{-1}A) \otimes I_n + O(z^{-1}),$$

which is identical to the behaviour of the underlying PDIRK.

On the other hand, the rates of convergence at the origin are different for Jacobi and Gauss-Seidel. From (6.7.9) and (6.7.10) it is seen that at the origin

$$\begin{aligned} Q_J(z) &= (E,0,0) + Z(DE, A-D, 0) + O(z^2) \\ Q_G(z) &= MZ + O(z^2), \end{aligned}$$

where $Z = \operatorname{diag}(z_1, \ldots, z_n)$ and

$$M = \begin{pmatrix} A-D & & & \\ H & A-D & & \\ & H & A-D & \\ & & \ddots & \ddots \end{pmatrix}, \quad H = E(A-D).$$

Clearly for $k < N$,

$$(Q_J(z))^k = C + O(z), \quad C \neq 0$$

where $\| C \|_\infty = 1$, and this leads to unacceptably slow convergence of the Jacobi iteration for the nonstiff components. Nevertheless, despite the fact that the stiff and nonstiff rates of convergence for the Gauss-Seidel iteration are satisfactory, this is not true for all $Re\,(z) \leq 0$. In fact, since the averaged rate of convergence is given by

$$R(N, k, z) := -\log(\sqrt[k]{\| Q(z)^k \|}),$$

this suggests attempting to choose D such that $R(N, k, z)$ is minimized over all $Re(z) \leq 0$, rather than choosing D so that $\rho(I - D^{-1}A)$ is minimized.

Van der Houwen et al. (1993b) report on the performance of the Gauss-Seidel step-parallel scheme where the underlying method is the four-stage Radau IIA corrector implemented with fixed stepsize in a sequential environment on a small set of test problems. For a given accuracy, speed-up factors in the range 3.5-5, in terms of the number of sequential steps, over a PDIRK implementation were obtained. However, it remains to be seen if this improvement translates when implemented in parallel in variable stepsize mode.

Operator splitting

This approach consists of splitting a general problem given by (2.1.1) into two equations

$$y'(x) = g(y(x)), \quad y'(x) = h(y(x)),$$

through the decomposition

$$f(y) = g(y) + h(y).$$

The hope here is that these equations will be individually easier to integrate than the original equation and can better reflect the structure of the underlying geometry (see Crouch and Grossmann (1993), for example). This approach can of course be generalized to spatially-dependent splittings of the form

$$y' = f(y) = \sum_{j=0}^{k} g_j(y) h_j(y).$$

In the case of linear problem (2.2.4) (with $g(x) \equiv 0$), a natural splitting for Qy is

$$g(y) = Cy, \quad h(y) = Dy$$

with $Q = C + D$. Hence, the solutions to the individual problems $y' = Cy$ and $y' = Dy$ corresponding to (2.1.1) are

$$e^{(x-x_0)C}y_1, \quad e^{(x-x_0)D}y_2,$$

where y_1 and y_2 are appropriate initial conditions.

The matrices C and D do not, in general, commute and it can be shown that

$$e^{(x-x_0)C}e^{(x-x_0)D} = e^{(x-x_0)Q+\frac{(x-x_0)^2}{2}[C,D]+\frac{(x-x_0)^3}{12}([C,[C,D]+[D,[D,C]])+\cdots},$$

where $[C, D]$ is the commutator $CD - DC$. Thus, for example,

$$e^{\frac{x-x_0}{2}C}e^{(x-x_0)D}e^{\frac{x-x_0}{2}C} = e^{(x-x_0)Q+O((x-x_0)^3)},$$

and it follows that

$$e^{\frac{x-x_0}{2}C}e^{(x-x_0)D}e^{\frac{x-x_0}{2}C}y_0 = e^{(x-x_0)Q}y_0 + O((x-x_0)^3). \qquad (6.7.11)$$

The computation of the left-hand side of (6.7.11) represents the solution of three different systems of equations and can be done, for example, by using a pipelining approach with three processors. The hope is that in the case of waveform relaxation, for example, convergence will take place more rapidly than for the original problem. Observe that the left-hand side of (6.7.11) approximates $e^{(x-x_0)Q}$ if the size of the window of integration is not too large. Note also that more accurate approximations than given by (6.7.11) can be obtained by increasing the number of splittings and hence the number of subproblems.

Finally, in this section a new approach due to Gear and Xuhai will be mentioned. Gear and Xuhai (1991) attempt to construct methods which exploit parallelism across the steps by noting that for quadrature problems and nonlinear systems efficient parallelism is readily achievable. For quadrature problems, integration can be performed in $O(\log(N))$ time on N processors and this is a lower limit on the speed of parallelism across the steps since N terms cannot be added in less time. On the other hand the nonlinear problem $f(x, y) = 0$ can be solved for N different values of x in $O(1)$ time on N processors. Furthermore, these two problems can be considered as limiting cases of a differential equation in which the stiffness changes from zero to infinity. This is seen by considering the singularly perturbed problem

$$y' = \frac{1}{\epsilon}f(x, y) + \phi(x),$$

so that $\epsilon \to \infty$ gives a quadrature problem, while $\epsilon \to 0$ gives the infinitely stiff problem $f(y, x) = 0$.

Thus Gear and Xuhai (1991) suggest using a blended combination of methods that are suitable for these two extremes, based on an analysis of the Jacobian of the system. The stiff components are computed by BDF methods and the nonstiff components by a quadrature formula such as Adams methods, but in such a way that much of the computation is performed in parallel. The process computes a sequence of discrete points $n = j+1, \ldots, j+N$ of the form

$$y_n^{(k+1)} = S_n y_{n-1}^{(k+1)} + G(y_{\{n-i\}}^{(k)}).$$

Here

$$S_n = (1 - h\theta M)^{-1},$$

where M is some approximation to the Jacobian, θ is chosen to obtain good stability and convergence properties and where G describes the blending of the two methods. However, preliminary results are not particularly encouraging because the convergence of the discrete waveforms can be very slow unless N is small and, in addition, M has to be a good approximation to the Jacobian.

6.8 Delay differential equations

Delay differential equations (DDEs) arise from the modelling process when there is a lagging effect and occur naturally in such disparate areas as control theory, population dynamics and bio-mathematics. The main difference between DDEs and ODEs is the presence of memory which may exist for the complete history of the solution or for only a finite history. Thus the analytical solution of DDEs may have derivative discontinuities which can propagate and affect the performance of a numerical method. Hence any robust DDE solver must attempt either to produce a continuous approximation to the solution (based perhaps on some modification of a dense-output facility associated with an ODE solver) or attempt to track the propagation points, associated with the discontinuities, in some way.

The general form of a system of DDEs is given by

$$y'(x) = f(x, y(x), y(\theta_1(x, y(x))), \ldots, y(\theta_k(x, y(x)))), \quad x \geq x_0, \tag{6.8.1}$$

where the **delay functions** $\theta_j(x, y(x))$ satisfy

$$\theta_j(x, y(x)) = x - \tau_j(x, y(x)) \leq x, \quad j = 1, \ldots, k. \tag{6.8.2}$$

The **lag functions** $(\tau_j(x, y(x)))$ are non-negative functions and are said to be **state-dependent** if they depend on y. The solution of (6.8.1) is required to satisfy an initial function $\psi(x)$ defined on some initial interval $[a, x_0]$.

Delay differential equations have a much richer structure than ODEs with smooth solutions or highly discontinuous solutions arising as the consequence of the lag functions (see the review article by Baker et al. (1994), for example). An important class of problems is one for which the lag function vanishes as x approaches some point x_v. For some classes of DDEs it is possible to adapt ODE techniques in a straight-forward manner.

Thus the general scalar DDE

$$y'(x) = f(x, y(x), y(x - \tau(x, y(x)))), \quad x \geq x_0, \tag{6.8.3}$$

can be solved as a sequence of systems of ODEs of increasing dimension (which may become unbounded) on successive intervals $I_k = [a_k, a_{k+1}]$. The dimension of the system on I_k is $k+1$ and the set of points $\{a_j\}$ satisfy

$$a_j - \theta(a_k, y(a_k)) = 0, \quad \text{for} \quad \text{some} \quad a_k > a_j \quad \text{with} \quad a_0 = x_0. \tag{6.8.4}$$

This method is sometimes called the **method of steps** and is the basis of the method of Bellman et al. (1965). Thus scalar problems with a constant delay can easily be solved using an ODE code with a dense-output facility.

In the case of non-constant delays the above approach can still be used provided some modifications are made. Thus suppose that $\tilde{y}(x)$ has been computed on some interval $[x_0, x_n]$ using a dense-output ODE solver with a sequence of stepsizes $\{h_j\}$. Then the next stage involves computing the solution on the interval $[x_n, x_{n+1}]$ of

$$y'(x) = f(x, y(x), \tilde{y}(x - \tau(x, y(x)))), \quad x \in [x_n, x_{n+1}]. \tag{6.8.5}$$

For a constant lag such that

$$h_n := x_{n+1} - x_n \leq \tau$$

the implementation is straight-forward. But if τ is small compared with a natural choice of the stepsize or if there is a vanishing lag then great care is needed for obtaining an efficient implementation.

This leads to the question of how to produce continuous approximations to the solution of an ODE on some interval. Clearly polynomial or spline interpolation can be used to "glue together" the discrete approximations produced by a discrete numerical method. However, the tendency now is to develop continuous extensions of the underlying discrete numerical method. The ease with which this can be done depends on the nature of the underlying method (multistep or one step) and on the stage order of the method.

For example, in the case of collocation methods, the collocating polynomial can be used as the continuous extension. Natural continuous extensions of Runge-Kutta methods for ODEs have been studied by Zennaro (1986), for example, and extended to DDEs by, for example, Paul (1992) and Neves and Thompson (1992). In these cases, of course, the numerical behaviour of the algorithms depends on the order, local error and stability properties of the dense output formulation.

As Baker et al. (1994) note "scalar DDEs possess a richer dynamical structure than scalar ODEs". Furthermore, there is a propagation of discontinuities which in the scalar case occur at those points $a = a_k$ for which

$$a_j - \theta(a, y(a)), \quad a > a_j$$

changes sign for some a_j. This of course can affect the implicit assumptions of smoothness made in stepsize control mechanisms for ODEs. It is known (Neves and Feldstein (1976)) that if a_k has even multiplicity there is no propagation of the discontinuity while if a_k has odd multiplicity the propagated discontinuity can be smoothed. Willé and Baker (1992) have analysed the behaviour of derivative discontinuities for systems of DDEs, which is influenced by the coupling of the components.

Baker et al. (1994) devote considerable time to discussing the appropriateness of various classes of numerical methods to different types of DDEs. Their conclusion is that linear multistep formulas are suitable for problems with smooth solutions, while codes based on Runge-Kutta methods are versatile and suitable for problems requiring frequent stepsize changes. However, if a general robust solver is required it seems clear that it is necessary to track the discontinuities by solving

$$a_j - \theta(a_k, \tilde{y}(a_k)) = 0 \tag{6.8.6}$$

for an a_k which has been propagated from a previously computed discontinuity a_j, given that $\tilde{y}$ is the dense-output approximation. Difficulties can arise in computing these zeros sufficiently accurately, and ensuring the correct multiplicity. Additional difficulties can arise if the zeros are in advance of the current interval.

For problems with vanishing lags, explicit Runge-Kutta methods as applied to (6.8.3) can become implicit unless at each step

$$0 < h_n \leq \tau(x, y(x)), \quad \forall x \in [x_n, x_n + h_n].$$

This of course can unnaturally restrict the stepsize. One way of overcoming this is to approximate the solution values on $[x_n, x_n + h_n]$ by some extrapolation (such as prediction) from the previous interval. However, Baker and Paul (1993) have proposed a class of iterated continuous explicit

Runge-Kutta methods (ICEs) for solving vanishing lag problems which are parallelizable.

In this approach, the underlying method for solving the scalar DDE

$$y'(x) = f(x, y(x), y(\theta(x, y(x)))) \tag{6.8.7}$$

is a continuous explicit Runge-Kutta method (CERK) which can be written in tableau form as

$$\begin{array}{c|c} c & A \\ \hline \theta & b^{\top}(\theta) \end{array} \tag{6.8.8}$$

This method is to be interpreted in the usual way, except that the update gives an approximation at $x_n + \theta h_n$ and the quadrature condition $B(w)$ now becomes

$$\sum_{i=1}^{s} b_i(\theta) c_i^{j-1} = \frac{\theta^j}{j}, \quad j = 1, \dots, w.$$

The parallel continuous extension for $[x_n, x_n + h_n]$ is then computed from a prediction-correction process which is very similar to that described in (5.5.6). Thus a prediction process based on (6.8.8) and a correction process based on l corrections gives, for $k = 1, \dots, l$,

$$\begin{aligned} \tilde{y}^{(0)}(x_{ni}) &= \tilde{y}(x_n) + h_n \sum_{j=1}^{i-1} a_{ij} f(x_{nj}, \tilde{y}^{(0)}(x_{nj}), \tilde{z}_n^{(0)}(\tilde{\theta}_{nj}^{(0)})) \\ \tilde{y}^{(k)}(x_n + \theta h_n) &= \tilde{y}(x_n) + h_n \sum_{j=1}^{s} b_j(\theta) f(x_{nj}, \tilde{y}^{(k-1)}(x_{nj}), \tilde{z}_n^{(k-1)}(\tilde{\theta}_{nj}^{(k-1)})), \end{aligned}$$

where

$$x_{nj} = x_n + c_j h_n, \quad j = 1, \dots, s$$

$$\tilde{z}_n^{(k)}(x) = \begin{cases} \tilde{y}(x), & x \le x_n \\ \tilde{z}_n^{(0)}(x), & x > x_n, \quad k = 0 \\ \tilde{y}^{(k)}(x), & x > x_n, \quad k \ge 1 \end{cases}$$

$$\tilde{y}(x) = \tilde{y}^{(l)}(x), \quad x_0 \le x \le x_n$$

$$\tilde{\theta}_{nj}^{(k)} = \theta(x_{nj}, z_n^{(k)}(x_{nj})).$$

This iterated method can be written as the $s(l+1)$-stage CERK method

$$\begin{array}{c|cccccc} c & A & & & & & \\ c & B & 0 & & & & \\ c & 0 & B & 0 & & & \\ \vdots & \vdots & & \ddots & \ddots & & \\ c & 0 & \cdots & 0 & B & 0 & \\ \hline 0 & 0 & \cdots & 0 & 0 & b^{\top}(\theta) & \end{array} \tag{6.8.9}$$

where $B = (b_j(c_i))_{i,j=1}^{s}$.

Baker and Paul (1993, 1994) have analysed the order and stability properties of ICEs. Due to the propagation of derivative discontinuities it is natural to require $c_1 = 0, c_s = 1$ and $c_j \in [0,1]$. Baker and Paul (1993) conclude that a suitable choice for A in a parallel implementation is

$$A = (c, 0, \ldots, 0). \tag{6.8.10}$$

As for the iterated Runge-Kutta methods described in section 5.5 it can be shown (Baker and Paul (1994)) that each correction increases the global convergence order of $\tilde{y}(x)$ to $y(x)$ by one until some upper bound is reached. This is illustrated in the following result given in Paul (1994).

Theorem 6.8.1. *If (6.8.7) is Lipschitz continuous in its second and third arguments and has sufficient continuous partial derivatives with $y^{(c-1)}(x)$ Lipschitz continuous, and if (6.8.8), with $A = 0$, has continuous quadrature order w then the order of (6.8.9) satisfies*

$$\sup_{x \in [x_0, T]} |y(x) - \tilde{y}(x)| = O(H^p), \quad p = \min\{l, w, r\}, \quad H = \max\{h_j\}.$$

For methods satisfying (6.8.9) and (6.8.10), $c_2, \ldots, c_{s-1}$ are free parameters and Paul (1994) has investigated various strategies for choosing the abscissae based on maximizing stability regions or minimizing the order defects. Paul (1994) also compares the efficiency of a fifth-order five-stage ICE with a fifth-order continuous version of the Dormand and Prince pair (1980).

Clearly much of the theory described in Chapter 5 can be extended to DDEs. The high order ICEs are undoubtedly efficient in a sequential environment at stringent tolerances but how efficient in a parallel environment still remains to be seen. In section 7.12 some consideration will be given to understanding how waveform relaxation techniques can be applied to DDEs. This approach has been considered by Bjørhus (1993) for DDEs and Jackiewicz et al. (1994) for functional differential systems of neutral type.

6.9 Applications to other differential systems

Much of the work described in Chapters 5 and 6 can be applied to other classes of problems such as second-kind Volterra equations, Volterra integro-differential equations and higher-order differential equations.

Volterra equations

For example, the general form of a Volterra integro-differential equation is given by

$$y' = f(x, y, \int_{x_0}^{x} k(x, s, y(s))ds), \quad y(x_0) = y_0. \tag{6.9.1}$$

Two possible ways of treating this problem is either to approximate the integral by a quadrature formula and then to integrate the differential equation or to define a new function

$$z(x,t) = \int_{x_0}^{t} k(x, s, y)ds$$

and to replace (6.9.1) by

$$y' = f(x, y, z(x, x)). \tag{6.9.2}$$

In the former approach the stability of the process depends on the nature of the quadrature formula (see Brunner and van der Houwen (1986), for example). The second approach described above is a more stable one and requires the application of some ODE solver to (6.9.2) where the values of $z(x, x)$ are obtained by solving the IVP

$$\frac{\partial z(x,t)}{\partial t} = k(x, t, y(t)), \quad z(x, x_0) = 0, \quad t \in [x_0, x].$$

Sommeijer et al. (1992) consider the stability of parallel block methods when using this second approach.

Implicit predictor-corrector techniques based on a Runge-Kutta corrector have also been applied to second-kind Volterra integral equations of the form

$$y(x) = g(x) + \int_{x_0}^{x} k(x, s, y(s))ds, \quad x_0 \le x \le T$$

by Crisci et al. (1991b). Method parameter selection in terms of accuracy and good stability is discussed in Crisci, Russo and Vecchio (1991a).

Parallelism can be exploited at two levels in this case. Firstly, in any s-stage Volterra Runge-Kutta implementation, s lag terms of the form

$$g(x_n + \theta_i h) + \int_{x_0}^{x_n} k(x_n + \theta_i h, s, y(s)), \quad i = 1, \ldots, s$$

have to be calculated and these expensive calculations can be evaluated independently on the available processors. A second level of parallelism can be exploited by performing diagonal iterations on the stage components (the analogue of PDIRK methods).

Preconditioning

Van der Houwen (1991) considers a very general multivalue approach which is sufficiently general to be suitable for many of the applications described in this section. It takes the form

$$\begin{aligned} Y &= U(Y_{n-k}, \ldots, Y_n) + h^\nu G(Y) \\ Y_{n+1} &= P(Y_{n-k}, \ldots, Y_n) + h^\nu Q(Y). \end{aligned} \tag{6.9.3}$$

Here Y and Y_n represent, respectively, vectors with s and r components which represent numerical approximations to the exact solution at various abscissae while G is a function such that

$$\frac{\partial G}{\partial Y} = M \otimes J,$$

where M is some $s \times s$ matrix and J is the Jacobian of the underlying problem evaluated at some point. The parameter ν reflects the order of the differential problem given by

$$y^{(\nu)} = f(y).$$

By writing the implicit relations which describe Y in (6.9.3) as

$$R_n(Y) := Y - U(Y_{n-k}, \ldots, Y_n) - h^\nu G(Y) = 0,$$

van der Houwen (1991) suggests solving this equation by l iterations of the form

$$(I - h^\nu(T \otimes J))Y^{(j+1)} = (I - h^\nu(T \otimes J))Y^{(j)} - P_{j+1}R_n(Y^{(j)}). \tag{6.9.4}$$

Here T is some $s \times s$ matrix which controls the convergence of (6.9.4) and P_j is an $sm \times sm$ matrix which represents a preconditioning of the residual. For example, the choice $P_j = I$ and $T = 0$ represents standard fixed-point iteration, while $P_j = I$ and $T = M$ represents a modified Newton approach. A similar approach has been considered by Burrage (1993d) and is described in section 5.11.

If

$$G(u) - G(v) = (M \otimes J)(u - v)$$

then the iteration error satisfies

$$Y^{(j+1)} - Y = Z_{j+1}(Y^{(j)} - Y), \quad j = 0, \ldots, l-1$$

where, for $j = 1, \ldots, l$,

$$Z_j = (I - h^\nu(T \otimes J))^{-1}(I - h^\nu(T \otimes J) - P_j(I - h^\nu(M \otimes J))). \tag{6.9.5}$$

In the case that $P_j = I$, a damping factor can be defined by

$$\alpha_1(\lambda) = h^\nu \lambda \rho((I - h^\nu \lambda T)^{-1}(M - T)) \tag{6.9.6}$$

where $\lambda \in \sigma(J)$, and it is clear that Z is $O(h^\nu)$ as $h \to 0$.

The convergence of (6.9.4) can be accelerated by choosing P differently so that if, for example,

$$P = I + h^\nu (M - T) \otimes J + O(h^{\nu+1})$$

then Z is $O(h^{2\nu})$ as $h \to 0$. In particular, with

$$P = (I - h^\nu (T \otimes J))^{-1} (I + h^\nu (M - 2T) \otimes J) \tag{6.9.7}$$

then the damping factor is given by

$$\alpha_2(\lambda) = h^{2\nu} \lambda^2 \rho((I - h^\nu \lambda T)^{-2} (M^2 - 2TM + T^2)). \tag{6.9.8}$$

In the nonstiff case, with $T = 0$, it is clear from (6.9.6) and (6.9.8) that in one iteration using (6.9.7) the same damping is achieved as with two iterations in the nonpreconditioned case. However, this is at the expense of requiring an evaluation of J and the evaluation of $(M \otimes J) R_n(Y^{(j)})$ at each iteration.

However, in the stiff case $T \neq 0$, α_2 is not necessarily the square of α_1. Van der Houwen (1991) suggests choosing T such that $\rho(Z)$ is minimized at $\lambda = \infty$. Since

$$Z = h^{2\nu} (I - h^\nu \lambda T)^{-2} (M^2 - 2TM + T^2),$$

where $\lambda \in \sigma(J)$, this is equivalent to choosing T which minimizes

$$\rho(T^{-2} M^2 - 2T^{-1} M + I).$$

With P chosen as in (6.9.7) then the implementation requires an initial LU factorization of $I - h^\nu (T \otimes J)$, which may be retained from a previous step, plus a series of forward and backward substitutions as well as an evaluation of $((M - 2T) \otimes J) R_n(Y^{(j)})$ at each iteration. Other ways in which to exploit parallelism are to choose $T = \tau I$ or to require T to be block lower triangular whose diagonal blocks are themselves diagonal.

One final approach considered by van der Houwen (1991) is based on Chebyshev preconditioning. By choosing

$$P_j = (I - h^\nu w_j (T \otimes I))(I - h^\nu w_j (M \otimes I))^{-1},$$

then good damping properties can be obtained by choosing the w_j based on where the eigenvalues of J lie. For example, if

$$\sigma(J) \subset [a, b]$$

then an optimal choice for the w_j is

$$w_j = \frac{1}{2} [(a - b) \cos\left(\frac{(2j+1)\pi}{l} \right) + a + b], \quad j = 1, \ldots, l.$$

Some numerical results suggest that the Chebyshev conditioning approach is an effective one.

Second-order problems

Finally, in this section some discussion is presented on the construction of parallel methods for solving second-order equations of the form

$$y'' = f(y), \quad y(x_0) = y_0, \quad y'(x_0) = y'_0. \qquad (6.9.9)$$

Clearly (6.9.9) can be converted to a first-order problem, but since (6.9.9) does not have a y' term, greater efficiency can be obtained by constructing special methods.

One popular class of methods is the so-called Runge-Kutta-Nyström (RKN) method which takes the form

$$\begin{aligned} Y_i &= y_n + c_i h y'_n + a_i h^2 f(y_n) + h^2 \sum_{j=1}^{s} a_{ij} f(Y_j), \quad i = 1, \ldots, s \\ y_{n+1} &= y_n + h y'_n + b_0 h^2 f(y_n) + h^2 \sum_{j=1}^{s} b_j f(Y_j) \qquad (6.9.10) \\ y'_{n+1} &= y'_n + h d_0 f(y_n) + h \sum_{j=1}^{s} d_j f(Y_j). \end{aligned}$$

This can be represented in a Butcher array as

$$\begin{array}{c|cc} 0 & 0 & 0 \\ c & a & A \\ \hline & b_0 & b^\top \\ & d_0 & d^\top \end{array}$$

The order of (6.9.10) can be studied in much the same way as the order of Runge-Kutta methods for first-order problems. Thus suppose $y(x_n) = y_n$ and $y'(x_n) = y'_n$ and let $Y(x_{n+1}) = (y(x_n + c_1 h), \ldots, y(x_n + c_s h))^\top$. Then (6.9.10) will be of order $w = \min\{w_1, w_2\}$ if

$$y(x_{n+1}) - y_{n+1} = O(h^{w_1+1}), \quad y'(x_{n+1}) - y'_{n+1} = O(h^{w_2+1}),$$

and stage order $q = \min\{w_1, w_2, w_3\}$ if, in addition,

$$Y(x_{n+1}) - e \otimes y_n - c \otimes h y'_n - h^2 (A \otimes I) F(Y(x_{n+1})) = O(h^{w_3+1}).$$

There are many important problems in structural mechanics which take the form (6.9.9) whose solution components have wide-ranging frequencies.

Usually only the lower harmonics are of interest and these correspond to the eigenvalues of $\frac{\partial f}{\partial y}$ close to the origin. Thus an appropriate method should damp the higher harmonics. The presence of these unwanted harmonics can cause an order reduction phenomenon very similar to that for first-order problems. Thus a stiffly accurate method can be introduced by requiring

$$c_s = 1, \quad b_0 = a_s, \quad b^\top = e_s^\top A.$$

In order to damp the high harmonics, the property of A-stability can be introduced as applied to the test problem

$$y'' = qy.$$

By letting $v_n = (y_n, hy'_n)^\top$ then for this problem

$$v_{n+1} = M(z)v_n, \quad z = qh^2$$

where

$$M(z) = \begin{pmatrix} 1 + zb^\top(I - Az)^{-1}e & 1 + zb^\top(I - Az)^{-1}c \\ zd^\top(I - Az)^{-1}e & 1 + zd^\top(I - Az)^{-1}c \end{pmatrix}.$$

Thus (6.9.10) will be A-stable if

$$\rho(M(z)) < 1, \quad \forall z \in (-\infty, 0).$$

For nonstiff problems in which the eigenvalues of the Jacobian are not too far from zero, implicit methods are not needed and van der Houwen et al. (1991) and Cong (1993b) have generalized the approach of van der Houwen and Sommeijer (1991) for first-order problems by performing fixed-point iterations based on the iteration of an implicit RKN corrector. If w is the order of the underlying corrector and l corrections are performed based on the trivial predictor then the order of the corrected method is $\min\{w, 2l{+}2\}$. This suggests a method which is twice as cheap as a standard implementation.

In constructing an appropriate corrector there are two approaches. The first approach involves converting (6.9.9) to first-order form and then using standard high order correctors such as Gauss or Radau IIA methods. This is called the indirect approach and allows the construction of an explicit RKN method of order $2s$ requiring s sequential function evaluations per integration step on s processors. The second approach, used by van der Houwen et al. (1991), involves the application of collocation techniques directly to (6.9.9). This generalizes the work of Kramarz (1980). Methods of order $2s$ with stage order $s + 1$ can be constructed, but these methods with an increased stage order cannot be A-stable. However, if the stage order is weakened to s then there do exist A-stable methods. One advantage of the direct approach is that for a given method $\rho(A_{\text{direct}})$ is at most 68% of $\rho(A_{\text{indirect}})$ for orders less than or equal to ten and this gives a faster convergence of the fixed point process.

For stiff problems, the diagonally implicit iteration approach of van der Houwen and Sommeijer (1991) can be extended to problems of the form (6.9.9) and this is done by van der Houwen et al. (1992a) and Cong (1993a). The splitting matrix D is chosen as in van der Houwen and Sommeijer (1992) by minimizing the spectral radius of the stage vector iteration matrix as the stiffness becomes infinite. A-stable diagonally iterated methods of orders 4, 6 and 8 can be constructed with, respectively, two, three and four iterations. These compare favourable with, for example, the sequential diagonally implicit RKN methods of Sharp et al. (1990) and Cooper and Sayfy (1979) which have orders 4 and 6 with, respectively, three and five sequential stages. Finally in this section the work of van der Houwen and Cong (1993) who construct parallel block predictor-corrector methods based on high-order predictors of Lagrangian or Hermitian interpolation can be mentioned.

Most of the approaches considered in Chapters 5 and 6 have offered small-scale parallelism. In the case of the predictor-corrector approach, the technique of constructing a theoretical framework which allows the analysis of the local truncation error of a general class of predictor-corrector methods and then the construction of certain predictor-corrector methods based on the minimization of the norm of the principal error coefficients and the error coefficients for the next order term, has enabled some improvement in efficiency to be obtained by allowing freedom in the choice of method parameters. Other approaches based on preconditioning of the residual also offers some improvements.

However, in spite of these improvements and although some authors have hinted and even stated that massive parallelism is possible a careful analysis of the local error analysis shows that at best there is only a moderate parallelism available through the block predictor-corrector approach. This is hardly surprising since a prediction step is for the most part an extrapolatory step so that if the blocksize is too large it takes a large number of corrections to produce an acceptable error.

While there is some scope of obtaining massive parallelism through the process of parallelism across the steps or by exploiting the structure associated with the solution of linear differential systems, this is unlikely to be a general-purpose approach. However, the deflation techniques described in Chapter 4 for linear and nonlinear systems do appear to be promising in giving good performance across the steps for solving the associated fixed-point problem. In addition, the concept of operator splitting and pipelining also appears promising. But much more analysis is needed here before a general robust approach becomes available. Consequently, in the last three chapters of this monograph attention is focused on an approach that does offer the likelihood of very good parallel performance through the concept of waveform relaxation.

7

WAVEFORM RELAXATION TECHNIQUES

The standard application of initial value problem codes such as VODE (Brown and Hindmarsh (1989)) or STRIDE (Butcher et al. (1979)) can become inefficient for large systems of equations where different state variables have widely differing time rates. This is due to the fact that the same method and stepsize scheme is used in an attempt to represent accurately all components to a specified vector of tolerances. Various approaches have been proposed to overcome this drawback, based on partitioning (Hairer (1981)) and multirate (Skelboe (1989)) techniques. The idea here is to decompose the system into subsystems with similar time rates which can then be treated by possibly different methods and time-step strategies, more or less independently of one another. Of course the difficulty here is in knowing (automatically) how to couple the components and the nature of the coupling (perhaps variable) between the subsystems.

Multirate methods exploit the fact that some differential systems can be partitioned into rapidly and slowly varying subsystems by integrating the slow subsystem with larger stepsizes than the fast subsystem. This involves an interpolation between the slow components and the fast components. Skelboe and Andersen (1989) have shown that the stability of such methods must be analysed by requiring at least one test equation for each different stepsize, while Skelboe (1989) notes that if there is only a one-way coupling between the subsystems then the stability properties of the multirate process is identical with those of the numerical methods applied to the two independent subsystems.

Skelboe (1991) has considered partitioned systems in which the couplings between subsystems are weak and each subsystem is monotonically stable in some maximum norm. Skelboe then solves such problems in a stable manner by a decoupled implicit Euler approach which essentially discretizes each subsystem independently of one another with information exchanged only after the completion of a step and not during the solution of the associated nonlinear algebraic systems.

This partitioning approach has been extended through the family of **waveform relaxation (WR) algorithms**, in which each subsystem is integrated independently over a number of iterative step sweeps with information from the other subsystems being input only at the end of each step sweep. WR algorithms were originally introduced by Lelarasmee (1982) and Lelarasmee et al. (1982) for solving problems arising from the modelling of metal oxide semiconductor digital circuits. The simulation of very large scale integrated (VLSI) circuits involves the solution of very large systems of stiff equations. Fortunately, for such problems the physicality of the problem (transistors are highly directional) allows tightly coupled nodes to be lumped together and this can be done automatically. In addition, Carlin and Vachoux (1984) have noted that any strong coupling between components occurs over short time intervals. As a consequence, WR techniques can perform extremely efficiently on problems arising from electrical network modelling (see White and Sangiovanni-Vincentelli (1987) and White et al. (1985), for example). The emphasis in this chapter, however, will be on the application of WR techniques to more general problems and to parallel architectures, and will deal only with the continuous case. The convergence behaviour of numerical methods when used in conjunction with waveform relaxation will be analysed in Chapter 8. Thus this chapter will cover the following material:

- section 7.1: an introduction to the Picard method and the shifted Picard method, with a discussion on their respective convergence behaviours;
- section 7.2: an introduction to waveform relaxation for differential systems based on Jacobi, Gauss-Seidel and SOR iteration with extensions to a more general setting;
- section 7.3: the generalization of the multisplitting algorithms used for linear systems and described in Chapter 4 to waveform relaxation algorithms for differential systems;
- section 7.4: linear convergence results based on the convolution operator and the Laplace transform of the underlying differential systems;
- section 7.5: comparison convergence results between the static (linear systems) case and the dynamic (linear differential systems) case;
- section 7.6: techniques for accelerating the convergence of WR algorithms based on linear acceleration and preconditioning;
- section 7.7: convergence and comparison results in a multisplitting setting;
- section 7.8: an introduction to multigrid waveform relaxation with convergence results for linear systems;
- section 7.9: general convergence results for nonlinear problems satisfying a one-sided Lipschitz condition;
- section 7.10: an introduction to the waveform Newton method and its associated convergence properties;

- section 7.11: a brief description of asynchronous waveform techniques;
- section 7.12: a description of dynamic iteration techniques for delay differential equations, hyperbolic partial differential equations, second-order problems and differential-algebraic systems;
- section 7.13: the generalization of the concept of left and right preconditioning to differential equations and the analysis of the convergence of a general preconditioning approach.

7.1 The Picard method

The basic idea of waveform relaxation is to solve a sequence of differential equations for solution sequences $y^{(1)}(x), \ldots, y^{(p)}(x)$, given an initial starting solution $y^{(0)}(x)$, over the interval of integration with the hope that this solution sequence converges to the solution $y(x)$. The simplest and most well-known example of this approach is the Picard method in which the sequence of differential equations of the form

$$y^{(k+1)\prime}(x) = f(x, y^{(k)}(x)), \quad y^{(k+1)}(x_0) = y_0, \quad x \in [x_0, x_f], \tag{7.1.1}$$

is solved (given that the underlying problem is defined as in (2.1.2)).

Since the solution to (7.1.1) is given by

$$y^{(k+1)}(x) = y_0 + \int_{x_0}^{x} f(\xi, y^{(k)}(\xi))d\xi, \tag{7.1.2}$$

Picard iteration allows the decoupling of the problem into m independent parallel quadrature problems. This is a very natural parallelism. In addition, different quadrature methods can be used, if appropriate, on different components of the system. Furthermore, if more than m processors are available, parallelism can also be exploited on each individual quadrature problem in terms of the function evaluations.

Such an approach is deemed to be "embarrassingly parallel". The only communication that is necessary is the updating of the waveforms among the processors. Unfortunately, the slow convergence of the iterations mitigates against the efficient use of this technique.

Theorem 7.1.1. *The application of the Picard method to the standard linear test problem given in (2.4.6) on the interval $[0, T]$ results in the following global error bound*

$$|y(x) - y^{(k)}(x)| \le \frac{(|q|x)^{k+1}}{(k+1)!}, \quad x \in [0, T], \quad q < 0. \tag{7.1.3}$$

Proof: From (2.4.6) and (7.1.1) it is easily seen that

$$y^{(k)}(x) = 1 + qx + \cdots + \frac{(qx)^k}{k!}$$

and the result follows from the fact that $y(x) = e^{qx}$ with $q < 0$. □

It is clear from this result that Picard iteration increases the order of accuracy by one at each iteration, and that there can be no convergence until

$$k \geq |q|T.$$

Hence if the interval of integration is long or the stiffness factor $|q|$ is large then an enormous number of iterations will be needed to get a suitably accurate answer. For the more general nonlinear problem, q can be replaced by L, the Lipschitz constant, for f. Hence the rate of convergence will be too slow for stiff problems while only a modest speed-up is likely for nonstiff problems.

One way to improve the convergence of Picard integration is by the technique of windowing in which the region of integration is split up into a series of windows and this iterative process then takes place on each window. The advantage of this approach is that a more accurate initial guess is available for each new window based, for example, on the computed solution at the end of the previous window.

Another way to improve the convergence behaviour is via the concept of the splitting of the right-hand side. This leads to the shifted Picard method (see Skeel (1989), for example) in which (7.1.1) is replaced by

$$y^{(k+1)\prime}(x) - M_k y^{(k+1)}(x) = f(x, y^{(k)}(x)) - M_k y^{(k)}(x). \qquad (7.1.4)$$

This approach can be considered to be a generalization of the (M, N) splitting of the (static) linear system $Qy = b$ described in section 4.5 and just as there is a difficulty in knowing how to choose M in that case, there are difficulties in knowing how to choose the window size and the splitting matrix M_k, automatically.

Yet another approach for improving the convergence of (7.1.1) is by a suitable acceleration possibly based on estimates of the extremal eigenvalues of the iteration operator. Thus defining the residual

$$r^{(k)}(x) = f(x, y^{(k)}(x)) - y^{(k)\prime}(x),$$

an accelerated Picard method could take the form

$$y^{(k+1)\prime}(x) = y^{(k)\prime}(x) + \alpha_k r^{(k)}(x). \qquad (7.1.5)$$

If (7.1.5) is applied to the test problem given by (2.2.4) on the range of integration [0,1], and denoting $\varepsilon^{(k)}(x) = y^{(k)}(x) - y(x)$, it can be shown that

$$\varepsilon^{(k+1)}(x) = [I_m - \alpha_k(I_m - Q\mathcal{L})]\varepsilon^{(k)}(x),$$

where $\mathcal{L}$ is the integral operator defined by $\mathcal{L}y(x) = \int_0^x y(s)ds$. Hence it is seen that

$$\varepsilon^{(k)}(x) = p_k(Q\mathcal{L})\varepsilon^{(0)}(x), \tag{7.1.6}$$

where $p_k(z)$ is a polynomial of degree k satisfying $p_k(1) = 1$. This approach is similar to the polynomial preconditioning techniques applied to the conjugate gradient method and discussed in section 4.8.

In the case that (2.2.4) is scalar with Q satisfying a Lipschitz condition

$$-L \leq Q \leq 0, \quad L \geq 1, \tag{7.1.7}$$

Skeel (1989) relates p_k to certain Chebyshev polynomials defined on $[-L, 0]$ and with

$$\|y\|_\infty := \max_{x\in[0,1]} |y(x)|,$$

shows

$$\|\varepsilon^{(k)}\|_\infty \leq \frac{L^{3/2}}{k} e^{(-k^2/L+O(1))}\|\varepsilon^{(0)}\|_\infty, \quad k = O(L^{3/4}).$$

Hence the number of iterations needed to obtain an order reduction of

$$\frac{\|\varepsilon^{(k)}\|_\infty}{\|\varepsilon^{(0)}\|_\infty} = O(\tau), \quad \tau \geq e^{-\sqrt{L}}$$

is

$$k \approx \sqrt{L \log(L/\tau)}, \tag{7.1.8}$$

which compares with

$$k \approx eL - \log(\tau\sqrt{L}) \tag{7.1.9}$$

in the non-accelerated case.

Thus for stiff problems, preconditioning reduces the number of iterations from approximately L to approximately $\sqrt{L \log(L)}$ on the interval $[0, 1]$. While this is a dramatic reduction in the number of iterations, this number is still too large for any practical stiff implementation.

In the case of the shifted Picard method introduced in (7.1.4), Skeel (1989) analyses the case where $M_k = m_k I$, where m_k is some real constant.

In this case the iteration process is based on a weighted quadrature of the form

$$y^{(k+1)}(x) - y^{(k)}(x) = \int_0^x e^{m_k(x-s)} r^{(k)}(s) ds.$$

For the model problem (2.2.4) it is easily seen that

$$\varepsilon^{(k+1)}(x) = (Q - m_k I)\mathcal{L}_k \varepsilon^{(k)}(x) \tag{7.1.10}$$

where $\mathcal{L}_k$ is the shifted integral operator defined by

$$\mathcal{L}_k y(x) = \int_0^x e^{m_k(x-s)} y(s) ds. \tag{7.1.11}$$

Hence, Skeel (1989) is able to show that for the scalar case satisfying (7.1.7) with $-L \leq x \leq 0$

$$\frac{\|\varepsilon^{(k)}\|_\infty}{\|\varepsilon^{(0)}\|_\infty} \leq |e^{[0,m_0,\cdots,m_{k-1}]}| \prod_{j=1}^{k} |(x - m_{j-1})|, \tag{7.1.12}$$

where the square bracket denotes a kth order divided difference. The second factor is minimized by choosing

$$m_j = \frac{L}{2}(\cos \frac{j+\frac{1}{2}}{k}\pi - 1), \quad j = 0, \ldots, k-1 \tag{7.1.13}$$

and Skeel (1989) is able to show that the product of the two factors in (7.1.12) is minimized locally for the choice in (7.1.13).

Skeel makes no attempt to compare this error bound with that obtained for the accelerated case and there are certainly difficulties in extending this analysis to nonlinear systems, although Skeel (1989) suggests that this analysis could provide a basis for choosing the shift parameters if the extremal eigenvalues of the Jacobian matrix can be estimated automatically.

7.2 Jacobi and Gauss-Seidel WR algorithms

The material in the previous section suggests that although the Picard method (and its accelerated and shifted variants) is a highly parallel method in that it decouples a problem of dimension m into m independent quadrature problems, the convergence of the waveform iterates is too slow for stiff problems to allow for an efficient parallel implementation. Consequently, in this section more general WR algorithms are introduced and their convergence properties studied.

In order to illustrate the very general nature of WR algorithms consider the two-dimensional nonlinear autonomous problem of the form

$$\begin{aligned} y_1' &= f_1(y_1, y_2), \quad y_1(x_0) = y_{10}, \\ y_2' &= f_2(y_1, y_2), \quad y_2(x_0) = y_{20}, \quad x \in [x_0, T]. \end{aligned} \tag{7.2.1}$$

One possible iterative scheme takes the form

$$\begin{aligned} y_1^{(k+1)'}(x) &= f_1(y_1^{(k+1)}(x), y_2^{(k)}(x)), \quad y_1^{(k+1)}(x_0) = y_{10} \\ y_2^{(k+1)'}(x) &= f_2(y_1^{(k)}(x), y_2^{(k+1)}(x)), \quad y_2^{(k+1)}(x_0) = y_{20}, \end{aligned} \tag{7.2.2}$$

so that at each level k, two single decoupled differential equations can be solved in parallel and independently of one another on the interval $[x_0, T]$. Communication between the two processors takes place only at the end of each iterate when the (possibly interpolated values) $y_1^{(k+1)}(x)$ and $y_2^{(k+1)}(x)$ need to be swapped between processors. The initial iterates $y_1^{(0)}(x)$ and $y_2^{(0)}(x)$ can be chosen freely but must satisfy the initial conditions. Such an iteration scheme is called the **Jacobi WR method** because of its obvious similarities with the Jacobi method for solving linear systems of equations.

There is of course a **Gauss-Seidel WR method** for (7.2.1) and this takes the form

$$\begin{aligned} y_1^{(k+1)'}(x) &= f_1(y_1^{(k+1)}(x), y_2^{(k)}(x)) \\ y_2^{(k+1)'}(x) &= f_2(y_1^{(k+1)}(x), y_2^{(k+1)}(x)). \end{aligned}$$

However, in this case there is no natural decoupling of the system since $y_2^{(k+1)}$ cannot be calculated until $y_1^{(k+1)}$ has been calculated and so this process is inherently sequential.

Pursuing the analogy with linear systems of equations even further, an **SOR WR method** can be defined for (7.2.1) which takes the form

$$\begin{aligned} z_1^{(k+1)'}(x) &= f_1(z_1^{(k+1)}(x), y_2^{(k)}(x)), & z_1^{(k+1)}(x_0) &= y_{10} \\ z_2^{(k+1)'}(x) &= f_2(z_1^{(k+1)}(x), z_2^{(k+1)}(x)), & z_2^{(k+1)}(x_0) &= y_{20} \\ y_1^{(k+1)}(x) &= wz_1^{(k+1)}(x) + (1-w)y_1^{(k)}(x) \\ y_2^{(k+1)}(x) &= wz_2^{(k+1)}(x) + (1-w)y_2^{(k)}(x). \end{aligned}$$

These three methods can be placed into a very general setting by again considering the idea of splitting functions. Thus for the non-autonomous problem of dimension m given by (2.1.2), the general form of a continuous-time waveform relaxation method is given by

$$
\begin{aligned}
z^{(k+1)'}(x) &= G(x, z^{(k+1)}(x), y^{(k+1)}(x), y^{(k)}(x)), \quad z^{(k+1)}(x_0) = y_0 \\
y^{(0)}(x_0) &= y_0 \qquad\qquad\qquad\qquad\qquad\qquad\qquad\qquad\qquad (7.2.3) \\
y^{(k+1)}(x) &= g(x, z^{(k+1)}(x), y^{(k)}(x)),
\end{aligned}
$$

where G and g are known as splitting functions and satisfy

$$
\begin{array}{ll}
G(x,y,y,y) = f(x,y), & G : [x_0, T] \times \mathbb{R}^m \times \mathbb{R}^m \times \mathbb{R}^m \to \mathbb{R}^m \\
g(x,y,y) = f(x,y), & g : [x_0, T] \times \mathbb{R}^m \times \mathbb{R}^m \to \mathbb{R}^m.
\end{array}
$$

A simpler formulation of (7.2.3) can be given which is a natural generalization of the Picard approach, namely

$$
y^{(k+1)'}(x) = F(x, y^{(k+1)}(x), y^{(k)}(x)), \quad y^{(k+1)}(x_0) = y_0, \qquad (7.2.4)
$$

where F satisfies

$$
F(x,y,y) = f(x,y), \quad F : [x_0, T] \times \mathbb{R}^m \times \mathbb{R}^m \to \mathbb{R}^m. \qquad (7.2.5)
$$

Denoting the m components of the function f in (2.1.2) by $f_1, \ldots, f_m$ and the m components of the kth iterate $y^{(k)}$ by $y_1^{(k)}, \ldots, y_m^{(k)}$ then Jacobi WR and Gauss-Seidel WR can be expressed in the form of (7.2.4) as, respectively,

$$
y_i^{(k+1)'} = f_i(x, y_1^{(k)}, \ldots, y_{i-1}^{(k)}, y_i^{(k+1)}, y_{i+1}^{(k)}, \ldots, y_m^{(k)}), \quad i = 1, \ldots, m
$$

and

$$
y_i^{(k+1)'} = f_i(x, y_1^{(k+1)}, \ldots, y_i^{(k+1)}, y_{i+1}^{(k)}, \ldots, y_m^{(k)}), \quad i = 1, \ldots, m.
$$

Similarly, SOR WR can be expressed in the form of (7.2.3) with

$$
\begin{aligned}
z_i^{(k+1)'} &= f_i(x, y_1^{(k+1)}, \ldots, y_{i-1}^{(k+1)}, z_i^{(k+1)}, y_{i+1}^{(k)}, \ldots, y_m^{(k)}), \quad i = 1, \ldots, m \\
y_i^{(k+1)} &= w z_i^{(k+1)} + (1-w) y_i^{(k)}, \quad i = 1, \ldots, m.
\end{aligned}
$$

The advantage of the Jacobi approach is that each component of the system can be solved independently in parallel, while a drawback is that a great deal of past information must be stored if m is large. Storage is not such a problem in the Gauss-Seidel case, but then there is no obvious decoupling of the systems. On the other hand it is well known that SOR methods can be parallelized by different orderings of the components (the so-called chequerboard effect). Fang (1991) has considered multicoloured implementations on sparse problems.

Furthermore, it is well known that for certain linear systems of equations, Gauss-Seidel iteration will converge approximately twice as fast as Jacobi iteration. This suggests that the convergence properties of these waveform relaxation algorithms must be closely studied to see if similar effects apply for differential equations. This will be done in section 7.5.

Another way of viewing Jacobi and Gauss-Seidel splittings is in terms of the function $F(x, v, w)$ defined in (7.2.4) and (7.2.5).

Definition 7.2.1. *A waveform relaxation scheme is said to be of Jacobi type if* $\frac{\partial F}{\partial v}|_{v=u,w=u} = \mathit{diag}\left(\frac{\partial f}{\partial u}\right)$.

Definition 7.2.2. *A waveform relaxation scheme is said to be of Gauss-Seidel type if* $\frac{\partial F}{\partial v}|_{v=u,w=u}$ *is a lower triangular matrix consisting of the lower triangular part of* $\frac{\partial f}{\partial u}$.

Of course these definitions only make sense if $\frac{\partial f}{\partial u}$ exists. The generalization to block Jacobi and block Gauss-Seidel WR schemes is now very simple in this framework as it is just a matter of allowing for block diagonal and block lower triangular matrices in the above definitions.

In general the block structured waveform approach may often prove to be more efficacious. This is due to the fact that many problems that arise from the modelling of physical phenomena have a natural coupling between blocks of components and if this coupling is not preserved in the waveform scheme then the convergence of the iterates is likely to be very slow. The difficulty is in knowing how to couple automatically.

As already mentioned, for some classes of problems the physicality of the situation can be exploited and this will often lead to an efficient waveform implementation (as is the case for circuit simulation). But for problems in which there is no obvious coupling between the components or, worse still, for problems in which the coupling changes (as in some biological and chemical problems), the efficacy of the waveform approach can be variable. However, one possible way of dealing with this problem is to allow the overlapping of components between differing subsystems.

7.3 Multisplitting WR algorithms

The idea of splitting a problem more than once was first introduced by O'Leary and White (1985) for solving linear systems of equations (see section 4.9), and has been extended by Pohl (1992) to waveform relaxation algorithms for ordinary differential equations. The generalization of Pohl allows the overlapping of components between different subsystems and although this overlapping introduces additional computational overheads (because the subsystems are bigger than they would otherwise be) the hope is that this approach would result in better convergence properties of the waveform iterates.

In order to illustrate this concept, consider a system of dimension 9 which has been decomposed into three subsystems each of dimension 3, then an overlap of 0, 1, 2 and 3 can be represented graphically as in Figure 7.1

Here the word overlap represents the same overlap in both directions. More sophisticated definitions can be given to allow differing overlaps in each direction and differing overlaps between subsystems. Note that in the this example, an overlap of 3 means that the second subsystem becomes the original system. Thus by controlling the size of the overlap the effects of strong coupling between subsystems can be ameliorated by allowing an overlap between the subsystems.

Of course there is still the difficulty in knowing what value(s) to choose for the overlap but numerical results presented in Chapter 9 and section 7.13 suggest that a modest amount of overlap is often better than none at all. This has been further studied by Jeltsch and Pohl (1991) and Frommer and Pohl (1993) for linear differential systems.

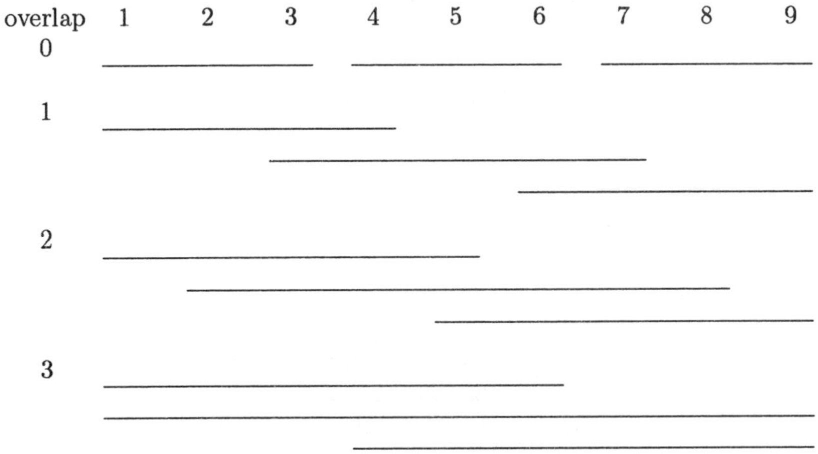

overlap 1 2 3 4 5 6 7 8 9

0

1

2

3

FIG. 7.1. overlapping of components

The other issue that has to be addressed when considering an overlapping approach is to decide what weighting should be given to the overlapping components. This is discussed in Jeltsch and Pohl (1991) in relation to the linear problem

$$y'(x) + Qy(x) = g(x), \tag{7.3.1}$$

and this is the formulation that will be used, rather than that of (2.2.4), for the rest of this chapter. Thus for an (M, N) splitting of the matrix Q, a general waveform algorithm is given by the iteration scheme

$$\begin{aligned} y^{(k+1)\prime}(x) + My^{(k+1)}(x) &= Ny^{(k)}(x) + g(x) \\ y^{(k+1)}(x_0) &= y_0. \end{aligned} \tag{7.3.2}$$

The extension to the multisplitting case has been given in Pohl (1992) and is completely analogous to the multisplitting case for linear systems described in section 4.9. Thus consider a multisplitting of Q into L splittings as in (4.9.4) and weighting matrices E_ℓ satisfying (4.9.6) and if $\tilde{y}_\ell^{(k+1)}(x)$ denotes the unknown components of the ℓth splitting at the $(k+1)$th iterate, (7.3.2) can be written as

$$\begin{aligned} \tilde{y}_\ell^{(k+1)\prime}(x) + M_\ell \tilde{y}_\ell^{(k+1)}(x) &= N_\ell y^{(k)}(x) + g(x), \quad \ell = 1, \ldots, L \\ \tilde{y}_\ell^{(k+1)}(x_0) &= y_0, \end{aligned} \tag{7.3.3}$$

with a new approximation computed by

$$y^{(k+1)}(x) = \sum_{\ell=1}^{L} E_\ell \tilde{y}_\ell^{(k+1)}(x). \tag{7.3.4}$$

Some numerical work by Frommer and Pohl (1993) suggests that the 0–1 weighting criterion is a reasonable one. In this case the overlapping components do not contribute in (7.3.3) to the formation of $y^{(k+1)}(x)$. Thus for the overlapping described by Figure 7.1

$$\begin{aligned} E_1 &= \text{diag}\ (1,1,1,0,0,0,0,0,0\} \\ E_2 &= \text{diag}\ (0,0,0,1,1,1,0,0,0\} \\ E_3 &= \text{diag}\ \{0,0,0,0,0,0,1,1,1\}, \end{aligned}$$

irrespective of the overlap.

7.4 The convergence of linear WR algorithms

The convergence properties of the sequence of iterates $y^{(1)}(x), y^{(2)}(x), \ldots$ of (7.3.2) have been investigated by Miekkala and Nevanlinna (1987a, 1987b) and Nevanlinna (1989, 1990). It was noted there that these iterates can be written in a fixed-point iterative form as

$$y^{(k+1)}(x) = Ky^{(k)}(x) + \phi(x), \tag{7.4.1}$$

where K is the convolution operator. Assuming, without loss of generality, that $x_0 = 0$ then K and the kernel function r are defined by

$$Ku(x) = \int_0^x r(x-s)u(s)ds, \quad r(x) = e^{-xM}N \tag{7.4.2}$$

and

$$\phi(x) = e^{-xM}y_0 + \int_0^x e^{(s-x)M}g(s)ds.$$

If $y(x)$ represents both the solution of (7.3.1) and the fixed-point formulation of (7.4.1), and if

$$\varepsilon^{(k)}(x) = y^{(k)}(x) - y(x),$$

then from (7.4.1)

$$\varepsilon^{(k+1)}(x) = K\varepsilon^{(k)}(x)$$

and hence

$$\varepsilon^{(k)}(x) = K^k\varepsilon^{(0)}(x), \tag{7.4.3}$$

where K^k denotes the kth fold convolution of K.

Clearly convergence will take place if $\rho(K) < 1$. In order to get realistic estimates of bounds on the error in the kth iterate, $\varepsilon^{(k)}(x)$, in terms of the initial error, the following definition is given.

Definition 7.4.1. *Let $\|\cdot\|$ be any fixed norm on $\mathbb{C}^m$. Then the maximum norm on the space of continuous $\mathbb{C}^m$-valued functions on $[0,T]$ is defined by*

$$\|u\|_T = \max_{x\in[0,T]} \|u(x)\|.$$

As a consequence, it can be shown that

$$\|K^k\|_T \le \int_0^x \| \tilde{r}^k(x) \| \, dx \tag{7.4.4}$$

where $\tilde{r}^k$ denotes the kth fold convolution, $r * r \ldots * r$ of r, with

$$r * u(x) = \int_0^x r(x-s)u(s)ds.$$

Now since $T < \infty$

$$\|r\|_T \leq C \tag{7.4.5}$$

with, for example $C = e^{T\|M\|}\|N\|$, then

$$\|\tilde{r}^k(x)\| \leq C \int_0^x \|\tilde{r}^{k-1}(s)\| ds, \quad x \in [0, T],$$

and an induction argument gives

$$\|\tilde{r}^k(x)\| \leq C \frac{(Cx)^{k-1}}{(k-1)!}, \quad x \in [0, T],$$

which on substitution into (7.4.4) gives the following result:

Theorem 7.4.2. *For the waveform relaxation algorithm given by (7.3.2) and with K and $r(x)$ defined as in (7.4.2) with r satisfying (7.4.5) the following is a bound on K:*

$$\|K^k\|_T \leq \frac{(CT)^k}{k!}. \tag{7.4.6}$$

Remark. Since

$$\lim_{k \to \infty} \frac{(CT)^k}{k!} = 0,$$

Theorem 7.4.2 implies convergence for all finite T.

The speed of convergence can be analysed by noting that the resolvent of K, $R_\lambda(K)$, satisfies

$$\begin{aligned} \|R_\lambda(K)\| &= \|(K - \lambda I)^{-1}\| \\ &\leq |\lambda| \, \|(I - \frac{1}{\lambda}K)^{-1}\|_T \\ &\leq |\lambda| \sum_{l=0}^{\infty} \frac{1}{|\lambda|^l} \|K^l\|_T \\ &\leq |\lambda| e^{CT/|\lambda|}. \end{aligned}$$

Hence for every fixed $T < \infty$ and nonzero λ, $R_\lambda(K)$ is bounded. The following definitions and results imply that the spectral radius of K is zero and hence convergence is superlinear.

Definition 7.4.3. *If Λ is a linear operator on a complex Banach space X then the resolvent set of Λ and the spectrum of Λ are defined, respectively, as*

$$\begin{aligned} R(\Lambda) &= \{\lambda : R_\lambda(\Lambda) \text{ exists and is bounded and defined on } X\} \\ \sigma(\Lambda) &= \mathbb{C} - R(\Lambda). \end{aligned}$$

Theorem 7.4.4. *If Λ is bounded,*

$$\sigma(\Lambda) = \{\lambda : R_\lambda(\Lambda) \text{ is unbounded or does not exist}\}$$

and the spectral radius of Λ is given by

$$\rho(\Lambda) = \sup_{\lambda \in \sigma(\Lambda)} |\lambda|.$$

For small values of T the bound in (7.4.6) is sharp. On the other hand it is seen that K is a contraction operator on the interval $[0, \frac{1}{C}]$ and when C is large, as is the case for stiff problems, this classical estimate implies that small windows must be used and this is unrealistic. This suggests that a different analysis for infinite domains of integration or stiff problems is needed. Since for stiff problems it is desirable that the convergence should be independent of the size of the Lipschitz constants, the stiff case should be treated by showing convergence independent of the interval of integration. In order to cope with large values for T, Miekkala and Nevanlinna (1987a) introduce an exponentially weighted norm and assume

$$\|r(x)\| \leq C\, e^{-\alpha x}, \quad 0 \leq x \leq T \leq \infty, \quad \alpha > 0. \tag{7.4.7}$$

Nevanlinna (1989) is then able to prove the following result, using a similar analysis to the above.

Theorem 7.4.5. *For the waveform relaxation algorithm given by (7.3.2) and with K and $r(x)$ defined as in (7.4.2) with r satisfying (7.4.7) then*

$$\|K^k\|_T \leq \left(\frac{C}{\alpha}\right)^k \frac{\Gamma_{\alpha T}(k)}{\Gamma(k)}, \tag{7.4.8}$$

where

$$\Gamma_\theta(s) = \int_0^\theta e^{-x} x^{s-1} dx$$

is the incomplete Γ-function.

Remark. Since $\Gamma_\theta(s) \to \Gamma(s)$ as $\theta \to \infty$, the iterations converge uniformly if

$$C \leq \alpha.$$

It is also worth noting that the spectral radius of K can be defined as

$$\rho(K) = \lim_{k\to\infty} \|K^k\|^{1/k}$$

which from (7.4.6) again implies that on finite windows for uniform norms

$$\rho(K) = 0$$

and this gives no information on the effects of splitting on convergence rates. However, by modifying Definition 7.4.1 to allow for weighted norms so that

$$\|u\|_T = \sup_{[0,T]} \| \, e^{-\theta x} u(x) \, \|, \quad \theta \geq 0, \tag{7.4.9}$$

the dependence of the convergence on the splitting becomes clearer.

Since

$$\|u\|_T \leq e^{\theta T} \sup_{x\geq 0} \| \, e^{-\theta x} u(x) \, \| = e^{\theta T} \, \| \, u \, \|_\infty,$$

it can be shown that

$$\|K^k\|_T \leq e^{\theta T} \, \| \, K^k \, \|_\theta,$$

where $\| \, K \, \|_\theta$ is the induced norm based on (7.4.9). Noting that

$$\rho_\theta(K) = \lim_{k\to\infty} \| \, K^k \, \|_\theta^{1/k}$$

then analogously to (7.4.3) it is natural to find θ such that

$$e^{\theta T} \rho_\theta(K)^k$$

is minimized.

In order to do this an estimate of $\rho_\theta(K)$ is needed and it is necessary that K be a bounded operator. Miekkala and Nevanlinna (1987a) have shown that this is the case if the eigenvalues of M have real part greater than $-\theta$, in which case

$$K(z) = (zI + M)^{-1} N \tag{7.4.10}$$

and

$$\rho_\theta(K) = \sup_{Re(z)\geq 0} \rho(K(z+\theta)) = \sup_{Re(z)\geq -\theta} \rho(K(z)). \tag{7.4.11}$$

This result is a consequence of a theorem by Paley and Wiener (1934) which states that $\lambda I - K$ is an invertible operator on $L^p(\mathbb{R}^+, \mathbb{C}^m) (1 \leq p \leq \infty)$ if and only if the matrices $\lambda I - K(z)$ are invertible for $Re(z) \geq 0$. Finally from the Maximum Modulus principle the following result can be stated:

Theorem 7.4.6. *If all eigenvalues of M have real part greater than $-\theta$ then*

$$\rho_\theta(K) = \max_{\xi \in \mathbb{R}} \rho(((i\xi + \theta)I + M)^{-1}N). \tag{7.4.12}$$

Remark. By taking the Laplace transforms of (7.3.2) it is easily seen that

$$s\hat{y}^{(k+1)}(s) + M\hat{y}^{(k+1)}(s) = N\hat{y}^{(k)}(s) + \hat{g}(s) + y_0$$

and hence

$$\hat{y}^{(k+1)}(s) = (sI + M)^{-1}N\hat{y}^{(k)}(s) + (sI + M)^{-1}(\hat{g}(s) + y_0). \tag{7.4.13}$$

By also taking the Laplace transform of the underlying differential equation given by (7.3.1) then

$$s\hat{y}(s) + Q\hat{y}(s) = \hat{g}(s) + y_0 \tag{7.4.14}$$

and substituting (7.4.14) into (7.4.13) gives, with $Q = M - N$,

$$\hat{y}^{(k+1)}(s) = (sI + M)^{-1}N\hat{y}^{(k)}(s) + (sI + M)^{-1}(sI + M - N)\hat{y}(s).$$

Upon simplification this gives

$$\hat{y}^{(k+1)}(s) - \hat{y}(s) = (sI + M)^{-1}N(\hat{y}^{(k)}(s) - \hat{y}(s)). \tag{7.4.15}$$

Now the function $(sI+M)^{-1}N$ can be interpreted as the Laplace transform of the kernel function $r(x) = e^{-xM}N$. Thus

$$K(s) = (sI + M)^{-1}N \tag{7.4.16}$$

will denote the transfer function of the convolution operator K. By comparing (7.4.16) with (7.4.3) it can be seen that the decay of K^k can in fact be accurately estimated by an analysis of $K(z)$. This technique of working with waveform relaxation Laplace transforms will prove to be significant when analysing the behaviour of discretized waveform relaxation methods when applied to linear problems.

Given a certain splitting it is natural to define the abscissae of w-convergence by

$$\theta_w = \inf\{\theta : \rho_\theta(K) < w\}.$$

The value when $w = 1$ is called the **convergence abscissa** of the splitting. Leimkuhler (1993) has shown that

$$\|K^n \epsilon^{(k)}\|_{T_w} \simeq e w^n \|\epsilon^{(0)}\|_{T_w}, \tag{7.4.17}$$

where $T_w = \frac{1}{\theta_w}$ is the window of convergence and the symbol "$\simeq$" means "approximately bounded by". Thus for $w = 1$, $T = \frac{1}{\theta_1}$ is an estimate of a window of stable convergence. Furthermore, the convergence abscissa can be used to compare the practical convergence properties of different splittings.

It has been shown in this section through Theorem 7.4.2 that it is possible to obtain convergence without any restrictions on Q or the splitting, but this is at the expense of having, possibly, a very small window. This of course is computationally inefficient. On the other hand in order to obtain convergence results on arbitrarily large windows certain structures have to be imposed on the underlying problem which are akin to dissipativeness. It is interesting that this analysis is akin to the classical linear stability theory of schemes for nonstiff and stiff differential equations.

7.5 Some convergence results

If z is set equal to zero in (7.4.10) then this effect can be considered as a static iteration of the linear system of equations $Mx = Nx + b$ which will converge if M is nonsingular and

$$\rho(M^{-1}N) < 1.$$

On the other hand as seen from Theorem 7.4.6 the dynamic iteration scheme (7.3.2), characterized by the splitting $Q = M - N$, converges if the eigenvalues of M have positive real part and

$$\rho(K) = \max_{\xi \in \mathbb{R}} \rho((i\xi I + M)^{-1}N) < 1.$$

It is clear that the convergence of the static iterations will always be at least as fast as the dynamic iterations, but in some cases the rates may be equal. Miekkala and Nevanlinna (1987a) have studied the convergence properties for a number of well-known dynamic iterative methods for various imposed structures on Q. A selection of these results is given below and the proofs are to be found in Miekkala and Nevanlinna (1987a).

Theorem 7.5.1. *If all eigenvalues λ_j of Q have positive real parts then the dynamic iteration scheme (7.3.2) with $M = \frac{1}{w}I$ converges if and only if*

$$w \in \left(0, 2\min_j \left(\frac{Re(\lambda_j)}{|\lambda_j|^2}\right)\right).$$

Corollary 7.5.2. *If Q has positive real eigenvalues $0 < \lambda_1 \le \cdots \le \lambda_m$ then the spectral radius of the dynamic iteration scheme (7.3.2) with $M = \frac{1}{w}I$ is minimized when $w = \frac{2}{\lambda_1+\lambda_m}$ in which case $\rho(K) = \frac{r-1}{r+1}, r = \frac{\lambda_m}{\lambda_1}$.*

Theorem 7.5.3. *For any H-matrix Q, (with $D = \text{diag}(Q) > 0$ and whose companion matrix is a nonsingular M-matrix) both the dynamic Jacobi over-relaxation method (characterized by $M = \frac{1}{w}D$) and dynamic SOR method will converge whenever*

$$0 < w < \frac{2}{1+\theta}, \quad \theta = \rho(|J|), \quad J = I - D^{-1}Q.$$

Here $J = I - D^{-1}Q$ represents the amplification matrix for the Jacobi method in the static case while $|Q|$ represents the matrix whose elements are $|q_{ij}|$. Furthermore, from the static case given in Theorem 4.5.12, it is known that $\theta < 1$ for $w \in (0, 1]$ and so the following corollary is proved.

Corollary 7.5.4. *Under the assumptions of Theorem 7.5.3, the dynamic Gauss-Seidel iteration scheme will converge.*

These two results show that there are remarkable similarities between the static and dynamic iteration. These similarities also hold for consistently ordered matrices, as will be seen in Theorems 7.5.5 and 7.5.6 (Miekkala and Nevanlinna (1987a)).

In what follows, the subscript D will be used to represent dynamic iteration, while the absence of a subscript will denote static iteration. In addition $\rho(G_D(w))$ will denote the spectral radius of dynamic SOR.

Theorem 7.5.5. *Let Q be either a consistently ordered matrix with constant positive diagonal or be a consistently ordered, real, symmetric and positive definite matrix. Then*

$$\begin{aligned} \rho(G_D(w)) &= 1 - w + \tfrac{1}{2}(wu)^2 + wu\sqrt{1 - w + \tfrac{1}{4}(wu)^2}, \quad w \le \gamma \\ \rho(G_D(w)) &\le 8(w-1)^2/(8(w-1) - (wu)^2), \quad w > \gamma, \end{aligned}$$

where $u = \rho(J), \gamma = \frac{4(2-\sqrt{4-3u^2})}{3u^2}$ and where equality in $\rho(G_D(w))$ holds if Q has constant positive diagonal. In addition, $\rho(G_D(w))$ takes its minimum at

$$w_{opt} = \frac{4}{4 - u^2}, \quad 0 < u < 1. \tag{7.5.1}$$

Theorem 7.5.6. *For real, symmetric, positive definite, consistently ordered matrices or consistently ordered matrices with positive constant diagonal*

$$\rho(G_D) = (\rho(J))^2 = \rho((J_D))^2 < 1. \tag{7.5.2}$$

Asymptotic estimates for the spectral norms of Jacobi and SOR were given in (4.5.9) in the static case for linear systems of the form (4.5.3). Although these estimates are the same for both static and dynamic Jacobi and static and dynamic Gauss-Seidel there is a substantial difference between $\rho(G_D(w_{\text{opt}}))$ and $\rho(G(w_{\text{opt}}))$. In the dynamic case Miekkala and Nevanlinna (1987a) have shown that

$$\rho(G_D(w_{\text{opt}})) = 1-2(\pi h)^2+O(h^4), \quad w_{\text{opt}} = \frac{4}{3}-\frac{4}{9}(\pi h)^2+O(h^4), \quad (7.5.3)$$

while from (4.5.9), $\rho(G(w_{\text{opt}})) = 1 - 2\pi h + O(h^2)$. (Note that these results change if waveform SOR is defined differently – see below). Thus, for the above class of problems, optimal dynamic SOR has a convergence rate which is only twice that of dynamic Gauss-Seidel which in turn has a convergence rate which is only twice that of dynamic Jacobi. For large problems these convergence rates are not acceptable and need to be accelerated for efficient implementation. Fortunately, the multigrid approach outlined for the static case in section 4.6 can be extended to the dynamic case and this will result in dramatic acceleration of the underlying dynamic iteration scheme (see section 7.8). However, as will be seen in the next two sections, there are other ways of improving the convergence behaviour of WR algorithms.

Before discussing this in more detail a new approach of Reichelt et al. (1993) should be mentioned which greatly accelerates SOR waveform relaxation. This is done by replacing the fixed relaxation parameter by an overrelaxation convolution with a time-dependent kernel. From problem (7.3.1) this leads to the convolution (CSOR) scheme

$$\begin{aligned} &y^{(k+1)'}(x) + Dy^{(k+1)}(x) + L\textstyle\int_0^x w(s)y^{(k+1)}(x-s)ds = \\ &y^{(k)'}(x) + Dy^{(k)}(x) - (U + D + \tfrac{d}{dx})\textstyle\int_0^x w(s)y^{(k)}(t-s)ds, \end{aligned} \quad (7.5.4)$$

where $Q = D + L + U$.

Note that this is not a generalization of dynamic SOR introduced by Miekkala and Nevanlinna (1987a) but rather a generalization of waveform SOR in which a derivative term appears on both the lefthand and righthand sides, i.e. the scheme

$$(\frac{d}{dx} + D + wL)y^{(k+1)}(x) = ((1-w)(\frac{d}{dx} + D) - wU)y^{(k)}(x). \quad (7.5.5)$$

With waveform SOR defined in this way, Reichelt et al. have shown that on a finite window, and for a given stepsize and relaxation parameter w, waveform SOR and static SOR have the same asymptotic convergence rates. Reichelt et al. have extended Theorem 7.5.5 to find an optimal parameter for WSOR, but in general this still does not dramatically improve convergence because the amplification matrix can be far from normal in which case the effective convergence rate is determined by the pseudo-spectra rather than the eigenvalue spectrum (Trefethen (1991)).

By using CSOR in conjunction with a fixed stepsize linear multistep implementation, Reichelt et al. have shown how to choose the $w(t)$ by using the theory of z-transforms. The z-transform of a sequence $w[m]$ is defined by

$$w(z) := \sum_{m=0}^{\infty} w[m]z^{-m} := Zw[m],$$

where $w[m] = w(mh)$.

By defining

$$D_z = D + \theta I$$

$$\theta = \frac{\sum_{j=0}^{k} \alpha_j z^{-j}}{h\sum_{j=0}^{k} \beta_j z^{-j}}$$

then the z-transform of the resulting discrete-time CSOR operator is given by

$$H_C(z) = (D_z + w(z)L)^{-1}((1 - w(z))D_z - w(z)U).$$

If K is the discrete CSOR operator defined by

$$\epsilon^{(k+1)}[m] = K\epsilon^{(k)}[m]$$

then Reichelt et al. have proved the following results.

Theorem 7.5.7. $\rho(K) = \max_{|z|\geq 1} \rho(H_C(z))$.

Theorem 7.5.8. *For a consistently ordered problem if, for a given* $z \in \mathbb{C}$, *the spectrum* $\mu(z)$ *of* $H_J(z) = \bar{D}(D - Q)$, $\bar{D} = (D + zI)^{-1}$, *is such that* $\mu(z) \subseteq [-\mu_1(z), \mu_1(z)]$ *with* $|\mu_1(z)| < 1$ *then* $\rho(H_C(z))$ *is minimized with*

$$w_{\text{opt}}(z) = \frac{2}{1 + \sqrt{1 - \mu_1(z)^2}}. \tag{7.5.6}$$

Theorem 7.5.9. *If* $w_{\text{opt}}(z)$ *in (7.5.6) is analytic then the corresponding sequence* $w_{\text{opt}}[m] = Z^{-1}w_{\text{opt}}(z)$ *is optimal for discrete CWOR and minimizes* $\rho(K)$.

Reichelt et al. have extended the CSOR algorithm to nonlinear problems by a waveform Newton approach and implemented this on semiconductor device problems on an Intel iPSC/860.

7.6 Linear acceleration of waveform iteration

Techniques for the acceleration of the convergence of the Picard method based on the work of Skeel (1989) have already been discussed earlier in this chapter. These techniques are based on either a polynomial preconditioning of the basic Picard iteration or by a dynamic shifting (see (7.1.4)) of the Picard method. This linear acceleration approach has been generalized by both Nevanlinna (1990) and Lubich (1992) to the linear waveform iteration algorithm (7.3.2) in which linear combinations of earlier iterates are taken.

It is initially troublesome that the theory of Skeel (1989) and some numerical results suggest that linear acceleration of waveform iteration can offer substantial improvements whereas the theoretical results of Nevanlinna (1990) suggest the reverse. Nevanlinna suggests that when $y^{(k+1)}$ is computed from (7.3.2) an acceleration can be computed by setting

$$\tilde{y}^{(k+1)} = (1 - w_{k+1})\tilde{y}^{(k)} + w_{k+1}y^{(k)}. \tag{7.6.1}$$

Thus in the terminology of section 7.4, where K is the convolution operator

$$Ku(x) = \int_0^x e^{-(x-s)M} Nu(s)ds,$$

it can be shown that

$$\tilde{\varepsilon}^{(k)}(x) = p_k(K)\tilde{\varepsilon}^{(0)}(x), \quad \tilde{\varepsilon}^{(k)} = \tilde{y}^{(k)} - y, \tag{7.6.2}$$

where

$$p_k(x) = \prod_{j=1}^{k}(1 - w_j + w_j x). \tag{7.6.3}$$

Nevanlinna (1989, 1990) addresses the question of how quickly $\rho_k(K)$ decays as $k \to \infty$ by generalizing the formulation of the spectral radius of K from

$$\rho(K) = \lim_{n\to\infty} \|K^n\|^{\frac{1}{n}} = \lim_{n\to\infty} \max_{w\in\sigma(K)} |w^n|^{\frac{1}{n}},$$

so that if $\rho(K) > 0$ then

$$\begin{aligned} \lim_{n\to\infty} \sup \|p_n(K)\|^{\frac{1}{n}} &= \lim_{n\to\infty} \sup \max_{w\in\sigma(K)} |p_n(w)|^{\frac{1}{n}} \\ \lim_{n\to\infty} \inf \|p_n(K)\|^{\frac{1}{n}} &= \lim_{n\to\infty} \inf \max_{w\in\sigma(K)} |p_n(w)|^{\frac{1}{n}}. \end{aligned}$$

By introducing the optimal reduction factor

$$\eta(K) = \inf \|p_n(K)\|^{\frac{1}{n}}$$

where the infimum is taken over all positive integers n and polynomials of degree less than $n+1$ satisfying $p_n(1) = 1$, Nevanlinna (1989) is able to estimate how much smaller $\eta(K)$ is compared with $\rho(K)$ and hence to obtain an estimate on the efficacy of linear acceleration.

Unfortunately, Nevanlinna's conclusion is that if

$$\rho(K) \sim 1 - \varepsilon + O(\varepsilon^2) \tag{7.6.4}$$

then

$$\eta(K) \sim 1 - c\varepsilon + O(\varepsilon^2)$$

and since (7.6.4) is exactly the behaviour expected for many waveform algorithms on large consistently ordered problems, it appears that linear acceleration does not offer a great deal.

Lubich (1992) was able to resolve these apparent discrepancies between the work of Skeel and that of Nevanlinna by noting that the taking of linear combinations of earlier iterates does not encompass all possible linear accelerations. Indeed Lubich (1992) suggests solving a modification of (7.3.2) namely

$$y^{(k+1)\prime} + My^{(k+1)} = N\tilde{y}^{(k)} + g, \quad y^{(k+1)}(0) = y_0 \tag{7.6.5}$$

then solving an additional differential equation

$$u^{(k+1)\prime} + \mu_{k+1}u^{(k+1)} = y^{(k+1)} - \tilde{y}^{(k)}, \quad u^{(k+1)}(0) = 0 \tag{7.6.6}$$

and then setting

$$\tilde{y}^{(k+1)} = (1 - w_{k+1})\tilde{y}^{(k)} + w_{k+1}y^{(k+1)} + \beta_{k+1}u^{(k+1)} \tag{7.6.7}$$

with $\beta_{k+1} = M - \mu_{k+1}I$.

The methods analysed by Nevanlinna have $\beta_{k+1} = 0$, so that the additional differential equation in (7.6.6) is not used (see (7.6.1)).

Lubich (1992) noted that this more general approach can also be obtained by a Chebyshev acceleration of the Laplace transformed Picard iteration. Now since the Laplace transform of the underlying differential equation (2.2.4) on $[0, T]$ is given by (7.4.14), and the Laplace transform of the Picard iteration is given by

$$s\hat{y}^{(k+1)}(s) + Q\hat{y}^{(k)}(s) = \hat{g}(s) + y_0,$$

then

$$\hat{y}^{(k+1)}(s) - \hat{y}(s) = -\frac{1}{s}Q(\hat{y}^{(k)}(s) - \hat{y}(s)). \tag{7.6.8}$$

Hence a polynomial acceleration of (7.6.8) gives

$$\hat{y}^{(k)}(s) - \hat{y}(s) = p_k(s, Q)(\hat{y}^{(0)}(s) - \hat{y}(s)), \tag{7.6.9}$$

where p_k is a polynomial of at most degree k and is dependent on Q.

It is of course well known that Chebyshev polynomial preconditioning is optimal over all polynomial preconditioning (see Hageman and Young (1981), for example). Thus if the eigenvalues of Q are known to lie in the interior of the ellipse in the complex (u, v)-plane given by

$$\left(\frac{u-d}{a}\right)^2 + \left(\frac{v}{b}\right)^2 \leq 1, \tag{7.6.10}$$

then Chebyshev polynomial preconditioning in the static case gives

$$p_k(s, z) = \frac{T_k\left((d-z)/c\right)}{T_k\left((d+s)/c\right)}, \quad c^2 = a^2 - b^2, \tag{7.6.11}$$

where T_k denotes the kth Chebyshev polynomial.

With $p_k(s, Q)$ in (7.6.9) chosen as in (7.6.11), Lubich shows that transforming (7.6.9) back into the time domain leads to the shifted Picard iteration schemes of Skeel (1989) given by

$$y^{(k+1)\prime} + \mu_{k+1} y^{(k+1)} = g - Qy^{(k)} + \mu_{k+1} y^{(k)}, \; y^{(k+1)}(0) = y_0, \; k = 0, \ldots, n-1$$

with

$$\mu_k = d - c\theta_k, \tag{7.6.12}$$

where θ_k denote the zeros of T_n, namely $\cos\left(\frac{2k+1}{2n}\pi\right)$.

Furthermore, from (7.6.9) and (7.6.11) it follows (with $\tau = 1/T$) that

$$\hat{y}^{(k)}(s) - \hat{y}(s) = \frac{T_k((d+\tau)/c)}{T_k((d+s)/c)} p_k(\tau, Q)(\hat{y}^{(0)}(s) - \hat{y}(s)).$$

Using (7.6.11), Lubich notes that the first factor on the right-hand side is just

$$\prod_{j=1}^{k} \left(\frac{\tau + \mu_k}{s + \mu_k}\right)$$

whose individual products are the Laplace transforms of

$$\phi_k(x) = (\tau + \mu_k) e^{-\mu_k x}.$$

Noting that convolution with this kernel represents a linear operator on the exponentially weighted space $S = L^p_\tau(\mathbb{R}^+)$ $(1 \leq p \leq \infty)$ consisting of functions v on $\mathbb{R}^+$ for which

$$\| v \|_S := \| e^{-\tau x} v(x) \|_{L^p(\mathbb{R}^+)}$$

is finite, and using the fact that $a > b$ implies $c \in \mathbb{R}$ and $\mu_k \in \mathbb{R}$ and

$$\int_0^\infty e^{-x/T}|\phi_k(x)|dx = 1,$$

Lubich (1992) shows

$$\| y^{(k)} - y \|_S \leq \| p_k(\tau, Q) \| \| y^{(0)} - y \|_S .$$

Finally, since

$$\| v \|_{L^p(0,T)} \leq e \| v \|_{L^p_\tau(0,T)},$$

Lubich (1992) proves

Theorem 7.6.1. *For Chebyshev Picard iteration, where the eigenvalues of the coefficient matrix satisfy (7.6.10) with $a > b$, the error satisfies for every $1 \leq p \leq \infty$ and $T > 0$*

$$\| y^{(k)} - y \|_{L^p(0,T)} \leq e \| p_k(\tau, Q) \| \| y^{(0)} - y \|_{L^p(0,T)}, \quad \tau = 1/T.$$

This brief but very elegant paper of Lubich is significant in that it unifies and resolves some of the apparent discrepancies between the work of Skeel and Nevanlinna and at the same time shows that polynomial preconditioning has the same effect in the dynamic waveform case as in the static case – and this is of course not unexpected.

7.7 The convergence of overlapping splittings

Multisplitting WR algorithms for linear systems were described in (7.3.3) and (7.3.4) and their convergence properties have been studied by Jeltsch and Pohl (1991) and Pohl (1992) and generalize the work of Nevanlinna (1989) in a very natural way.

By introducing for each subsystem

$$r_l(x) = e^{-xM_l} N_l, \quad l = 1, \ldots, L,$$

where $r_l(x)$ is the kernel of the linear integral operator K_l defined by

$$K_l u(x) = \int_0^x r_l(x-s)u(s)ds, \quad l = 1, \ldots, L$$

and

$$\phi_l(x) = e^{-xM_l} y_0 + \int_0^x e^{(s-x)M_l} g(s)ds, \quad l = 1, \ldots, L,$$

the solution $\tilde{y}_l(x)$ of subsystem l in (7.3.3) can be written as in (7.4.1) by

$$\tilde{y}_l^{(k+1)}(x) = K_l\tilde{y}_l^{(k)}(x) + \phi_l(x), \quad l = 1, \ldots, L.$$

Jeltsch and Pohl (1991) introduce

$$r(x) = \sum_{l=1}^{L} E_l r_l(x) \tag{7.7.1}$$

$$Ku(x) = \sum_{l=1}^{L} E_l K_l u(x) \tag{7.7.2}$$

$$\phi(x) = \sum_{l=1}^{L} E_l \phi_l(x), \tag{7.7.3}$$

so that $y^{(k+1)}(x)$ computed by (7.3.4) satisfies

$$y^{(k+1)}(x) = Ky^{(k)}(x) + \phi(x).$$

This formulation allows the extension of Nevanlinna's bound described by Theorem 7.4.2, by assuming (as in (7.4.5)) that there exist constants C_l such that

$$\| r_l \|_T \leq C_l, \quad l = 1, \ldots, L \tag{7.7.4}$$

so that

$$\| r(x) \|_T \leq C := \max_{l=1,\ldots,L} C_l \tag{7.7.5}$$

and hence

$$\| K^k \|_T \leq \frac{(CT)^k}{k!}. \tag{7.7.6}$$

As in the single-splitting case studied by Nevanlinna this shows that the multisplitting algorithm converges on the interval $[0, 1/C]$ without restriction on Q or the splitting. The significant point here is that $C = \max_{l=1,\ldots,L} C_l$ may be larger than C for the single-splitting case, so that by increasing the number of subsystems the width of the window may decrease in this analysis.

The stiff case for the multisplitting algorithm can also be treated in a similar fashion to the analysis of Nevanlinna. It is easy to show that

$$K(z) = \sum_{l=1}^{L} E_l(zI + M_l)^{-1}N_l \tag{7.7.7}$$

and

$$\rho(K) = \sup_{Re(z)\geq 0} \rho(K(z)). \tag{7.7.8}$$

By the use of three lemmas, the first of which can be found in Frommer (1990) and the last two in Pohl (1992), Jeltsch and Pohl (1991) prove Theorem 7.7.4.

Lemma 7.7.1. *If $Q \geq 0$ and $u \geq 0 \in \mathbb{R}^m$ then $Qu < u$ implies $\rho(Q) < 1$.*

Lemma 7.7.2. *$\rho(Q) \leq \rho(|Q|)$ for Q real or complex.*

Lemma 7.7.3. *If Q is real and an M-matrix then*

$$|(zI + Q)^{-1}| \leq \langle zI + Q\rangle^{-1}, \quad \forall z \in \mathbb{C}^+,$$

where $\langle Q\rangle$ denotes the comparison matrix of Q.

Theorem 7.7.4. *If for $l = 1, \ldots, L$, (M_l, N_l) is an M-splitting of the M-matrix Q, then $\rho(K) < 1$, where $\rho(K)$ is given by (7.7.8).*

The question of whether overlapping the components can improve the convergence rate of the waveform iterates when compared with the non-overlapping case is partially addressed in Frommer and Pohl (1993). In the static case there is the result of Elsner (1989) which states

Theorem 7.7.5. *If Q is real and nonsingular with $Q^{-1} \geq 0$ and $Q = M_1 - N_1 = M_2 - N_2$ are two regular or weak regular splittings of Q such that $M_1^{-1} \leq M_2^{-1}$ then $\rho(M_2^{-1}N_2) \leq \rho(M_1^{-1}N_1)$.*

Frommer and Pohl (1993) have generalized this result in the static case to a multisplitting of $Q = M_l - N_l, l = 1, \ldots, L$, with weighting matrices E_L based on the 0-1 weighting.

Theorem 7.7.6. *Let (M_l, N_l) and $(\tilde{M}_l, \tilde{N}_l)$, $l = 1, \ldots, L$, be two multisplittings of Q using the same weighting matrices in which each individual splitting is weak regular and the splitting $Q = M - N$ with $M = (\sum_{l=1}^{L} E_l M_l^{-1})^{-1}$ is regular. Then*

$$\rho(\sum_{l=1}^{L} E_l \tilde{M}_l^{-1}\tilde{N}_l) \leq \rho(\sum_{l=1}^{L} E_l M_l^{-1} N_l)$$

provided

$$M_l^{-1} \leq \tilde{M}_l^{-1}, \quad l = 1, \ldots, L.$$

Remark. This result can be shown to hold for the case when either (M_l, N_l) correspond to non-overlapping block Jacobi splittings and $(\tilde{M}_l, \tilde{N}_l)$ correspond to overlapping block Jacobi splittings or when (M_l, N_l) correspond to non-overlapping block Gauss-Seidel splittings and $(\tilde{M}_l, \tilde{N}_l)$ correspond to overlapping block Gauss-Seidel splittings provided the set of weighting matrices, which are the same for each comparison and do not overlap.

Unfortunately, there is as yet no corresponding result for the dynamic case, although Frommer and Pohl (1993) do provide a comparison result on a finite window using either the implicit Euler method or the trapezoidal rule in the discretization process. This, and other results, on the behaviour of the discretized waveform process are presented in Chapter 8.

7.8 Multigrid waveform techniques

It was shown in section 7.5 that differential systems that arise from the semi-discretization of parabolic partial differential equations of heat conduction type by central differencing based on an equidistant mesh length of $h = \frac{1}{N+1}$ can be solved by various waveform relaxation techniques. In particular, the well-known convergence rates for the static Gauss-Seidel and static Jacobi methods are preserved in the dynamic case. Thus

$$\begin{aligned} \rho(G_D) &\approx 1 - \pi^2 h^2 + O(h^4) \\ \rho(J_D) &\approx 1 - \tfrac{1}{2}\pi^2 h^2 + O(h^4). \end{aligned}$$

A consequence of this is that if a reasonable accuracy is required h must be small and many iterations are required to attain convergence. Even a dynamic successive over-relaxation approach does not substantially improve the convergence rates, although the convolution approach by Reichelt et al. (1993) does appear promising.

Of course one way to overcome this difficulty is to apply a discretization immediately after the semi-discretization. This then leads to a system of linear equations at each integration point and at this stage the standard iterative techniques can be accelerated by multigrid methods (see section 4.6).

However, it was Lubich and Ostermann (1987) who noted that the multigrid method can be applied directly in space to the evolution equation as described by a parabolic partial differential equation. As will be seen this leads to a system of stiff ordinary differential equations in which each equation corresponds to a node on a spatial grid. In principle each equation or block of equations can be discretized independently of one another with differing stepsizes. As Lubich and Ostermann (1987) note, this approach can be considered as a waveform relaxation and the case in which the same temporal stepsize is used for all equations reduces to a method of Hackbusch (1984) proposed for parabolic problems.

Lubich and Ostermann (1987) restrict their convergence analysis to two-dimensional linear parabolic problems, but note that there is no limitation in extending this multigrid waveform approach to nonlinear problems. Thus if a semi-discretization is applied to the linear elliptic differential operator with a grid meshsize of h, a linear initial value problem of the form (2.2.4) is obtained while in the case of a nonlinear problem, the semi-discretization leads to a nonlinear system of ordinary differential equations of the form

$$y'(x) = Q(y) + g(x), \quad y(x_0) = y_0, \tag{7.8.1}$$

where Q is the discretized, possibly nonlinear, spatial operator.

As mentioned in section 4.6, the efficacy of the multigrid technique in the static case is that it can dramatically accelerate the convergence of the underlying iterative method. This acceleration is obtained by the interplay of fine-grid smoothing and course-grid correction on a nested sequence of grids. The smoothing is achieved by the application of one or more relaxations of a standard relaxation algorithm such as Jacobi or Gauss-Seidel, while the coarse -grid correction is obtained by projecting a current approximation onto a coarser grid, calculating a residual, relaxing on the residual and interpolating the correction back to the fine grid. In the static case the net effect is to annihilate high-frequency components of the error and for the coarse-grid correction to reduce low-frequency, or smooth components of the error.

In order to describe the **multigrid waveform relaxation algorithm** (MWR) the approach described in Vandewalle and Roose (1989), Vandewalle and Piessens (1991) and Roose and Vandewalle (1991), which can be applied to the nonlinear problems (7.8.1), is presented. Let $G_k, k = 0, 1, \ldots, r$, be a hierarchy of grids in which G_0 represents the coarsest grid and G_r the finest grid, and let the corresponding differential equation on grid k be

$$y^{(k)\prime}(x) = Q_k(y^{(k)}) + g^{(k)}, \quad y^{(k)}(x_0) = y_0^{(k)}, \tag{7.8.2}$$

where $y^{(k)}$ represents the solution on grid level k. A solution on G_r is found by iteratively applying the following algorithm to an initial approximation of $y^{(r)}$. The algorithm is a slight modification of that found in Roose and Vandewalle (1991).

ALGORITHM : $mwr(r, g^{(r)}, y^{(r)})$

if $r = 0$ solve $y^{(0)\prime} = Q_0(y^{(0)}) + g^{(0)}$;
else
  perform ν_r waveform relaxation presmoothing steps on $y^{(r)}$;
  project $y^{(r)}$ onto G_{r-1} by
    $\tilde{y}^{(r-1)} := I_r^{r-1} y^{(r)}$;
  calculate the coarse problem right-hand side by
    $g^{(r-1)} = \tilde{y}^{(r-1)\prime} - Q_{r-1}(\tilde{y}^{(r-1)}) - I_r^{r-1}(y^{(r)\prime} - Q_r(y^{(r)}) - g^{(r)})$;
  solve on G_{r-1}
    $y^{(r-1)\prime} = Q_{r-1}(y^{(r-1)}) + g^{(r-1)}, y^{(r-1)}(x_0) = \tilde{y}^{(r-1)}(x_0)$;
  repeat γ_r times
    $mwr(r-1, g^{(r-1)}, y^{(r-1)}), \quad y^{(r-1)} = \tilde{y}^{(r-1)}$;
  interpolate the correction to the fine grid by
    $y^{(r)} := y^{(r)} + I_{r-1}^r (y^{(r-1)} - \tilde{y}^{(r-1)})$;
  perform μ_r waveform relaxation post-smoothing steps on $y^{(r)}$;
endif

Remarks.

1. The smoothing process consists of the application a number of times of any standard waveform relaxation algorithm such as Jacobi, Gauss-Seidel, block Jacobi or block Gauss-Seidel waveform. Just as in the static case, parallelism can be introduced for Gauss-Seidel by allowing a red-black ordering of the components. The number of presmoothing and post-smoothing iterations at each level can be constant or variable.

2. The waveform relaxation restriction operators I_k^{k-1} represent a projection of the current approximation $y^{(k)}$ on G_k onto the coarse grid G_{k-1}. These operators are natural extensions of those used in the static case but operate on functions instead of scalar values. Thus the one-dimensional full-weighting restriction operator can be written as

$$\tilde{y}_I(x) = \frac{1}{4}(y_{i+1}(x) + 2y_i(x) + y_{i-1}(x)) \tag{7.8.3}$$

where $\tilde{y}_I$ and y_i are functions at corresponding grid points on the coarse grid and on the fine grid, respectively.

3. The operator I_{k-1}^k represents an interpolation from the coarse grid G_{k-1} to the fine grid G_k. Both the restriction and interpolation operators can be calculated using exactly those formulae used in the elliptic case with the proviso that they operate on functions.

V-cycle , W-cycle and full multigrid (FM) cycle implementations can be used just as in the static (elliptic) case. The advantage of the FM case is that it avoids the difficulty of obtaining good and cheap initial approximations to the solution by proceeding from the coarsest grid to the finest grid.

Rather than present a convergence analysis for the full multigrid waveform case, the approach followed by Hackbusch (1985) (for the elliptic case) and Lubich and Ostermann (1987) (for the parabolic case) will be adopted. Consequently, only a two-grid analysis will be given for linear systems of the form given by (7.3.1).

As in Lubich and Ostermann (1987), H will denote the coarse-grid meshsize, Q_H will denote the coarse-grid discretization of the elliptic operator, and I_h^H and I_H^h will denote the restriction from the fine grid to the coarse grid and the interpolation from the coarse grid to the fine grid, respectively.

Let $S(z)$ be the transfer function of the convolution operator associated with the smoothing waveform. Then a simple generalization of the analysis given in sections 4.6 and 7.4 gives for the two-grid iteration

$$\varepsilon^{(k+1)} = MG\varepsilon^{(k)}, \quad \varepsilon^{(k)} = y^{(k)} - y, \tag{7.8.4}$$

where MG is the convolution operator defined by

$$MGu(x) = \int_0^x m(x-s)u(s)ds, \tag{7.8.5}$$

and $m(x)$ is the associated multigrid kernel, whose Laplace transform is given by

$$MG(z) = S(z)^{\mu_1}(I - I_H^h(zI+Q_H)^{-1}I_h^H(zI+Q_h))S(z)^{\nu_1}, \tag{7.8.6}$$

with ν_1 presmoothing and μ_1 post-smoothing iterations.

By writing

$$\begin{aligned} MG(z) &= S(z)^{\mu_1}CG(z)S(z)^{\nu_1}, \\ CG(z) &= I - I_H^h(zI+Q_H)^{-1}I_h^H(zI+Q_h), \end{aligned} \tag{7.8.7}$$

then $CG(0)$ represents the coarse-grid correction for the corresponding static problem $Q_h y_h = g_h$ considered in (4.6.3). By making certain assumptions on Q_H, Q_h and the restriction and interpolation operators, $CG(z)$ can be related to $CG(0)$ in a very simple manner. The following result appears in Ta'asan and Zhang (1994).

Theorem 7.8.1. *If $I_h^H I_H^h = I$, Q_H and Q_h are symmetric positive definite and $\sigma(I_H^h Q_H^{-1} I_h^H) \geq 0$ then*

$$CG(z) = (I + zI_H^h Q_H^{-1} I_h^H) CG(0) \tag{7.8.8}$$

with

$$\max_{Re(z)\geq 0} \|(I + zI_H^h Q_H^{-1} I_h^H)^{-1}\|_2 = 1. \tag{7.8.9}$$

Proof: Under the assumption $I_h^H I_H^h = I$, then

$$(I + zI_H^h Q_H^{-1} I_h^H) I_H^h = I_H^h Q_H^{-1} (z + Q_H).$$

The assumptions of the theorem allow inverses to be taken and (4.6.3) leads to (7.8.8). In addition, (7.8.9) is a direct consequence of the definition of the Euclidean matrix norm and the fact that the minimum eigenvalue of $(I + iyI_H^h Q_H^{-1} I_h^H)(I + iyI_H^h Q_H^{-1} I_h^H)^*$ for real y is reached at $y = 0$. □

This result implies that the Euclidean norm of the coarse-grid corrector for the MWR algorithm is as good as the one for the steady-state multigrid iteration. Ta'asan and Zhang (1994) conjecture that the assumption on the restriction and interpolation operators either holds or is violated only by a small perturbation. For this latter case, it is conjectured that Theorem 7.8.1 will hold with a small variation.

Using Theorem 7.8.1, Ta'asan and Zhang (1994) are able to prove a number of convergence results using the fact that in an L^2 setting

$$||MG|| = \max_{Re(z)\geq 0} ||MG(z)||, \tag{7.8.10}$$

where $||\cdot||$ is the operator norm for $L^2(\mathbb{R}^+, \mathbb{C}^m)$ and $||\cdot||$ the induced Euclidean matrix norm. These results relate the convergence of the MWR approach to the corresponding static approach, where the terminology $CG(0), S(0)$ and $MG(0)$ will represent the coarse-grid correction, smoothing and two-grid iteration operators for the static case.

Theorem 7.8.2. *Given the assumptions of Theorem 7.8.1, diag $(Q_h) = d(I + \theta)$, $\theta =$ diag $(\theta_1, \ldots, \theta_m)$ with $|\theta_i| \leq c < 1$, and $\nu_1 = 0$, $\nu = \nu_1 + \mu_1$, then the two-grid damped Jacobi waveform relaxation operator satisfies*

$$||MG|| \leq ||MG(0)|| + ||CG(0)||\; ||S(0)||^\nu p(c/(1-c)), \tag{7.8.11}$$

where

$$p(x) = \sum_{j=1}^{\nu} \binom{\nu}{j} \alpha_{j\nu} x^j, \tag{7.8.12}$$

with

$$\alpha_{j\nu} = \begin{cases} \frac{j^{j/2}(\nu-j)^{(\nu-j)/2}}{\nu^{\nu/2}} & , \quad 1 \le j < \nu \\ 1 & , \quad j = \nu. \end{cases} \qquad (7.8.13)$$

Proof: For the weighted Jacobi method

$$M = \frac{1}{\omega}\text{diag}\,(Q_h), \quad S(z) = (zM^{-1} + I)^{-1}S(0), \quad S(0) = M^{-1}N.$$

In the case that diag $(Q_h) = d(I + \theta)$ then

$$S(z) = \frac{1}{\beta + 1}S(0) + E(\beta)S(0), \quad \beta = \omega z/d,$$

where

$$E(z) = \frac{z}{z+1}\,\text{diag}\left(\frac{\theta_1}{z+1+\theta_1}, \ldots, \frac{\theta_m}{z+1+\theta_m}\right).$$

By introducing the operator $P(z, \nu)$ defined as

$$P(z,\nu) = (\frac{1}{z+1}S(0) + E(z)S(0))^{\nu} - \frac{1}{(z+1)^{\nu}}S^{\nu}(0), \qquad (7.8.14)$$

it can be seen that

$$S^{\nu}(z) = \frac{1}{(\beta+1)^{\nu}}S^{\nu}(0) + P(\beta, \nu). \qquad (7.8.15)$$

From (7.8.7), (7.8.10) and (7.8.14)

$$||MG|| \le \max_{Re(z)\ge 0} ||\frac{1}{(\beta+1)^{\nu}}CG(z)S^{\nu}(0)|| + \max_{Re(z)\ge 0} ||CG(z)P(\beta,\nu)||$$

which from Theorem 7.8.1 implies

$$||MG|| \le ||MG(0)|| + ||CG(0)|| \max_{Re(\beta)\ge 0} ||P(\beta, \nu)||.$$

The result follows from bounding $||P(z, \nu)||$ by

$$||P(z,\nu)|| \le ||S(0)||^{\nu} \sum_{j=1}^{\nu} \binom{\nu}{j} \left|\frac{z^j}{(z+1)^{\nu}}\right| \left|\frac{c}{z+1-c}\right|^{j}. \quad \square$$

Remarks.
1. In the case that $\text{diag}(Q_h) = dI$

$$||MG|| = ||MG(0)||.$$

Furthermore, if the variable coefficients on the diagonal of Q_h are slowly varying ($0 < c << 1$) then

$$||MG|| = ||MG(0)|| + O(\varepsilon),$$

so that convergence of the damped Jacobi MWR algorithm is perturbed from the static iteration by the same order.
2. In the case that M and N are simultaneously diagonalizable, $\nu_2 = 0$ and the assumptions of Theorem 7.8.1 hold, then a commutability property and (7.8.8) and (7.8.9) imply

$$||MG|| \leq \max_{Re(z)\geq 0} ||(zM^{-1} + I)^{-\nu}|| \; ||MG(0)||. \tag{7.8.16}$$

For some linear partial differential equations which are semi-discretized by a simple difference scheme, the matrices M and N are simultaneously diagonizable by a unitary matrix and, in addition, the eigenvalues of M are known analytically. In this case accurate bounds can be obtained for (7.8.16). For example, Ta'asan and Zhang (1994) show that if the lexicographic Gauss-Seidel MWR algorithm is applied to the two-dimensional diffusion equation

$$u_t - \Delta u + u = 0$$

with periodic boundary condition using central differencing then the bound in (7.8.16) satisfies

$$\max_{Re(z)\geq 0} ||(zM^{-1} + I)^{-\nu}|| \leq (1.1547)^{\nu}.$$

This seems to suggest that the convergence rate of the MWR algorithm slowly worsens in terms of its static counterpart as the number of smoothings increases.
3. The $\alpha_{j\nu}$ given in (7.8.12) and (7.8.13) are in fact defined by

$$\alpha_{j\nu} = \max_{Re(z)\geq 0} \left| \frac{z^j}{(z+1)^{\nu}} \right|.$$

Quantities such as these are crucial in obtaining estimates on the convergence rates of MWR algorithms, and arise, for example, in the work of Lubich and Ostermann (1987).

Lubich and Ostermann (1987) have obtained bounds for $\rho(MG)$ in the case of linear systems of the form (7.3.1) which arise from the semi-discretization, based on central differencing, of the one-dimensional equation with Dirichlet boundary conditions on the unit square. For such a problem $Q_h = \frac{1}{h^2}(-1,2,-1)$, $h = \frac{1}{m+1}$. These bounds are given in the following result.

Theorem 7.8.3. *If a two-grid MWR process solves the one-dimensional semi-discretized heat equation, in which $H = 2h$, interpolation is piecewise linear and restriction is the standard one-dimensional full weighting given in (7.8.3), then the iteration operator on $L^p(\mathbb{R}^+,\mathbb{C}^m)$ $(1 \le p \le \infty)$ satisfies*

$$\rho(MG) \le \frac{1}{2}\sqrt{\eta_0(2\nu-1)}, \quad \nu = \nu_1 + \mu_1 \ge 1, \quad \eta_0(x) = \frac{x^x}{(x+1)^{x+1}},$$

for the red-black Gauss-Seidel smoother, and

$$\rho(MG) \le \mu^\nu, \quad \mu = \max\{|1 - \frac{\omega}{2}|, |1 - 2\omega|\},$$

for the damped Jacobi smoother.

Remarks.

1. The two-grid spectral norm based on static Gauss-Seidel iteration is $\eta_0(\nu)$ where ν is the total number of smoothing iterations (see Hackbusch (1985)). Thus the convergence properties for the dynamic case are only marginally worse than the static case (for $\nu > 1$). The values for the two-grid spectral norms for Gauss-Seidel and weighted Jacobi ($\omega = \frac{2}{3}$) for $\nu \le 4$ are given in Table 7.1.

Table 7.1 *two-grid spectral norms*

ν	GS static	GS dynamic	WJ static	WJ dynamic
1	0.25	0.25	0.6666	0.6666
2	0.148	0.1624	0.4444	0.4444
3	0.105	0.1294	0.2963	0.2963
4	0.082	0.1108	0.1975	0.1975

2. A non-optimal bound on $\| MG \|$ for a Gauss-Seidel smoother (see Lubich and Ostermann (1987)) is given by

$$\| MG \| \le \frac{1}{\sqrt{2}}\sqrt{\eta_0(2\nu-2)}.$$

3. The bound for the damped Jacobi smoother in the static case was given in Stüben and Trottenberg (1982) and is the same as that for the dynamic case. The values for the spectral norm for $\nu \leq 4$ for the weighted Jacobi smoother with $\omega = \frac{2}{3}$ are again given in Table 7.1.

The proof of Theorem 7.8.3 relies on knowing analytic expressions for the eigenvalues and eigenvectors of Q_h. However, these are only usually known for simple linear problems, and depend on the boundary conditions, the semi-discretization mesh, and the domain of definition in very complicated ways. Thus an analysis of the convergence rates of MWR algorithms based on an eigenvector approach does not appear to be a robust or easily generalizable one. Indeed, Lubich and Ostermann (1987) were unable to generalize Theorem 7.8.3 to the semi-discretized two-dimensional heat equation defined on the unit square using this eigenvector approach. However, they were able to study the smoothing effects for this problem by performing an orthogonal projection onto the space of high frequencies. This represents a generalization of the approach given in Stüben and Trottenberg (1982) for the static case.

Ta'asan and Zhang (1994), however, have suggested a much more general approach that significantly extends the work of Lubich and Ostermann. This is based on first taking the Laplace transform of (7.3.2) which leads to

$$(zI + Q)\hat{y}(z) = \hat{g}(z) + y_0, \quad Re(z) \geq 0 \tag{7.8.17}$$

and then applying a Fourier analysis to (7.8.17) to obtain convergence rates for the MWR algorithm.

Thus for the two-grid iteration, the Laplace transform of (7.8.4) leads to

$$\hat{\varepsilon}^{(k+1)}(z) = MG(z)\hat{\varepsilon}^{(k)}(z), \quad Re(z) \geq 0. \tag{7.8.18}$$

Suppose now that (7.8.17) arises from some r dimensional partial differential equation defined on an infinite space Ω of the form

$$u_t + Lu = g, \quad (t, x) \in (0, T] \times \Omega. \tag{7.8.19}$$

Then there exists a set, $X(\theta)$, of 2^r Fourier modes each mode being an infinite dimensional vector determined by the grid points of the semi-discretization operator. Here θ is a vector consisting of 2^r components with each component itself consisting of r components with $|\theta_1|_\infty \leq \frac{\pi}{2}$ and $|\theta_j|_\infty \leq \pi$, $j = 2, \ldots, 2^r$, where the r components of $\theta_j - \theta_1 \neq 0$ are either 0 or π.

There exists a matrix of dimension 2^r (the symbol of $MG(z)$ denoted by $MG_1(\theta, z)$) such that

$$MG(z)X(\theta) = X(\theta)MG_1(\theta, z)$$

and hence

$$\rho(MG(z)) = \sup_{\theta} \rho(MG_1(\theta, z)). \tag{7.8.20}$$

Thus convergence rates can be obtained by computing the spectral radius of a matrix of dimension 2^r. Ta'asan and Zhang (1994) do this for the two-dimensional heat equation defined on an infinite space in which L is approximated on an equidistant grid by the standard five-point difference operator. Because the space spanned by the Fourier mode is invariant under each of the operators in (7.8.6), $MG_1(\theta, z)$ can be computed in terms of the matrix symbols for each of the operators I_H^h, I_h^H and S. For the damped Jacobi and red-black Gauss-Seidel relaxation these matrix symbols are exactly those matrices computed by Lubich and Ostermann (1987) for which $S(z)$ is similar to the restriction under which $S(z)$ leaves the subspace spanned by the space of high frequencies invariant.

Ta'asan and Zhang (1994) also computes the matrix symbols for a bilinear interpolation operator and the restriction operator given by $I_h^H = (I_H^h)^\top$. The computed values for $\rho(MG)$ based on (7.8.20) are very similar to those given in Theorem 7.8.3 for the one-dimensional case.

It seems that this approach based on Fourier-Laplace analysis and the use of Fourier modes to estimate convergence rates of MWR algorithms is a very general one and can be applied to high-dimensional problems, general domains and non-constant coefficient problems.

7.9 Nonlinear convergence

A reasonably comprehensive analysis of the convergence rates of waveform iteration as applied to linear problems has been given in the previous sections because in this case it is possible to see how the splitting affects convergence. In the nonlinear case the analysis is more difficult and currently there are less comprehensive results in comparison with the linear case.

The first significant work in this area was due to Lelarasmee (1982) and then extended in a survey paper by White et al. (1985) and a monograph by White and Sangiovanni-Vincentelli (1987). These studies consider the more general problem

$$c(y(x), u(x))y'(x) = f(y(x), u(x)), \quad y(0) = y_0, \quad x \in [0, T], \tag{7.9.1}$$

where $y \in \mathbb{R}^m$ and $u \in \mathbb{R}^r$ are piecewise continuous. Here $c(y, u)^{-1}$ is assumed to exist and is uniformly bounded with respect to both y and u, while f is assumed to be globally Lipschitz continuous with respect to y for all $u \in \mathbb{R}^r$. These conditions guarantee a unique solution to (7.9.1).

It was Lelarasmee (1982) who first proved that waveform iteration based on either Gauss-Seidel or Jacobi smoothing applied to (7.9.1) will converge if $c(y, u)$ is diagonally dominant and independent of y. This result was generalized by White et al. (1985) in the following result:

Theorem 7.9.1. *If in addition to the assumptions associated with (7.9.1), $c(y, u)$ is strictly dominant for all $y(x) \in \mathbb{R}^m$ and $u(x) \in \mathbb{R}^r$ and is Lipschitz continuous with respect to $y(x)$ for all $u(x)$ then both the Gauss-Seidel and Jacobi waveform algorithms converge uniformly on all bounded intervals* $[0, T]$.

A proof of this result will not be given here, merely an outline. This outline will be used later in this section to extend some of these results.

Using the formulation (7.2.4) the autonomous version of the splitting function $F(y, z, u)$ is introduced along with a splitting function $C(y, z, u)$, where

$$F(y, y, u) = f(y, u), \quad C(y, y, u) = c(y, u).$$

Thus a natural generalization of (7.2.4) to (7.9.1) gives, with $y^{(k+1)}(0) = y_0$,

$$C(y^{(k+1)}, y^{(k)}, u)y^{(k+1)\prime} = F(y^{(k+1)}, y^{(k)}, u). \tag{7.9.2}$$

Suppose it is assumed that L_1 and L_2 are the Lipschitz constants on F with respect to the first and second arguments, respectively. Then

$$\begin{aligned} \|F(y_1, z, u) - F(y_2, z, u)\| &\leq L_1\|y_1 - y_2\|^2, \ \forall y_1, y_2, z \in \mathbb{R}^m, \ \forall u \in \mathbb{R}^r \\ \|F(y, z_1, u) - F(y, z_2, u)\| &\leq L_2\|z_1 - z_2\|^2, \ \forall y, z_1, z_2 \in \mathbb{R}^m, \ \forall u \in \mathbb{R}^r. \end{aligned}$$

In the case of Gauss-Seidel smoothing, $C(y^{(k+1)}, y^{(k)}, u)$ can be split into diagonal (D_{k+1}), lower triangular (L_{k+1}) and upper triangular (U_{k+1}) parts so that (7.9.2) becomes

$$y^{(k+1)\prime} = (L_{k+1} + D_{k+1})^{-1}(F(y^{(k+1)}, y^{(k)}, u) - U_{k+1}y^{(k)\prime}). \tag{7.9.3}$$

Now because $c(y, u)$ is uniformly diagonally dominant with respect to y, there exist four positive finite numbers which represent Lipschitz constants of $(L_{k+1} + D_{k+1})^{-1}$ and $(L_{k+1} + D_{k+1})^{-1}U_{k+1}$ with respect to its first and second arguments. Using these bounds as well as bounds on $\| y^{(k)\prime} \|$ and $\| F(y^{(k+1)}, y^{(k)}, u) \|$, White et al. (1985) show that for all positive integers k and j

$$\begin{aligned} \| y^{(k+1)\prime} - y^{(j+1)\prime} \| \leq \gamma \| y^{(k)\prime} - y^{(j)\prime} \| \ &+ \ \alpha_1 \| y^{(k+1)} - y^{(j+1)} \| \\ &+ \ \alpha_2 \| y^{(k)} - y^{(j)} \|, \end{aligned} \tag{7.9.4}$$

where $\gamma < 1$ and α_1 and α_2 depend, respectively, on L_1 and L_2 and other bounds.

White et al. (1985) then show that if (7.9.4) holds then there exists a norm $\| \cdot \|$ defined by $\| u \| = \max_{[0,T]} e^{-\theta x} \| u(x) \|, \theta > 0$ and a positive number $w < 1$ such that

$$\begin{aligned} \| y^{(k+1)\prime} - y^{(j+1)\prime} \| &\leq w \| y^{(k)\prime} - y^{(j)\prime} \| + \alpha_1 \| y^{(k+1)}(0) - y^{(j+1)}(0) \| \\ &+ \alpha_2 \| y^{(k)}(0) - y^{(j)}(0) \| . \end{aligned} \tag{7.9.5}$$

Since $y^{(k+1)}(0) = y^{(j+1)}(0), \quad \forall\, k$ and j, this implies

$$\| y^{(k+1)\prime} - y^{(j+1)\prime} \| < \| y^{(k)\prime} - y^{(j)\prime} \|$$

and hence $\{y^{(k)\prime}\}$ converges to the unique solution, y', of (7.9.1). Finally, the sequence $\{y^k\}$ converges to y because integration on $[0, T]$ which maps $y'(x)$ to $y(x)$ is a bounded continuous function.

Of course the proof described here also applies for Jacobi smoothing. Unfortunately, this approach is limiting in two respects:

(i) it only applies for Gauss-Seidel and Jacobi smoothing and not more general splitting functions;

(ii) no error bounds on $\varepsilon^{(k)}(x)$ are given in terms of $\varepsilon^{(0)}(x)$.

These limitations will be resolved later in this section.

While et al. (1985) noted that the waveform relaxation approach described as above is a stationary one in that the equations that define the iteration process do not change with the iterations. Lelarasmee (1982) was the first to allow the iterations to vary and studied their convergence properties. White et al. (1985) also noted that when a numerical method is used to solve the underlying differential equation, the solution obtained is inexact. On the other hand, since a discrete approximation can be interpreted as an exact solution of a perturbed differential system, then discrete waveform iteration can be interpreted as a nonstationary process. The behaviour of discrete waveform is discussed in depth in Chapter 8, but one of the advantages of the analysis given in White et al. is that nonstationary waveform iteration converges as a direct consequence of the contraction mapping established for the continuous waveform process. Their basic result is that any discrete method will not affect the convergence of the continuous waveform provided the errors in approximating the differential equation are driven to zero.

In order now to resolve the issues addressed in (i) and (ii) above a general convergence theorem is given which gives realistic bounds on the growth of errors in terms of the initial error.

Theorem 7.9.2. *Consider the general differential equation (2.1.1) defined on $[0,T]$ and assume there is a splitting characterized by the function $F(y,z)$ where F satisfies*

$$F(y,y) = f(y), \quad F : \mathbb{R}^m \times \mathbb{R}^m \to \mathbb{R}^m,$$

where F is Lipschitz continuous with respect to both components. Then the dynamic iteration scheme

$$y^{(k+1)\prime}(x) = F(y^{(k+1)}(x), y^{(k)}(x)), \quad y^{(k+1)}(0) = y_0 \tag{7.9.6}$$

converges uniformly on all finite intervals $[0,T]$ with

$$||\varepsilon^{(k)}||_T \le \frac{(L_2T)^k}{k!} e^{L_1T} ||\varepsilon^{(0)}||_T, \quad \varepsilon^{(k)}(x) = y^{(k)}(x) - y(x),$$

where L_1 and L_2 are the Lipschitz constants associated with the first and second arguments of F, respectively.

Proof: Let L_1 and L_2 be the Lipschitz constants on F with respect to the first and second components, respectively. Then (7.9.6) gives

$$\varepsilon^{(k+1)\prime} = y^{(k+1)\prime}(x) - y'(x) = F(y^{(k+1)}(x), y^{(k)}(x)) - F(y(x), y(x)).$$

Thus

$$\begin{aligned} \langle \varepsilon^{(k+1)'}, \varepsilon^{(k+1)} \rangle &= \langle F(y^{(k+1)}, y^{(k)}) - F(y, y^{(k)}), \varepsilon^{(k+1)} \rangle \\ &\quad + \langle F(y, y^{(k)}) - F(y,y), \varepsilon^{(k+1)} \rangle, \end{aligned}$$

and by the Lipschitz continuity of F in its first and second arguments

$$\langle \varepsilon^{(k+1)'}, \varepsilon^{(k+1)} \rangle \le L_1 ||\varepsilon^{(k+1)}||^2 + L_2 ||\varepsilon^{(k)}|| \, ||\varepsilon^{(k+1)}||. \tag{7.9.7}$$

But

$$\langle \varepsilon^{(k+1)'}, \varepsilon^{(k+1)} \rangle = \frac{1}{2} \frac{d}{dx} ||\varepsilon^{(k+1)}||^2 = ||\varepsilon^{(k+1)}|| \frac{d}{dx} ||\varepsilon^{(k+1)}||. \tag{7.9.8}$$

Hence, under the assumption that $||\varepsilon^{(k+1)}(x)|| \ne 0$, (7.9.7) and (7.9.8) imply

$$\frac{d}{dx} ||\varepsilon^{(k+1)}|| \le L_1 ||\varepsilon^{(k+1)}|| + L_2 ||\varepsilon^{(k)}||, \tag{7.9.9}$$

or with $v^{(k)}(x) :=\| \varepsilon^{(k)}(x) \|$ (a positive function),

$$v^{(k+1)\prime}(x) \le L_1 v^{(k+1)}(x) + L_2 v^{(k)}(x), \quad v^{(k+1)}(0) = 0. \tag{7.9.10}$$

Consequently,

$$v^{(k+1)}(x) \leq L_2 e^{L_1 x} \int_0^x e^{-L_1 s} v^{(k)}(s) ds. \tag{7.9.11}$$

Thus from Definition 7.4.1 and (7.9.11) (with $k = 0$)

$$v^{(1)}(x) \leq L_2 e^{L_1 x} \int_0^x e^{-L_1 s} ds ||v^{(0)}||_x$$

and so, since $L_1 \geq 0, L_2 \geq 0$,

$$v^{(1)}(x) \leq \left(\frac{L_2}{L_1}\right)(e^{L_1 x} - 1)||v^{(0)}||_x. \tag{7.9.12}$$

Substituting (7.9.12) into (7.9.11), then with $k = 1$

$$v^{(2)}(x) \leq \frac{L_2^2}{L_1} e^{L_1 x} \int_0^x (1 - e^{-L_1 s}) ds ||v^{(0)}||_x$$

or

$$v^{(2)}(x) \leq \left(\frac{L_2}{L_1}\right)^2 (1 - e^{L_1 x}(1 - L_1 x))||v^{(0)}||_x. \tag{7.9.13}$$

Hence by induction it is easy to verify that

$$v^{(k)}(x) \leq \left(\frac{L_2}{L_1}\right)^k (-1)^k (1 - e^{L_1 x} \sum_{j=0}^{k-1} \frac{(-L_1 x)^j}{j!})||v^{(0)}||_x. \tag{7.9.14}$$

Thus

$$||\varepsilon^{(k)}||_T \leq \left(\frac{L_2}{-L_1}\right)^k \alpha_k(T)||\varepsilon^{(0)}||_T,$$

where

$$\alpha_k(T) = e^{L_1 T} \sum_{j=k}^{\infty} \frac{(-L_1 T)^j}{j!}.$$

Hence

$$|\alpha_k(T)| \leq e^{L_1 T} \frac{(+L_1 T)^k}{k!}$$

and the result is proved. □

Theorem 7.9.3. *If in (7.9.6) the two-sided Lipschitz constant of F with respect to its first argument is replaced by a one-sided Lipschitz condition*

$$\langle F(y_1, z) - F(y_2, z), y_1 - y_2 \rangle \leq l_1 \parallel y_1 - y_2 \parallel^2, \quad \forall y_1, y_2, z \in \mathbb{R}^m$$

then

$$||\varepsilon^{(k)}||_T \leq \left(\frac{L_2}{-l_1}\right)^k \alpha_k(T)||\varepsilon^{(0)}||_T, \quad \alpha_k(T) = e^{l_1 T} \sum_{j=k}^{\infty} \frac{(-l_1 T)^j}{j!}. \quad (7.9.15)$$

Proof: In the case of a one-sided Lipschitz condition, a similar analysis that lead to (7.9.9) holds but with L_1 replaced by l_1 and the proof follows directly from Theorem 7.9.2. □

Remarks.

1. Theorem 7.9.2 generalizes the linear convergence result of Theorem 7.4.2 to the nonlinear case.
2. Theorem 7.9.3 gives a more useful bound than that of Theorem 7.9.2 which is pessimistic. In particular, it shows how a splitting characterized by a one-sided Lipschitz constant for the implicit part and a two-sided Lipschitz constant for the input part affects the waveform convergence.
3. If $l < 0$, so that a dissipative system is solved at each iteration then

$$0 < \alpha_{k+1}(T) < \alpha_k(T) < 1$$

and if, in addition, $-l > L_2$ then the waveform iteration is a contraction mapping on all bounded intervals $[0, T]$.
4. On the other hand, if $l > 0$ then $|\alpha_k(T)| < e^{lT}$ and if, in addition, $l > L_2$ then the initial error never grows more than by e^{lT} and after the initial iteration phase superconvergence takes place.
5. Theorem 7.9.3 suggests that the splitting function F should always be chosen so that $\frac{L_2}{|l_1|}$ is as small as possible, so that the coupling between successive components is as weak as possible. This ratio will henceforth be known as the **coupling ratio** and will be denoted by R.
6. Results similar to Theorems 7.9.2 and 7.9.3 are given in Chartier (1993) and Bjørhus (1992).
7. A special case of Theorem 7.9.3 arises when $f(y) = Qy$ and $Q = M - N$ is a splitting. In this case

$$||\varepsilon^{(k)}||_T \leq \left(\frac{\parallel N \parallel}{-\mu[M]}\right)^k \alpha_k(T)||\varepsilon^{(0)}||_T, \quad (7.9.16)$$

where $\mu[M]$ is the logarithmic norm of M.

In order to show how a system can be decomposed into subsystems in order to obtain reasonable convergence rates for the waveform iterations, Theorem 7.9.3 needs to be extended to allow for blocks of components. This is done in the following result, which follows from Definition 7.4.1 and Theorem 7.9.3.

Theorem 7.9.4. *If (2.1.1) can be split into p (possibly overlapping) subsystems with splitting functions $F_1, \ldots, F_p$, and if l_j $(j = 1, \ldots, p)$ and L_j $(j = 1, \ldots, p)$ denote, respectively, the one-sided and two-sided Lipschitz constants for the first and second arguments of F_j then*

$$||\varepsilon^{(k)}||_T \leq R^k \hat{\alpha}_k(T) ||\varepsilon^{(0)}||_T, \tag{7.9.17}$$

with

$$R = \max_{j=1,\ldots,p} \{R_j\}, \quad R_j = -L_j/l_j$$

$$\hat{\alpha}_k(T) = \max_{j=1,\ldots,p} \{e^{l_j T} \sum_{i=k}^{\infty} \frac{(-l_j T)^i}{i!}\}.$$

Again R should be made as small as possible. Of course, in general, estimates of L and l may be difficult to find and then more heuristic approaches are needed in order to see how to split the system and group the components. But as will be seen in the next chapter, for certain classes of dissipative problems arising from parabolic partial differential equations by the method of lines and simple splittings (such as Jacobi), estimates for L and l can be found.

7.10 Waveform Newton iteration

Both White et al. (1985) and White and Sangiovanni-Vincentelli (1987) noted that there is a generalization of the classical Newton-Raphson method for solving nonlinear systems of equations to a function space equivalent (called **waveform Newton iteration**).

Consider the general system of equations described in (7.9.1). These can be written in the form

$$F(y) := c(y, u)y' - f(y, u) = 0, \quad y(0) = y_0. \tag{7.10.1}$$

If the Newton-Raphson method is applied to (7.10.1) this gives

$$y^{(k+1)} = y^{(k)} - J_F^{-1}(y^{(k)})F(y^{(k)}), \tag{7.10.2}$$

where $J_F(y)$ is the Frechet derivative of F with respect to y defined by

$$\lim_{h\to 0} \frac{1}{\| h \|} \| F(y+h) - F(y) - J_F(y)h \| = 0. \tag{7.10.3}$$

Now, from (7.10.1)

$$\begin{aligned} F(y+h) - F(y) &= c(y+h,u)(y+h)' - c(y,u)y' - f(y+h,u) + f(y,u) \\ &= c(y,u)h' + \frac{\partial c(y,u)}{\partial y}(h)y' - \frac{\partial f(y,u)}{\partial y}(h) + O(\| h \|^2), \end{aligned}$$

while (7.10.3) implies (from the above)

$$J_F(y)h = c(y,u)h' + \frac{\partial c(y,u)}{\partial y}(h)y' - \frac{\partial f(y,u)}{\partial y}h. \tag{7.10.4}$$

Substitution of (7.10.4) into (7.10.2) gives the waveform Newton method for (7.9.1), namely

$$\begin{aligned} c(y^{(k)},u)y^{(k+1)\prime} &+ \frac{\partial c(y^{(k)},u)}{\partial y}(y^{(k+1)} - y^{(k)})y^{(k)\prime} \\ &= f(y^{(k)},u) + \frac{\partial f(y^{(k)},u)}{\partial y}(y^{(k+1)} - y^{(k)}). \end{aligned}$$

In the case c is the identity matrix, the waveform Newton for (7.9.1) is

$$y^{(k+1)\prime} = f(y^{(k)},u) + \frac{\partial f(y^{(k)},u)}{\partial y}(y^{(k+1)} - y^{(k)}). \tag{7.10.5}$$

By writing (7.10.5) in the form

$$y^{(k+1)'} - J_k y^{(k+1)} = f(y^{(k)}) - J_k y^{(k)} \tag{7.10.6}$$

it can be seen that waveform Newton is just an example of a more general splitting. In the case of waveform Newton, $J_k = \frac{\partial f(y^{(k)})}{\partial y}$.

Using a similar approach to that described in section 7.9, White et al. (1985) and White and Sangiovanni-Vincentelli (1987) have proved the following result.

Theorem 7.10.1. *If $\frac{\partial c(y,u)}{\partial y}$ is Lipschitz continuous with respect to y, for all u and f is continuous differentiable then waveform Newton converges uniformly.*

Remarks.

1. Although waveform Newton represents the function space equivalent of the classical Newton-Raphson method, there is an important difference

between the two – namely that waveform Newton converges quadratically for any initial guess.
2. Waveform Newton can be combined with waveform relaxation so that the nonlinear waveform iterations are reduced to a time-varying linear system which can then be solved by efficient linear system techniques such as spectral methods or cyclic reduction.

The analysis given in White et al. (1985) that leads to Theorem 7.10.1 again offers no error bounds on $\varepsilon^{(k)}$ and does not demonstrate the quadratic convergence. An alternative analysis is presented below.

Theorem 7.10.2. *Consider (2.1.1) defined on* $[0,T]$. *Then waveform Newton converges quadratically on all finite intervals* $[0,T]$ *with*

$$\|\varepsilon^{(k)}\|_T \leq \left(\frac{\mu_2}{-2\mu_1}\right)^{2^k-1} \|\varepsilon^{(0)}\|_T^{2^k},$$

where $\mu_1 < 0$ *and* μ_2 *are the logarithmic norms of* $\frac{\partial f}{\partial y}$ *and* $\frac{\partial^2 f}{\partial y^2}$, *respectively.*

Proof: Using (7.10.6) then with $\varepsilon^{(k)} = y^{(k)} - y$, where $y' = f(y)$, expand $f(y^{(k)}) = f(y+\varepsilon^{(k)})$ about y to give

$$f(y+\varepsilon^{(k)}) = f(y) + \frac{\partial f}{\partial y}\varepsilon^{(k)} + \frac{1}{2}\frac{\partial^2 f}{\partial y^2}(\varepsilon^{(k)})^2 + O(\varepsilon^{(k)})^3. \qquad (7.10.7)$$

Hence using a similar analysis that led to (7.9.9) and (7.9.8), (7.10.6) and (7.10.7) give

$$\frac{d}{dx} \| \varepsilon^{(k+1)} \| \leq \mu_1 \| \varepsilon^{(k+1)} \| + \frac{1}{2}\mu_2 \| \varepsilon^{(k)} \|^2, \qquad (7.10.8)$$

where μ_1 and μ_2 are logarithmic norms for $\frac{\partial f}{\partial y}$ and $\frac{\partial^2 f}{\partial y^2}$ for all $y \in \mathbb{R}^m$.

As in Theorem 7.9.2 let $v_k = \| \varepsilon^{(k)} \|$. Then (7.10.8) can be written as

$$v^{(k+1)\prime} \leq \mu_1 v^{(k+1)} + \frac{1}{2}\mu_2 (v^{(k)})^2, \quad v^{(k+1)}(0) = 0. \qquad (7.10.9)$$

Solving (7.10.9) gives

$$\begin{aligned} v^{(k+1)}(x) &\leq \tfrac{1}{2}\mu_2 e^{\mu_1 x}\int_0^x e^{-\mu_1 s}(v^{(k)}(s))^2 ds \\ &\leq \tfrac{1}{2}\mu_2 e^{\mu_1 x}\int_0^x e^{-\mu_1 s} ds \|v^{(k)}\|_x^2 \end{aligned}$$

and so

$$v^{(k+1)}(x) \le \left(\frac{\mu_2}{-2\mu_1}\right)(1 - e^{\mu_1 x})||v^{(k)}||_x^2.$$

Hence

$$||\varepsilon^{(k+1)}||_T \le \left(\frac{\mu_2}{-2\mu_1}\right)(1 - e^{\mu_1 T})||\varepsilon^{(k)}||_T^2, \qquad (7.10.10)$$

which implies quadratic convergence.

Now repeatedly solving (7.10.9) for $k = 0, 1, \ldots$, it is easily shown that

$$v^{(k)}(x) \le \left(\frac{\mu_2}{-2\mu_1}\right)^{2^k - 1} ||v^{(0)}||_x^{2^k} \alpha_k(x), \qquad (7.10.11)$$

where for $\mu_1 < 0$

$$|\alpha_k(x)| \le 1.$$

Hence from (7.10.11)

$$||\varepsilon^{(k)}||_T \le \left(\frac{\mu_2}{-2\mu_1}\right)^{2^k - 1} ||\varepsilon^{(0)}||_T^{2^k},$$

as required. □

7.11 Asynchronous techniques

It is known that in some instances, asynchronous iteration can offer substantial improvement over synchronous iteration because there is minimal synchronization between cooperating processes (see section 4.9). Aslam and Gear (1989) have applied asynchronous techniques to Picard-like iteration schemes for solving ordinary differential equation systems and implemented this approach in a shared-memory environment. This work has been extended by Aslam (1990) who has considered asynchronous waveform relaxation techniques.

Bertsekas and Tsitsiklis (1989) and Mitra (1985, 1987) have developed waveform relaxation schemes based on asynchronous iterative techniques for solving systems of ordinary differential equations in which there is little time spent in processor synchronization and thus the idle time of the processors is negligible. This approach is essentially independent of the number of processors or the architecture and the subsystems can be relaxed in an arbitrary order by the parallel processors operating asynchronously. Both Bertsekas and Tsitsiklis (1989) and Mitra (1987) consider the convergence behaviour of this asynchronous waveform approach on the linear problem

$$y'(x) + D(x)y(x) = B(x)y(x) + g(x),$$

where D is block diagonal.

Convergence on finite windows is proved in the usual way based on the determination of a two-sided Lipschitz constant while uniform convergence on infinite windows requires additional assumptions based on a modified diagonal dominance property. In the case that D and B are independent of x and the blocks associated with D are of dimension one, Bertsekas and Tsitsiklis (1989) show that these additional assumptions are of the form

$$D > 0, \quad \rho(D^{-1}|B|) < 1.$$

In the nonlinear case, assuming certain dominance conditions on the differential system, Mitra (1987) shows that the asynchronous waveforms converge geometrically given uniformly bounded delays in the updating of information and that a certain nonstarvation condition holds.

7.12 Other applications

Delay differential equations

The concept of dynamic iteration has been extended to hyperbolic partial differential equations by Bjørhus (1992), to delay differential equations by Bjørhus (1993), and to differential-algebraic systems by Jackiewicz and Kwapisz (1994). In the case of the general delay equation

$$y'(x) = f(y(x), y(\theta(x))), \quad y(0) = y_0, \quad \theta(0) = 0, \quad x \in [0, T]. \quad (7.12.1)$$

Bjørhus (1992) introduces a generalization of the Picard method given in (7.1.1) namely

$$y^{(k+1)'}(x) = f(y^{(k+1)}(x), y^{(k)}(\theta(x))), \quad y^{(k+1)}(0) = y_0. \quad (7.12.2)$$

An examination of the convergence properties of (7.12.2) proceeds in much the same way as described in the proof of Theorem 7.9.2. Thus, under the assumption that f is Lipschitz continuous (with Lipschitz constant L_2) in the second argument and f satisfies a one-sided Lipschitz condition (with constant l_1) with respect to the first argument, and under the assumption that

$$\theta(x) \leq x, \quad \theta'(x) > 0,$$

Bjørhus shows that

$$\frac{d}{dx} \parallel \varepsilon^{(k+1)}(x) \parallel \leq l_1 \parallel \varepsilon^{(k+1)}(x) \parallel + L_2 \parallel \varepsilon^{(k)}(\theta(x)) \parallel . \quad (7.12.3)$$

Note here the similarity with (7.9.9) and also note that there may be a set of points for which $\parallel \varepsilon^{(k+1)} \parallel$ is not differentiable.

With $\theta_0(x) = x$ and $\theta_k(x) = \theta \circ \ldots \circ \theta(x)$ (k times), Bjørhus (1993) uses an application of Gronwall's lemma to give

$$\| \varepsilon^{(k+1)}(x) \| \leq L_2 \int_0^x \| \varepsilon^{(k)}(\theta(s)) \| e^{l_1(x-s)} ds,$$

and hence proves

Theorem 7.12.1. *For the dynamic iteration scheme (7.12.2) satisfying the Lipschitz conditions described previously,*

$$||\varepsilon^{(k)}||_T \leq \frac{(L_2T)^k}{k!} e^{\mu T} \prod_{j=0}^{k-2} |\theta'|_{\theta_j(T)}^{k-1-j} \, ||\varepsilon^{(0)}||_T, \qquad (7.12.4)$$

where

$$\mu = \max\{0, l_1\}.$$

Remarks.
1. If θ' is bounded by a constant, this result proves convergence on all finite windows.
2. If $\theta(x) = x$, this result is identical with that given in Theorem 7.9.2.
3. Because μ appears in (7.12.4) rather than l_1, information is lost if the system is dissipative and so this bound may be unduly pessimistic.

Bjørhus (1993) introduces a more general splitting to (7.12.2) to allow the possibility of integrating different subsystems independently. In the autonomous form this leads to the iteration scheme

$$y^{(k+1)'}(x) = F(y^{(k+1)}(x), y^{(k)}(x), y^{(k)}(\theta(x))), \quad y^{(k+1)}(0) = y_0, \quad (7.12.5)$$

where

$$F(y, y, y(\theta)) = f(y, y(\theta)),$$

which is similar to the splitting described in (7.2.4) for the non-retarded case.

Under the assumption that F satisfies a one-sided Lipschitz condition (l_1) in the first component and is Lipschitz continuous in the second and third components (L_2 and L_3), a result similar to Theorem 7.9.3 can be proved.

Theorem 7.12.2. *For the dynamic iteration scheme (7.12.5) satisfying the Lipschitz conditions described previously*

$$||\varepsilon^{(k)}||_T \leq \left(\frac{L_2 + L_3}{-\mu}\right)^k \tilde{\alpha}_k(T) \, ||\varepsilon^{(0)}||_T$$

where

$$\tilde{\alpha}_k(T) = e^{\mu T} \sum_{j=k}^{\infty} \frac{(-\mu T)^j}{j!}, \qquad \mu = \max\{0, l_1\}.$$

Note that there is an important difference between this result and that of Theorem 7.9.3, since $\mu \geq 0$ and in this case the property

$$0 < \tilde{\alpha}_{k+1}(T) < \tilde{\alpha}_k(T) < 1$$

does not necessarily hold.

Hyperbolic problems

In the solution of time-dependent PDEs, waveform techniques have only been considered after the spatial discretization of the PDE to an ODE system. However, dynamic iteration techniques can be applied both before and after the spatial discretization of the PDE. In the PDE context this amounts to decoupling the equations through a domain decomposition approach over the time interval of integration, in which values that couple adjacent subdomains are input from the previous sweep. Bjørhus (1992) calls this process an **outer dynamic iteration** (ODI).

Bjørhus (1992) applies this idea to hyperbolic PDEs of the form

$$\begin{array}{rcllcl} (\frac{\partial}{\partial t} + L)y & = & f\,, & (x,t) & \in & \Omega \times [0,T] \\ By & = & b\,, & (x,t) & \in & \partial\Omega \times [0,T] \\ y & = & g\,, & (x,t) & \in & \Omega \times \{0\}. \end{array} \tag{7.12.6}$$

The domain Ω is then split into subdomains $\Omega_i, i = 1, \ldots, L$, where $\cup_i \overline{\Omega}_i = \overline{\Omega}$. The set of neighbours N_i of Ω_i is defined as

$$j \in N_i \quad \text{iff} \quad \overline{\Omega}_j \cap \overline{\Omega}_i \neq 0, \qquad j \neq i,$$

so that (7.12.6) can be written as L subproblems

$$\begin{array}{rcllcl} (\frac{\partial}{\partial t} + L)y_i & = & f\,, & (x,t) & \in & \Omega_i \times [0,T] \\ M(y_i, y_j) & = & 0\,, & (x,t) & \in & \partial\Omega_i \cap \overline{\Omega}_j \times [0,T],\ \forall j \in N_i \\ By_i & = & b\,, & (x,t) & \in & \partial\Omega_i \cap \partial\Omega \times [0,T] \\ y_i & = & g\,, & (x,t) & \in & \Omega_i \times \{0\}. \end{array} \tag{7.12.7}$$

Here M is an interface operator which couples the solutions on different domains. Bjørhus (1992) then decouples the problems by suppressing the coupling within M and this leads to the outer dynamic iteration

$$\begin{aligned}
(\tfrac{\partial}{\partial t} + L)y_i^{(k+1)} &= f\,, & (x,t) &\in \Omega_i \times [0,T] \\
M(y_i^{(k+1)}, y_j^{(k)}) &= 0\,, & (x,t) &\in \partial\Omega_i \cap \overline{\Omega}_j \times [0,T],\ \forall j \in N_i \\
By_i^{(k+1)} &= b\,, & (x,t) &\in \partial\Omega_i \cap \partial\Omega \times [0,T] \\
y_i^{(k+1)} &= g\,, & (x,t) &\in \Omega_i \times \{0\},\quad y_i^{(0)} = g|_{\Omega_i}.
\end{aligned}$$

This decouples the original problem (7.12.6) into independent subsystems which can be solved in parallel on some appropriate time window for each sweep k.

These independent subsystems can now be spatially discretized to produce a set of ordinary differential equations and standard waveform techniques can then be applied at this level. Bjørhus terms this the **inner dynamic iteration** (IDI) process. Thus the outer process affects only the couplings to other domains applied before spatial discretization, while the inner process affects only the couplings within the domain after spatial discretization.

Bjørhus (1992) formulates this approach for quasilinear hyperbolic PDEs and proves some uniqueness and convergence results and analyses in detail the behaviour of this process on the one-dimensional linear hyperbolic initial boundary value problem with constant coefficients, namely

$$y_t + Ay_x = f, \quad y \in \mathbb{R}^m. \tag{7.12.8}$$

In particular, Bjørhus shows that the outer dynamic iteration process applied to (7.12.8) yields the exact solution in L iterations, where L is the number of decompositions.

Bjørhus also compares the efficiency of this approach compared with the single domain approach on a system of m equations in a d-dimensional space using a collocation approach with ν collocation points in each direction. If in the ODI approach μ subdivisions are used in each direction then Bjørhus estimates a speed-up of μ if a constant-stepsize explicit Runge-Kutta method is used and a speed-up of approximately μ^d if an implicit Runge-Kutta method is used, ignoring communication effects.

Second-order problems

Bjørhus (1990) has also considered the extension of dynamic iteration techniques to higher derivative problems. In particular, Bjørhus (1990) has studied the second-order problem

$$y'' + Qy = 0, \quad y(0) = y_0, \quad y'(0) = y_0' \tag{7.12.9}$$

and the dynamic iteration scheme

$$y^{(k+1)''} + My^{(k+1)} = Ny^{(k)}, \quad y^{(k+1)}(0) = y_0, \quad y^{(k+1)'}(0) = y_0'. \tag{7.12.10}$$

On finite intervals, Bjørhus remarks that it is a straightforward generalization of the work of Miekkala and Nevanlinna (1987a) and Nevanlinna

(1990) to show superlinear convergence of (7.12.10) independently of the splitting.

In the case of infinite intervals these generalizations are not so clear. A number of approaches are possible. One approach is to rewrite (7.12.9), and hence (7.12.10), as a system of first-order equations and then consider splittings of the matrix

$$\tilde{Q} = \begin{pmatrix} 0 & -I \\ Q & 0 \end{pmatrix}.$$

However, in this case all the eigenvalues of $\tilde{Q}$ will be strictly imaginary and the convergence theory of first-order waveforms requires eigenvalues with positive real part. This difficulty can be resolved by considering a small perturbation of (7.12.9), but this is not a natural approach.

Instead, Bjørhus (1990) examines the second order dynamic iteration scheme directly. In this case the dynamic iteration scheme can be written in the form (7.4.1) where now

$$\begin{aligned} Ku(x) &= \int_0^x M^{-\frac{1}{2}} \sin(M^{\frac{1}{2}}(x-s))Nu(s)ds \\ \phi(x) &= M^{-\frac{1}{2}} \sin(M^{\frac{1}{2}}x)y_0' + \cos(M^{\frac{1}{2}}x)y_0. \end{aligned} \tag{7.12.11}$$

Under the assumptions that both M and Q are diagonalizable, under the same similarity transformation S, with real positive eigenvalues and letting Λ be the diagonal matrix consisting of the eigenvalues of M, Bjørhus (1990) proves the analogue of Theorem 7.4.2, where here

$$C = ||\Lambda^{-\frac{1}{2}}||\,||N||\,||S||\,||S^{-1}||, \tag{7.12.12}$$

and convergence takes place on the window $[0, 1/C]$.

In order to treat the infinite window case, the notation of Bjørhus (1990) will be used.

Definition 7.12.3. *X_T will denote the set of all functions $f : [0,T] \to \mathbb{C}^m$ such that each component of $f \in C[0,T]$, the space of continuous functions on $[0,T]$.*

If T is infinite in Definition 7.12.3, then the appropriate space will be denoted by X_∞. If this space is equipped with the norm

$$||y||_\infty = \sup_{x \geq 0} ||y(x)||, \tag{7.12.13}$$

then K, as given in (7.12.11), is an unbounded operator. Hence an expression for the spectral radius of K (see (7.4.12)) cannot be found using the techniques of Miekkala and Nevanlinna (1987a), as these techniques rely on the convolution operator being a bounded linear operator.

In order to ensure that K is a bounded operator on some appropriate space, Bjørhus (1990) applies dynamic iteration to the perturbed system

$$y'' - 2\epsilon y' + Qy = 0, \quad \epsilon < 0. \tag{7.12.14}$$

Thus given a $\delta > 0$ and a $T < 0$ there exists an $\epsilon < 0$ such that the difference between the solutions of (7.12.9) and (7.12.14) is bounded by δ on $[0, T]$.

Hence a dynamic iteration scheme applied to (7.12.14) gives

$$y^{(k+1)''} - 2\epsilon y^{(k+1)'} + My^{(k+1)} = Ny^{(k)}. \tag{7.12.15}$$

This can be written in the form of (7.4.1) with

$$\begin{aligned} K_\epsilon u(x) &= \int_0^x e^{\epsilon(x-s)} R^{-\frac{1}{2}} \sin(R^{\frac{1}{2}}(x-s)) N u(s) ds \\ \phi_\epsilon(x) &= e^{\epsilon x}(R^{-\frac{1}{2}} \sin(R^{\frac{1}{2}} x)(y_0' - \epsilon y_0) + \cos(R^{\frac{1}{2}} x) y_0) \\ R &= M - \epsilon^2 I. \end{aligned}$$

Now, in order to show that K_ϵ is a bounded operator for $\epsilon < 0$, Bjørhus (1990) constructs a new space $\bar{X}_\infty \subset X_\infty$, which is the set of all functions f for which both f and f' are elements of X_∞ with norm given by

$$||y||_{\infty,\epsilon} = \max\{||y||_\infty, ||y'||_\infty\},$$

where $||y||_\infty$ is given by (7.12.13).

The techniques of Miekkala and Nevanlinna (1987a) can now be applied to K_ϵ on the space $\bar{X}_\infty$, to compute the spectral radii of a bounded operator. Thus Bjørhus (1990) proves the analogue of Theorem 7.4.6.

Theorem 7.12.4. *Let Q and M be diagonalizable with positive real eigenvalues. Then on $\bar{X}_\infty$*

$$\rho(K_\epsilon) = \sup_{z \in \partial\Omega_\epsilon} \rho((M + zI)^{-1} N),$$

where $\partial\Omega_\epsilon$ is described by the curve $\bar{z}(s)$ given by

$$\bar{z}(s) = \epsilon^2 \left(\frac{e^{is}}{\cos^2(s/2)} - 1 \right), \quad -\pi < s < \pi.$$

A similar result can be proved by using an exponentially weighted version of $\bar{X}_\infty$, denoted by $\bar{X}_{\infty,\theta}$. This will be equipped with the norm

$$||y||_{\infty,\theta} = \sup_{x \geq 0} ||e^{-\theta x} y(x)||, \quad \theta > 0.$$

Theorem 7.12.5. *Let Q and M be diagonalizable with positive real eigenvalues. Then on $\bar{X}_{\infty,\theta}$*

$$\rho(K) = \sup_{z \in \partial\Omega} \rho((zI + M)^{-1}N),$$

where $\partial\Omega$ is described by the curve $\bar{z}(s)$ given by

$$\bar{z}(s) = \frac{\theta^2}{\cos^2(s/2)} e^{is}, \quad -\pi < s < \pi.$$

Differential algebraic systems

Jackiewicz and Kwapisz (1994) analyse the convergence of Picard iterations and more general WR iterations for differential-algebraic systems which may contain delay terms. It is proved that these iterations are convergent if $\rho(A) < 1$ where A is a 2×2 nonnegative matrix whose coefficients depend on Lipschitz constants of the problem under consideration. It is also shown that under certain conditions the convergence of WR iterations is always faster than the convergence of Picard iterations.

7.13 Preconditioning

As has already been noted, waveform relaxation can be very effective if the waveform iterates converge quickly. Unfortunately, in many cases convergence can be very slow, as is the case of Picard iteration applied to stiff problems on moderately sized windows. This situation is of course no different to the static case. Thus, for example, if Jacobi or Gauss-Seidel iteration is applied to a linear system (4.4.1) arising from the semi-discretization of an elliptic equation, convergence slows as the problem becomes more stiff (by, for example, refining the mesh). One way of improving this convergence is by preconditioning, as described in sections 4.5 and 4.8.

Consider now the linear system

$$Qy = b. \tag{7.13.1}$$

If some matrix P is applied to both sides of this equation this gives

$$\bar{Q}y = \bar{b}, \quad \bar{Q} = PQ, \quad \bar{b} = Pb. \tag{7.13.2}$$

By choosing P appropriately, so that the eigenvalues of $\bar{Q}$ are better behaved, the convergence properties of iterative techniques can be much improved. This is called **preconditioning on the left.**

Similarly, if R denotes a nonsingular matrix then (7.13.1) can be written as

$$\hat{Q}z = b, \quad \hat{Q} = QR^{-1}, \quad y = R^{-1}z,$$

and R is again chosen to obtain better behaviour for the spectrum of $\hat{Q}$. This is called **preconditioning on the right**.

Preconditioning on the left and on the right can be combined to give

$$\tilde{Q}z = \tilde{b}, \quad \tilde{b} = Pb, \quad \tilde{Q} = PQR^{-1}, \quad y = R^{-1}z. \tag{7.13.3}$$

These concepts can be generalized in a very general way to the dynamic (differential equation) case. The generalization will only be done here for linear differential systems but it is equally appropriate for nonlinear systems.

Consider the linear system of equations

$$y' + Qy = g(x), \quad y(x_0) = y_0. \tag{7.13.4}$$

By applying some matrix P to both sides of (7.13.4) a linear system given by

$$Py' + \bar{Q}y = Pg(x), \quad \bar{Q} = PQ$$

is formed. Let P and $\bar{Q}$ be split by

$$P = M_p - N_p, \quad \bar{Q} = M_q - N_q; \tag{7.13.5}$$

then this leads to a waveform iteration of the form

$$M_p y^{(k+1)'} + M_q y^{(k+1)} = N_p y^{(k)'} + N_q y^{(k)} + Pg, \; y^{(k+1)}(x_0) = y_0. \tag{7.13.6}$$

This can be written as fixed-point iteration of the form (7.4.1) where

$$\begin{aligned} Ky(x) &= A_1 y(x) + \int_{x_0}^{x} e^{M_p^{-1}M_q(s-x)} A_2 y(s)ds \\ \phi(x) &= e^{M_p^{-1}M_q(x_0-x)}(I - A_1)y_0 \\ &\quad + \int_{x_0}^{x} e^{M_p^{-1}M_q(s-x)} M_p^{-1}Pg(s)ds \\ A_1 &= M_p^{-1}N_p, \quad A_2 = M_p^{-1}(N_q - M_q M_p^{-1}N_p). \end{aligned} \tag{7.13.7}$$

Here it is assumed that M_p is nonsingular, otherwise (7.13.6) is a differential algebraic equation.

It is easy to show that K is bounded on X_T with norm given in Definition 7.4.1. In particular,

$$||Ky - Kz||_T \leq (||A_1|| + ||A_2||T)||y - z||_T.$$

Thus if

$$||M_p^{-1} N_p|| < 1 \tag{7.13.8}$$

it is possible to find an interval $[0, T_c]$ for which K is a contraction mapping.

In the case of infinitely long windows it is possible to obtain the analogue of Theorem 7.4.6. This is given in the following result.

Theorem 7.13.1. *If* $\det(M_p + \lambda M_q) \neq 0, \quad Re(\lambda) \geq 0$ *then*

$$\rho(K) = \max_{\xi \in \mathbb{R}} \rho((i\xi M_p + M_q)^{-1}(i\xi N_p + N_q)). \tag{7.13.9}$$

Remarks.
1. The assumption in Theorem 7.13.1 guarantees that K is a bounded operator on X_∞.
2. The proof of Theorem 7.13.1 was first given in Spilling (1993). In fact Spilling (1993) considered dynamic iterations of the form (7.13.6) to (7.13.4) where Q represents a pseudospectral differentiation matrix. The **pseudospectral method** is a technique used to approximate the spatial derivatives in a partial differential equation by constructing a global interpolant through certain collocation points and then differentiating the interpolant at every point to give a **pseudospectral differentiation matrix.**

The pseudospectral approximation to the approximate solution is represented as a linear combination of basis functions and the coefficients of this linear combination are known as the spectrum. The advantage of this approach is that high accuracy can be obtained from a relatively coarse discretization. However, Q in (7.13.4) is dense and is strongly sensitive to small perturbations in its entries.

Spilling (1993) uses dynamic iteration and the pseudospectral method to solve the one-dimensional hyperbolic equation

$$y_t + f(x, t, y)y_x = 0.$$

In the case of Chebyshev polynomial basis functions and collocation points based on the extrema of Chebyshev polynomials, Spilling constructs a specific preconditioner P with an explicitly known inverse. A discretized waveform implementation is given based on the order 2 method

$$\begin{array}{c|cc} 0 & 0 & 0 \\ \frac{1}{2} & \frac{1}{2} & 0 \\ \hline & 0 & 1 \end{array}$$

Various splittings of $\bar{Q}$ and P are considered by Spilling.

Having shown how to precondition on the left, the technique of preconditioning on the right is now described. Much of this work appears in Burrage et al. (1995). Consider again (7.13.4), and assume that Q can be split as

$$Q = C + D. \tag{7.13.10}$$

By making the transformation

$$z(x) = e^{D(x-x_0)}y(x), \tag{7.13.11}$$

it is easily seen that (7.13.4) can be written as

$$z'(x) + B(x - x_0)z(x) = e^{D(x-x_0)}g(x), \quad z(x_0) = y_0, \tag{7.13.12}$$

where

$$B(x) = e^{Dx}Ce^{-Dx}. \tag{7.13.13}$$

At first glance this transformation only complicates the problem. However, the idea here is to choose D in such a way that D absorbs most of the stiffness of Q leaving a "nonstiff" matrix C. Waveform relaxation can then be applied to (7.13.12) with the hope that convergence will take place more rapidly as the convergence rate will depend on the size of the Lipschitz constant for C rather than Q.

Choosing the splitting of the matrix $B(x)$

$$B(x) = M - N(x), \tag{7.13.14}$$

where M is a constant matrix determined by the desired iteration scheme (Jacobi, Gauss-Seidel, etc.) leads to the process

$$z^{(k+1)'}(x) + Mz^{(k+1)}(x) = N(x - x_0)z^{(k)}(x) + e^{D(x-x_0)}g(x). \tag{7.13.15}$$

Putting

$$Kz(x) \quad = \quad e^{-Mx}\int_{x_0}^{x} e^{Ms}N(s - x_0)z(s)ds$$

$$\phi(x) \quad = \quad e^{M(x_0-x)}z_0 + e^{-Mx}\int_{x_0}^{x} e^{Ms}e^{D(s-x_0)}g(s)ds,$$

(7.13.15) can be rewritten in the form

$$z^{(k+1)}(t) = Kz^{(k)}(t) + \phi(t), \quad z^{(k+1)}(t_0) = z_0. \tag{7.13.16}$$

Observe that the computation of the iterations (7.13.15) involves the extra cost of evaluating $e^{D(x_\mu - x_0)}Ce^{-D(x_\mu - x_0)}$ for the discrete set of points

$$x_0 < x_1 < \cdots < x_N = T,$$

which correspond to the discretization of the convolution integral defining K on the given window $[x_0, T]$. However, at the expense of extra storage space, this need only be computed once and reused for all iterations assuming that subsequent discretization grids are a subset of the discretization grid for the first sweep. Moreover, at the completion of the iteration process, the quantities

$$e^{-D(x_\mu - x_0)}z^{(\nu)}(x_\mu)$$

must also be computed, where ν is the index which corresponds to the accepted iterate $z^{(\nu)}$. Efficient techniques for doing this based on Krylov techniques are described in Sidje (1994) and Gallopoulos and Saad (1989). On the other hand, for some iterative schemes (block Jacobi, for example), the matrix D will have such a form that the computation of e^{Dt_μ} and e^{-Dt_μ} will be trivial, if Q has a block-tridiagonal structure (see Burrage et al. (1995)).

In the case that C and D commute (equivalently, Q and D commute), then (7.13.13) and (7.13.14) imply

$$B(x) = Q - D.$$

Hence the solution to (7.13.12), with $g(x) \equiv 0$, can be written as

$$e^{(Q-D)(x-x_0)}e^{D(x-x_0)}y_0$$

and waveform relaxation applied to (7.13.12) represents the computation of the first factor. This effect bears some relation to the concept of operator splitting which was discussed in section 6.7.

By letting

$$\varepsilon^{(k)}(x) = z(x) - z^{(k)}(x)$$

and subtracting from the equivalent form of (7.13.16) for the exact solution z, gives the error equation

$$\varepsilon^{(k+1)}(x) = e^{-Mx}\int_{x_0}^{x} e^{Ms}N(s - x_0)\varepsilon^{(k)}(s)ds. \tag{7.13.17}$$

To study the behavior of $\varepsilon^{(k)}$ the following lemmas are needed.

Lemma 7.13.1. *The following expansion holds*

$$e^{D(x-x_0)}Ce^{-D(x-x_0)} = \sum_{i=0}^{\infty} \Delta_i \frac{(x-x_0)^i}{i!}, \tag{7.13.18}$$

where the matrices Δ_i *are defined by*

$$\Delta_i = \sum_{j=0}^{i} \binom{i}{j} D^{i-j}C(-D)^j. \tag{7.13.19}$$

Proof. Expanding $e^{D(x-x_0)}$ and $e^{-D(x-x_0)}$ into Taylor series around $x = x_0$ and comparing the corresponding terms in (7.13.18) yields (7.13.19). □

Lemma 7.13.2. *The matrices* Δ_i, $i = 0, 1, \ldots$, *satisfy the recurrence relation*

$$\Delta_{i+1} = D\Delta_i - \Delta_i D,$$

with $\Delta_0 = C$.

Proof.

$$\begin{aligned}
\Delta_{i+1} &= \sum_{j=0}^{i+1} \binom{i+1}{j} D^{i+1-j}C(-D)^j \\
&= D^{i+1}C + \sum_{j=1}^{i} \left(\binom{i}{j} + \binom{i}{j-1} \right) D^{i+1-j}C(-D)^j + C(-D)^{i+1} \\
&= D^{i+1}C + \sum_{j=1}^{i} \binom{i}{j} D^{i+1-j}C(-D)^j \\
&\quad + \sum_{j=0}^{i-1} \binom{i}{j} D^{i-j}C(-D)^{j+1} + C(-D)^{i+1} \\
&= D\sum_{j=0}^{i} \binom{i}{j} D^{i-j}C(-D)^j - \sum_{j=0}^{i} \binom{i}{j} D^{i-j}C(-D)^j D \\
&= D\Delta i - \Delta_i D. \quad \square
\end{aligned}$$

Remark. Observe that if C and D commute then $\Delta_i = 0$ for $i \geq 1$.

Using (7.13.18) the equation (7.13.17) can be rewritten in the form

$$\varepsilon^{(k+1)}(x) = e^{-Mx} \int_{x_0}^{x} e^{Ms} \left(M - Q + D + \sum_{i=1}^{\infty} \Delta_i \frac{(s-x_0)^i}{i!} \right) \varepsilon^{(k)}(s)ds.$$

D will be chosen to annihilate the constant term $M - Q + D$ hoping that this will speed up the convergence to zero of the error $\varepsilon^{(k)}$ as $k \to \infty$. That this is indeed the case follows from the theorem given below.

Theorem 7.13.3. *Assume that* $M - Q + D = 0$, $||e^{-Mx}\Delta_i|| \leq A_i$ *and that* $||\varepsilon^{(0)}||_T \leq 1$. *Then*

$$||\varepsilon^{(\nu)}||_T \leq \sum_{i=1}^{\infty} \frac{A_i^\nu (T - x_0)^{\nu(i+1)}}{((i+1)!)^\nu \nu!}, \quad \nu = 0, 1, \ldots. \tag{7.13.20}$$

Proof. Define $\varepsilon_i^{(\nu)}(x)$ by

$$\varepsilon_i^{(\nu+1)}(x) = -\frac{1}{i!}\int_{x_0}^{x} e^{-M(x-s)}\Delta_i (s - x_0)^i \varepsilon_i^{(\nu)}(s)ds.$$

Then

$$||\varepsilon_i^{(\nu+1)}(x)|| \leq \frac{1}{i!}\int_{x_0}^{x} A_i (s - x_0)^i ||\varepsilon_i^{(\nu)}(s)||ds,$$

and since $||\varepsilon_i^{(0)}(s)|| \leq 1$ an easy induction argument leads to

$$||\varepsilon_i^{(\nu)}(x)|| \leq \frac{A_i^\nu (x - x_0)^{\nu(i+1)}}{((i+1)!)^\nu \nu!}.$$

It follows from the superposition principle that

$$\varepsilon^{(\nu)}(x) = \sum_{i=1}^{\infty} \varepsilon_i^{(\nu)}(x)$$

which leads to (7.13.20). □

In the case that there is no preconditioning the bound on the right hand side of (7.13.20) can be written as

$$\psi_N(x, \nu) = \frac{A_0^\nu (T - x_0)^\nu}{\nu!}. \tag{7.13.21}$$

On the other hand, preconditioning gives a bound which consists of an infinite number of terms. However, if the A_i do not grow (as is the case for waveform relaxation block Jacobi scheme) then a reasonable approximation to this bound is

$$\psi_P(x, \nu) = \frac{A_1^\nu (T - x_0)^{2\nu}}{2^\nu \nu!}, \tag{7.13.22}$$

as long as $T - x_0$ is not too large.

For the test problem arising from the one-dimensional heat equation, a waveform relaxation block Jacobi method leads to the same bound for A_0 and A_1, which in the case of the infinity norm is $||A_0||_\infty = ||A_1||_\infty = 1$. Furthermore, for this problem (7.13.22) is an accurate bound if $T - x_0 \leq$

Table 7.2 *predicted iteration counts*

tolerance	10^{-4}	10^{-8}	10^{-4}	10^{-8}	10^{-4}	10^{-8}	10^{-4}	10^{-8}
x	0.25	0.25	0.5	0.5	1.0	1.0	2.0	2.0
ν_N	6	9	7	11	10	14	14	14
ν_P	3	5	5	7	7	11	19	19

1. The table below gives the number of iterations required to attain an accuracy of 10^{-4} and 10^{-8} for (7.1.3.21) and (7.13.22) when $T - x_0$ is, respectively, 0.25, 0.5, 1, and 2 and $||A_0||_\infty = ||A_1||_\infty = 1$.

These numbers predicted by the above theory are in good agreement with the experimental numbers obtained (see Burrage et al. (1995)). As can be seen, for windows of small length, approximately only half the number of iterations are required in the preconditioned case when compared with the non-preconditioned case.

In some cases it may be advantageous from the point of view of parallel processing to consider more general preconditioning with a function $R(x)$, where $R(x)$ is polynomial or rational approximation to the matrix exponential e^{Dx}. Thus by letting $z(x) = R(x)y(x)$, (2.2.4) can be rewritten in the form

$$z'(x) + Mz(x) = P(x)z(x) + R(x)g(x),$$

where

$$P(x) = M - (R(x)Q - R'(x))R^{-1}(x).$$

Assume that

$$\begin{aligned} R(x) &= I + C_1 x + C_2 x^2 + \cdots \\ P(x) &= \Delta_0 + \Delta_1 x + \Delta_2 x^2 + \cdots . \end{aligned}$$

then comparing the corresponding terms in the equation

$$P(x)R(x) = MR(x) - R(x)Q - R'(x),$$

gives the following relationship between C_i and Δ_i:

$$C_{i+1} = \frac{1}{i+1}\left(C_i Q - MC_i + \sum_{j=0}^{i} \Delta_j C_{i-j}\right),$$

or

$$\Delta_i = (i+1)C_{i+1} + MC_i - C_i Q - \sum_{j=0}^{i-1} \Delta_j C_{i-j}.$$

Choosing $\Delta_0 = 0$, $C_1 = Q - M = D$ and $C_i = 0$ for $i \geq 2$ corresponds to approximating the exponential e^{Dx} by the polynomial $R(x) = I + Dx$ of the first degree which can be easily evaluated in parallel. In this case

$$\Delta_1 = MD - DQ,$$

and

$$\Delta_{i+1} = (-1)^i \Delta_1 D^i, \quad i = 1, 2, \ldots.$$

Hence here D should be chosen, if possible, so that $D^2 = 0$. Similarly, it is easy to obtain relevant expressions when the function $R(x)$ has the form

$$R(x) = (I + D_1 x + D_2 x^2 + \ldots)^{-1}(I + C_1 x + C_2 x^2 + \ldots).$$

The hope here is that D_j and C_j can be chosen in such a way that $\Delta_1 = \Delta_2 = \ldots = \Delta_k = 0$ for some $k \geq 1$. In this case the leading term of $||\varepsilon^{(\nu)}||_T$ is

$$\frac{A_{k+1}^{\nu}(T - x_0)^{\nu(k+2)}}{((k+2)!)^{\nu}\nu!}$$

and this will be a realistic bound on the error if $T - x_0$ is not too large.

Various preconditioned and non-preconditioned waveform relaxation techniques have been implemented on linear systems of differential equations arising from the semi-discretization of the heat equation in either one or two dimensions (see Burrage et al. (1995)). In the case of the one-dimensional heat equation, overlapping block Jacobi WR is very effective in comparison with the corresponding nonoverlapping scheme. Hence preconditioning does not lead to a significant improvement over the overlapping implementation. On the other hand, preconditioning is much more effective on the two-dimensional equation (Burrage et al. (1995)).

8

DISCRETE WAVEFORM METHODS

The previous chapter dealt with the issue of convergence rates for continuous waveform algorithms when applied to either linear or nonlinear problems. Techniques for choosing appropriate splittings were formulated based on estimating the coupling between components or subsystems by an analysis of the one-sided and two-sided Lipschitz constants associated with the splitting. However, so far nothing substantial has been proved about the behaviour of numerical methods when they are used to solve the iteration waveforms.

As noted in Chapter 7, both White et al. (1985) and White and Sangiovanni-Vincentelli (1987) observed that when a discrete approximation is applied to a continuous waveform algorithm the discrete approximations can be interpreted as the exact solution of a perturbed system which changes with each iteration. This perturbed system is called a nonstationary waveform relaxation algorithm.

White et al. (1985) and White and Sangiovanni-Vincentelli (1987) are then able to show that the nonstationary algorithms converge as a direct consequence of the contraction mapping property associated with the original waveform relaxation algorithm in which, as the number of nonstationary iterates goes to infinity, the nonstationary map approaches the stationary map. A consequence of this result is that the discretized iterates will converge as long as the global discretization errors in the numerical approximation are driven to zero.

However, as in the continuous waveform case, these results do not offer any bounds on the convergence rates of the discretized waveform iterates and the norm used to prove the contraction mapping is problem dependent. In addition, it is not clear what is meant by driving the numerical approximations to zero – in other words, how accurately should the iteration equations be solved in order to guarantee convergence within a specified tolerance? Finally, this approach does not allow the convergence rates of different integration methods to be compared.

These issues and others will be further discussed in this chapter, which covers the following material:

- section 8.1: an in-depth analysis of the convergence behaviour of the Euler and implicit Euler methods in conjunction with Gauss-Seidel or

Jacobi relaxation on a linear problem with variable coupling;

- section 8.2: the convergence behaviour of linear multistep methods in conjunction with general relaxation splittings for linear systems;
- section 8.3: the relationship between the discretization operator of an algebraically stable Runge-Kutta method and the continuous waveform operator on linear systems;
- section 8.4: the generalizations of some of these discretization results to multisplitting algorithms;
- section 8.5: the convergence behaviour of discretized WR algorithms on nonlinear problems and the importance of mesh refinement;
- section 8.6: implementation and convergence aspects of multistage WR methods in a variable-stepsize setting;
- section 8.7: the introduction of a new class of waveform relaxation Runge-Kutta methods based on Jacobi or SOR relaxation which break the order barrier of one for unconditionally contractive Runge-Kutta methods in the maximum norm.

8.1 An example

Consider the two-dimensional linear problem

$$y' = Qy, \quad y(0) = y_0, \quad Q = \begin{pmatrix} -1 & \lambda \\ -\lambda & -1 \end{pmatrix}, \quad \lambda > 0. \tag{8.1.1}$$

If the implicit Euler method is applied to this problem with constant stepsize h then

$$(I - hQ)y_{n+1} = y_n \tag{8.1.2}$$

and

$$y_N = (I - hQ)^{-N} y_0. \tag{8.1.3}$$

Now since the eigenvalues of Q are $-1 \pm i\lambda$, the numerical approximations computed from (8.1.3) will remain bounded from step to step for all positive values of h.

On the other hand if the Euler method is applied to this problem then

$$y_N = (I + hQ)^N y_0$$

and since the eigenvalues of $I + hQ$ are $(1 - h) \pm ih\lambda$, the approximations will remain bounded only if $h < \frac{2}{1+\lambda^2}$. Suppose now that the implicit Euler method is used in conjunction with a Gauss-Seidel waveform with the same constant stepsizes for both components. Then the waveform iterates can be written as

$$\begin{aligned} y_1^{(k+1)\prime} &= -y_1^{(k+1)} + \lambda y_2^{(k)}, & y_1^{(k+1)}(0) &= y_{10} \\ y_2^{(k+1)\prime} &= -\lambda y_1^{(k+1)} - y_2^{(k+1)}, & y_2^{(k+1)}(0) &= y_{20}, \end{aligned}$$

and an application of the implicit Euler method with $z = \lambda h$ gives, for steps $n = 0, \dots, N-1$

$$\begin{aligned} y_{1,n+1}^{(k+1)} &= \tfrac{1}{1+h} y_{1,n}^{(k+1)} + \tfrac{z}{1+h} y_{2,n+1}^{(k)} \\ y_{2,n+1}^{(k+1)} &= \tfrac{1}{1+h} y_{2,n}^{(k+1)} - \tfrac{z}{1+h} y_{1,n+1}^{(k+1)}. \end{aligned}$$

Let $Y_1^{(k+1)} = (y_{1,1}^{(k+1)}, \dots, y_{1,N}^{(k+1)})^\top, Y_2^{(k+1)} = (y_{2,1}^{(k+1)}, \dots, y_{2,N}^{(k+1)})^\top$; then this can be written in matrix form as

$$QY_1^{(k+1)} = h\lambda Y_2^{(k)} + e \otimes y_{10}, \quad QY_2^{(k+1)} = -h\lambda Y_1^{(k+1)} + e \otimes y_{20},$$

where Q is the $N \times N$ bidiagonal matrix $(-1, 1+h, 0)$.

Hence

$$Y_2^{(k+1)} = -(\lambda h Q^{-1})^2 Y_2^{(k)} - \lambda h Q^{-1}(e \otimes y_{10}) + e \otimes y_{20},$$

and this will converge only if $\lambda h \rho(Q^{-1}) < 1$ or if

$$\frac{\lambda h}{1+h} < 1, \quad \lambda > 0$$

or

$$h < \frac{1}{\lambda - 1}, \quad \lambda > 1. \tag{8.1.4}$$

A similar bound can be found for Jacobi waveform.

On the other hand, if the Euler method is used in conjunction with Gauss-Seidel iteration the following difference scheme is obtained, for steps $n = 0, \dots, N-1$,

$$\begin{aligned} y_{1,n+1}^{(k+1)} &= (1-h) y_{1,n}^{(k+1)} + h\lambda y_{2,n}^{(k)} \\ y_{2,n+1}^{(k+1)} &= -h\lambda y_{1,n}^{(k)} + (1-h) y_{2,n}^{(k+1)}. \end{aligned}$$

This can be written in matrix form as

$$Q_1 Y_1^{(k+1)} = h\lambda Q_2 Y_2^{(k)} + e \otimes y_{10}, \quad Q_1 Y_2^{(k+1)} = -h\lambda Q_2 Y_1^{(k)} + e \otimes y_{20},$$

where Q_1 and Q_2 are the $N \times N$ bidiagonal matrices given by $(h-1, 1, 0)$ and $(1, 0, 0)$, respectively.

Hence

$$Y_2^{(k+2)} = -(h\lambda Q_1^{-1}Q_2)^2 Y_2^{(k)} - h\lambda Q_1^{-1}Q_2Q_1^{-1}(e \otimes y_{10}) + Q_1^{-1}(e \otimes y_{20}),$$

and since $\sigma(Q_1^{-1}Q_2) = \{0\}$, convergence takes place for all positive values of h.

Thus the reverse of the usual situation when implicit or explicit methods are applied to a stiff problem occurs. In the case of the discretized explicit waveform there is no restriction at all on the stepsize to guarantee convergence, while in the case of the implicit method there is a severe stepsize restriction. In fact by considering what happens when the Gauss-Seidel method is applied to the linear system defined in (8.1.2), it can be shown that the spectral radius of the amplification matrix is given by

$$\frac{\lambda h}{1+h}$$

so that convergence will only take place in the static case if

$$h < \frac{1}{\lambda - 1}, \quad \lambda > 1$$

which is the same as the bound in (8.1.4).

This situation is typical of discretized waveform algorithms and has been studied in some depth by White and Sangiovanni-Vincentelli (1987). In particular, they have proved the following results:

Theorem 8.1.1. *If a linear system of differential equations is discretized identically in each component by a Jacobi or Gauss-Seidel waveform relaxation algorithm used in conjunction with a consistent and zero-stable linear multistep method, then the iterations will converge for any initial guess and stepsize sequence if and only if the corresponding relaxation algorithms applied to the ensuing linear system of equations converges for the same method and stepsize sequence.*

Theorem 8.1.2. *If, in addition to the assumptions of Theorem 8.1.1, the waveform relaxation iteration equations are solved by a consistent, zero-stable linear multistep method with a fixed stepsize h, then $\exists H > 0$ such that the sequence $\{y_n^{(k)}\}$ generated by a Gauss-Seidel or Jacobi discretized waveform algorithm will converge for all $0 < h < H$.*

Theorem 8.1.3. *The discretized waveform methods of Theorem 8.1.2 will converge if different but fixed stepsizes are used independently for each subsystem given that the points produced by the numerical method are interpolated linearly.*

Remarks.

1. Odeh and Zein (1983) and White et al. (1985) have shown that the waveform iterates will converge for all positive values of h when solved by the discretized Gauss-Seidel or Jacobi waveform relaxation algorithms used in conjunction with an explicit linear multistep method. On the other hand, explicit methods have, in general, small regions of absolute stability to ensure the boundedness of numerical solutions. In the example described in (8.1.1), the stability constraint for the explicit method is more severe than the waveform constant for the implicit method on the stepsize for all values of the parameter λ. Of course this will not always hold as the constraints are problem and method dependent. In general, however, a waveform relaxation implementation will allow the use of larger stepsizes than a direct explicit implementation (see White et al. (1985) for a fuller description of this).

2. Theorems 8.1.1 and 8.1.2 provide no information on the nature of the waveform stepsize restriction as a function of the waveform algorithm, the method or the problem. In the case of the previous example, the restriction on the stepsize of the implicit Euler, Gauss-Seidel discretized waveform method is of the form $h < 1/L$ where L is a Lipschitz constant for the problem. If this is to hold in general then discretized waveform algorithms would not prove efficacious because the stepsize would be restricted to that needed for an explicit method. This will be investigated later in this chapter, but note that if Q_{21} in (8.1.1) is replaced by $c > 0$ then the eigenvalues of Q are $-1 \pm i\sqrt{c\lambda}$ and the stepsize restriction in (8.1.4) becomes

$$\frac{h}{1+h} < \frac{1}{\sqrt{c\lambda}}$$

or

$$h < \frac{1}{\sqrt{c\lambda}-1}, \quad c\lambda > 1. \tag{8.1.5}$$

White et al. (1985) note that the restriction on the stepsize is, in general, smaller if the system is loosely coupled. For example, if $c \leq \frac{1}{\lambda}$ in the above, there is no restriction on h.

It is well known that differential equation systems that arise from the modelling of integrated circuits have couplings between components which can be strong but which usually exist only for short time intervals. Consequently, waveform relaxation algorithms have proved to be efficient for these problems because small stepsizes are only needed for short windows when the coupling is strong. For the modification of (8.1.1) described above, a splitting based on Gauss-Seidel relaxation leads to a one-sided Lipschitz constant for the first argument given by

$$l = \lambda - 1$$

while a two-sided Lipschitz constant for the second argument is

$$L = c.$$

The coupling ratio (R) is thus

$$R = \frac{c}{\lambda - 1}$$

and for λ large, the coupling ratio gets smaller as c becomes smaller and the restriction on h becomes accordingly less and less severe.

3. The above comments apply for a constant-stepsize implementation. One of the potential advantages of the waveform approach is that it allows for the subsystems to be solved independently with different stepsize controls. However, interpolation schemes are needed to interpolate the discrete sequence of points produced by the integration methods for each subsystem in order to provide waveform iterates for subsequent iterations. The choice of the interpolation scheme is crucial if the convergence properties of the waveform iterates are to be preserved (see White and Sangiovanni-Vincentelli (1987)).

The proof of Theorem 8.1.3 is given in the above monograph, and is based on viewing the multirate discretized waveform method (which allows for different stepsize sequences for each subsystem) along with the linear interpolation operator as a map of continuous functions on $[0, T]$ with an underlying b-norm defined by

$$\| y \|_b = \max_{[0,T]} e^{-bx} \max_i | I_x(y_i(x_j^i)) | .$$

Here I_x denotes an interpolation operator defined on the sequence of discrete integration points x_j^i for each component equation $i = 1, \ldots, m$. Again no comparisons are able to be made between individual methods and White and Sangiovanni-Vincentelli (1987) suggest that this approach may not be generalizable to other interpolation schemes, although they do observe that Theorem 8.1.3 needs no restriction on the ratio of the stepsizes from one system to the next.

8.2 Linear multistep WR methods

In this section an analysis of the convergence behaviour of the class of linear multistep methods is presented when used in conjunction with waveform relaxation algorithms for solving linear problems of the form (7.3.2) with the same stepsize and same method for each component. It will be seen that just as the Laplace transform plays a crucial role in analysing the convergence properties of continuous waveform algorithms, so will the discrete Laplace transform play a crucial role in the discretized waveform case.

Consider first the application of an arbitrary k-step linear multistep method given by (2.6.1) with constant stepsize h to the linear waveform equation

$$y^{(\nu+1)\prime} + My^{(\nu+1)} = Ny^{(\nu)} + g(x), \quad y^{(\nu)}(0) = y_0. \tag{8.2.1}$$

This leads to the following discretization

$$\begin{aligned}\frac{1}{h}(y_{n+1}^{(\nu+1)} - \sum_{j=1}^{k} \alpha_j y_{n+1-j}^{(\nu+1)}) \quad &+ \quad M\sum_{j=0}^{k} \beta_j y_{n+1-j}^{(\nu+1)} \\ &= \quad \sum_{j=0}^{k} \beta_j (Ny_{n+1-j}^{(\nu)} + g_{n+1-j}).\end{aligned} \tag{8.2.2}$$

Let $\tilde{y}^{(\nu)} = \{\tilde{y}_n^{(\nu)}\}_{n=0}^{\infty}$ and $\tilde{y} = \{y_n\}_{n=0}^{\infty}$ denote the discretized approximation vector at the νth iterate and the discretized approximation vector for the underlying differential equation (7.3.1), respectively. Then the error vector given by

$$\varepsilon^{(\nu)} = \tilde{y}^{(\nu)} - \tilde{y} = \{y_n^{(\nu)} - y_n\}_{n=0}^{\infty} \tag{8.2.3}$$

will satisfy (8.2.2) with $g \equiv 0$.

Hence, analogously to the continuous case

$$\varepsilon^{(\nu+1)} = K_h \varepsilon^{(\nu)} \tag{8.2.4}$$

where K_h denotes the discrete convolution operator.

Now given a set of discrete values $\{u_j\}_{j=0}^{\infty}$, the **discrete convolution operator** is defined by

$$K_h u_j = \sum_{i=0}^{j} k_{j-i} u_i, \quad j \geq 0$$

and $\{k_j\}_{j=0}^{\infty}$ can be interpreted as the **kernel** of the discrete convolution operator. This kernel can be calculated from the **discrete Laplace transform** $K_h(w)$ given by

$$K_h(w) = \sum_{j=0}^{\infty} k_j w^j. \tag{8.2.5}$$

Now from (8.2.4) and (8.2.5) it is easily seen that

$$K_h(w) \quad = \quad (\frac{1}{h}I + \sum_{j=0}^{k} \beta_j w^j M - \frac{1}{h}\sum_{j=1}^{k} \alpha_j w^j I)^{-1} \sum_{j=0}^{k} \beta_j w^j N$$

$$= \left(\frac{1}{h}\left(\frac{1-\sum_{j=1}^{k}\alpha_j w^j}{\sum_{j=0}^{k}\beta_j w^j}\right)I + M\right)^{-1} N$$

$$= \left(\frac{1}{h}\frac{\rho(w)}{\sigma(w)}I + M\right)^{-1} N, \tag{8.2.6}$$

where $\rho(w)$ and $\sigma(w)$ are given in (3.3.5). Hence from (7.4.10)

$$K_h(w) = K\left(\frac{1}{h}\frac{\rho(w)}{\sigma(w)}\right), \tag{8.2.7}$$

where $K(w)$ denotes the continuous Laplace transform, and where it is assumed that $\rho(w)$ and $\sigma(w)$ have no common factors.

Example 8.2.1. The discrete Laplace transforms (in terms of the continuous Laplace transform K) for both the trapezoidal rule and the family of k-step BDF methods are, respectively,

$$K\left(\frac{2}{h}\frac{1-w}{1+w}\right), \quad K\left(\frac{1}{h}(1+\sum_{j=1}^{k}\frac{1}{j}(1-w)^j)\right).$$

Since a k-step linear multistep method requires k initial approximations to commence the implementation, the assumption has been made in the above analysis that there is no iterating of the k given starting values which are denoted by negative subscripts. Thus it is assumed that $\varepsilon_j^{(\nu)} = 0, \quad j = -1, \ldots, -k$. In fact Miekkala and Nevanlinna (1987a) have given a different analysis to the one presented above which does not rely on the use of the discrete Laplace transform and have analysed the effect of the initial errors on the rate of convergence of the iterates.

It should be also noted that $K_h(w)$ can be written as

$$K_h(w) = (\frac{1}{h}\rho(w)I + \sigma(w)M)^{-1}\sigma(w)N$$

and that K_h will be bounded in $l_p \quad (1 \le p \le \infty)$ (see Miekkala and Nevanlinna (1987a), for example) if $K_h(w)$ is bounded for $|w| \ge 1$, or equivalently, whenever

$$\det\,(\rho(w)I + h\sigma(w)M) \neq 0 \text{ for } |w| \ge 1.$$

Suppose there exists w_1 with $|\, w_1 \,| \ge 1$ such that

$$\det\,(\rho(w_1)I + h\sigma(w_1)M) = 0.$$

Then $\sigma(w_1) \neq 0$ since ρ and σ have no common factors, and so

$$\det\left(\frac{\rho(w_1)}{\sigma(w_1)}I + hM\right) = 0.$$

Hence

$$\frac{\rho(w_1)}{\sigma(w_1)} \in \sigma(-hM). \tag{8.2.8}$$

Now with the stability region of a linear multistep method defined as

$$S = \{z \in \mathbb{C} : \rho(w) - z\sigma(w) \text{ satisfies the root condition}\},$$

(8.2.8) implies

$$\sigma(-hM) \not\subset \operatorname{int}(S),$$

and thus the following result is proved:

Theorem 8.2.1. *If $\sigma(-hM) \subset \operatorname{int}(S)$ then K_h is bounded in l_p.*

Both Miekkala and Nevanlinna (1987a) (using the Wiener lemma) and Lubich and Ostermann (1987) (using the discrete version of the Paley-Wiener theorem (Zygmund (1959)) note that

$$\rho(K_h) = \max_{|w|\geq 1} \rho(K_h(w)) = \max_{|w|\geq 1} \rho(K(\frac{1}{h}\frac{\rho(w)}{\sigma(w)})). \tag{8.2.9}$$

By also noting (see Miekkala and Nevanlinna (1987a)) that

$$\mathbb{C} - \operatorname{int}(S) = \{\frac{\rho(w)}{\sigma(w)} : |w| \geq 1\}$$

then

Theorem 8.2.2. *If $\sigma(-hM) \subset \operatorname{int}(S)$ then*

$$\rho(K_h) = \sup\{\rho(K(z)) : hz \in \mathbb{C} - \operatorname{int}(S)\}$$

A consequence of these results is the following:

Theorem 8.2.3. *If K is the integral operator on $L^p(\mathbb{R}^+, \mathbb{C}^m)\,(1 \leq p \leq \infty)$ associated with the continuous waveform case and K_h is the discretized operator associated with the application of an $A(\alpha)$-stable linear multistep method with constant stepsize h then*

$$\rho(K_h) \leq \max_{z \in S_\alpha} \rho(K(z)), \quad S_\alpha = \{z : |\arg(z)| \leq \pi - \alpha\} \cup \{0\}. \tag{8.2.10}$$

Theorem 8.2.4. *If in Theorem 8.2.3, a linear multistep method is A-stable then*

$$\rho(K_h) \leq \rho(K) = \max_{Re(z)\geq 0} \rho(K(z)). \tag{8.2.11}$$

Remarks.

1. A similar approach to the one outlined above has been given for discretized multigrid waveform algorithms based on Gauss-Seidel dynamic iteration and $A(\alpha)$-stable linear multistep methods by Lubich and Ostermann (1987).
2. Theorems 8.2.3 and 8.2.4 represent a generalization of Theorem 8.1.1 (due to White and Sangiovanni-Vincentelli (1987)) to the stiff case (or infinitely long windows) and show that if a bound can be placed on the convergence rate of the continuous waveform algorithm, then, under appropriate stability considerations, the discretized waveform will converge at a faster rate.
3. Lubich and Ostermann (1987) have remarked on the importance of a numerical method being A-stable or $A(\alpha)$-stable for large values of α within a WR scheme. They note that for the two-level dynamic multigrid algorithm using Gauss-Seidel smoothing the quantity

$$C_\alpha = \max_{z \in S_\alpha} \left\{ \frac{z}{(1+z)^{2\nu}} \right\}$$

will determine the convergence rate if the discretization is performed with an $A(\alpha)$-stable linear multistep method. The following table for BDF methods of orders 1 to 6 is given in Lubich and Ostermann (1987).

Table 8.1 *BDF convergence rates*

order	α	$\nu : 1$	2	3	4
1, 2	90°	0.5	0.325	0.259	0.222
3	88°	0.518	0.345	0.280	0.243
4	73°	0.707	0.6	0.585	0.596
5	51°	1.349	2	3.173	5.133
6	18°	10.2	105	1101	11512

Although these figures only represent upper bounds on the convergence rates, they do suggest that lower-order $A(\alpha)$-stable methods are preferred over higher-order methods within WR schemes.

4. By assuming that the initial values satisfy

$$\| y_n^{(\nu)} - y_n \| \leq \gamma \theta^\nu, \quad n < k, \quad \theta \leq \rho(K_h),$$

Miekkala and Nevanlinna (1987a) have shown that $\rho(K_h)$ will always determine the convergence rate of the iterates given that

$$\sigma(-hM) \cup \sigma(-hQ) \subset \text{ int } (S).$$

Miekkala and Nevanlinna (1987a) have presented a number of other useful results which are consequences of the analysis outlined above. Before stating these results the following definitions are given:

Definition 8.2.5. *A linear multistep method is said to be strongly stable if all roots of ρ are less than one in modulus apart from the consistency root which equals one.*

Definition 8.2.6. *A linear multistep method has order of amplitude fitting equal to p if the principal root $w_1(z)$ of $\rho(w) - z\sigma(w) = 0$ satisfies $\mid w_1(it) \mid -1 = O(t^{p+1})$ for small real t.*

Theorem 8.2.7. *If a linear multistep method is strongly stable with order of amplitude fitting equal to p then there exists $H > 0$ such that*

$$\rho(K_h) = \rho(K) + O(h^p), \quad \forall h \leq H.$$

Theorem 8.2.8. *If the corresponding static iteration converges and $\infty \in \text{int}(S)$, then there exists H such that the discretized dynamic iteration converges for all $h \geq H$ and*

$$\rho(K_h) = \rho(M^{-1}N) + O(1/h) \text{ as } h \to \infty.$$

Theorem 8.2.9. *If the discretized dynamic iteration converges then $\sigma(-hQ) \subset \text{ int } (S)$.*

A consequence of Theorem 8.2.8 is that

$$\rho(K_h) \geq \rho(M^{-1}N) \tag{8.2.12}$$

so that discretized dynamic iteration can never converge faster than the static iteration. Theorem 8.2.9 is significant in that if Q represents a stiff problem then the iterations will not converge if M is a nonstiff matrix. Hence a nonstiff solver would be inappropriate as the basis of a WR scheme.

8.3 Runge-Kutta WR methods

Consider the s-stage Runge-Kutta method for solving (2.1.1) given by

$$\begin{aligned} Y_{n,i} &= y_n + h\sum_{j=1}^{s} a_{ij} f(Y_{n,j}), \quad i = 1, \dots, s \\ y_{n+1} &= y_n + h\sum_{j=1}^{s} b_j f(Y_{n,j}). \end{aligned}$$

If this is applied to the linear waveform equation (8.2.1), then with

$$\begin{aligned} \varepsilon_{n,j}^{(\nu)} &= Y_{n,j}^{(\nu)} - Y_{n,j}, \quad j = 1, \dots, s \\ \varepsilon_{n+1}^{(\nu)} &= y_{n+1}^{(\nu)} - y_{n+1} \end{aligned}$$

defining the waveform errors of the internal approximations and the updated approximation, respectively, a recursive system of equations for these errors can be obtained (for $i = 1, \dots, s$) of the form

$$\begin{aligned} \varepsilon_{n,i}^{(\nu+1)} &= \varepsilon_n^{(\nu+1)} + h\sum_{j=1}^{s} a_{ij}(-M\varepsilon_{n,j}^{(\nu+1)} + N\varepsilon_{n,j}^{(\nu)}) \\ \varepsilon_{n+1}^{(\nu+1)} &= \varepsilon_n^{(\nu+1)} + h\sum_{j=1}^{s} b_j(-M\varepsilon_{n,j}^{(\nu+1)} + N\varepsilon_{n,j}^{(\nu)}), \end{aligned} \tag{8.3.1}$$

where $\varepsilon_0^{(\nu+1)} = 0$.

Let

$$E_n = (\varepsilon_{n1}, \dots, \varepsilon_{ns})^\top$$

and introduce the generating power series

$$\begin{aligned} E(w) &= \sum_{j=0}^{\infty} E_j w^j \\ \varepsilon(w) &= \sum_{j=0}^{\infty} \varepsilon_j w^j; \end{aligned}$$

then from (8.3.1)

$$E^{(\nu+1)}(w) = e \otimes \varepsilon^{(\nu+1)}(w) - hA \otimes (ME^{(\nu+1)}(w) - NE^{(\nu)}(w)) \tag{8.3.2}$$

$$\frac{1}{w}\varepsilon^{(\nu+1)}(w) = \varepsilon^{(\nu+1)}(w) - hb^\top \otimes (ME^{(\nu+1)}(w) - NE^{(\nu)}(w)). \tag{8.3.3}$$

Solving for $\varepsilon^{(\nu+1)}(w)$ in (8.3.3) and substituting into (8.3.2) gives after some simplification

$$(I + h\mathcal{A} \otimes M)E^{(\nu+1)}(w) = h\mathcal{A} \otimes NE^{(\nu)}(w),$$

where

$$\mathcal{A} = A + \frac{w}{1-w} eb^\top. \tag{8.3.4}$$

Hence

$$E^{(\nu+1)}(w) = (\frac{1}{h}I + \mathcal{A} \otimes M)^{-1}\mathcal{A} \otimes NE^{(\nu)}(w), \tag{8.3.5}$$

assuming the inverse exists, which it will if $Re(\sigma(M)) \geq 0$ and the eigenvalues of $\mathcal{A}$ have nonnegative real part for $\mid w \mid \leq 1$ with $w \neq 1$ (see below).

Now assuming that $\mathcal{A}^{-1}$ exists, (8.3.5) and (8.3.3) imply

$$\varepsilon^{(\nu+1)}(w) = \frac{w}{1-w} b^{\top} \mathcal{A}^{-1} \otimes E^{(\nu+1)}(w). \tag{8.3.6}$$

Using the Shermann-Morrison formula for rank-one updates

$$\mathcal{A}^{-1} = A^{-1} - \frac{\frac{w}{1-w}(A^{-1}e)(b^{\top}A^{-1})}{1 + \frac{w}{1-w} b^{\top} A^{-1} e} \tag{8.3.7}$$

and hence with

$$\theta = b^{\top} A^{-1} e, \tag{8.3.8}$$

(8.3.6), (8.3.7) and (8.3.8) give

$$\varepsilon^{(\nu+1)}(w) = \frac{w}{1 + w(\theta - 1)} b^{\top} A^{-1} \otimes E^{(\nu+1)}(w). \tag{8.3.9}$$

Note that in the case of a stiffly accurate method,

$$b^{\top} A^{-1} = e_s^{\top}$$

and

$$\varepsilon^{(\nu+1)}(w) = w e_s^{\top} \otimes E^{(\nu+1)}(w). \tag{8.3.10}$$

As in (8.2.4), let K_h be the discrete convolution operator such that

$$E^{(\nu+1)} = K_h E^{(\nu)}, \quad E^{(\nu)} = \{E_n^{(\nu)}\}_{n=0}^{\infty} \tag{8.3.11}$$

where, given a set of discrete values $\{u_j\}_{j=0}^{\infty}$,

$$K_h u_j = \sum_{i=0}^{j} k_{j-i} u_i, \quad j \geq 0$$

and $\{k_j\}_{j=0}^{\infty}$ represents the kernel of K_h.

In this case, (8.3.5) and (8.3.11) give

$$\begin{aligned} K_h(w) &= \left(\frac{1}{h} I + \mathcal{A} \otimes M\right)^{-1} \mathcal{A} \otimes N. \\ &= \left(\frac{1}{h} \mathcal{A}^{-1} \otimes I + I \otimes M\right)^{-1} I \otimes N \\ &= K\left(\frac{1}{h} \mathcal{A}^{-1}\right), \end{aligned} \tag{8.3.12}$$

where $K(z)$ represents the continuous Laplace transform for the splitting satisfying $K(z) = (zI + M)^{-1} N$ with corresponding operator K satisfying

$$\rho(K) = \max_{Re(z) \geq 0} \rho(K(z)). \tag{8.3.13}$$

Here K can be interpreted as an operator on $L^p(\mathbb{R}^+, \mathbb{C}^m)$ while K_h is an operator on $l^p(\mathbb{N}, \mathbb{C}^{ms})$ for $1 \leq p \leq \infty$.

For a vector, $u = (u_1^\top, ..., u_s^\top)$, $u_i \in \mathbb{C}^m$, a norm can be placed on $\mathbb{C}^{ms}$ of the form

$$\| u \|_b = \sum_{j=1}^{s} b_j \| u_j \|^2,$$

where $\| \cdot \|^2$ is some inner product norm on $\mathbb{C}^m$, assuming that

$$b_j > 0, \quad j = 1, ..., s. \tag{8.3.14}$$

(This of course holds only for the case $p = 2$, but it is sufficient to consider this case without loss of generality.)

In fact from (8.3.4)

$$\mathcal{A} = \hat{A} + \frac{1}{2}\frac{1+w}{1-w} eb^\top, \quad \hat{A} = A - \frac{1}{2} eb^\top. \tag{8.3.15}$$

Now by defining an inner product on $\mathbb{C}^s$ by

$$\langle u, v \rangle = \sum_{i=1}^{s} b_i u_i \bar{v}_i, \quad u, v \in \mathbb{C}^s$$

it is clear that

$$\langle eb^\top v, v \rangle = \left| \sum_{j=1}^{s} b_j v_j \right|^2$$

and

$$\langle (\hat{A} + \hat{A}^\top) v, v \rangle = \sum_{i,j=1}^{s} (b_i a_{ij} + b_j a_{ji} - b_i b_j) v_i \bar{v}_j.$$

Now the assumption of algebraic stability implies that $eb^\top$ is positive semi-definite and $\hat{A} + \hat{A}^\top$ is positive semi-definite. This together with the fact that

$$Re\left(\frac{1+w}{1-w}\right) \geq 0, \text{ for } |w| \leq 1, \quad w \neq 1$$

implies from (8.3.15)

$$Re\langle \mathcal{A} v, v \rangle \geq 0, \text{ for } v \in \mathbb{C}^s \text{ and } |w| \leq 1, \quad w \neq 1,$$

and the eigenvalues of $\mathcal{A}$ have nonnegative real part.

Hence, analogously to the proof of Theorems 8.2.2 and 8.2.3, (8.3.13) implies

$$\rho(K_h) = \sup_{|w| \leq 1} \rho\left(K\left(\frac{1}{h}\mathcal{A}^{-1}\right)\right) \leq \sup_{Re(z) \geq 0} \rho(K(z)) = \rho(K)$$

and the following result holds.

Theorem 8.3.1. *For any algebraically stable Runge-Kutta method with positive stepsize h both $\rho(K_h) \leq \rho(K)$ and $\| K_h \| \leq \| K \|$ hold.*

Remarks.

1. This result generalizes a result presented in Lubich and Ostermann (1987) for the discretized multigrid waveform algorithm based on a Gauss-Seidel splitting to arbitrary splittings.
2. Theorem 8.3.1 can be generalized to more general classes of methods such as multivalue methods under the assumption that these methods are algebraically stable (see Burrage and Butcher (1980)). Note the consistency of this result and the corresponding result for A-stable linear multistep methods when it is remembered that A-stability and algebraic stability are equivalent for linear multistep methods.

If an s-stage Runge-Kutta method is applied to (8.2.1) then it can be seen that

$$Y_n^{(\nu+1)} = R_h Y_n^{(\nu)} + g,$$

where $Y_n^{(\nu+1)} = (Y_{n,1}^{(\nu+1)}, \dots, Y_{n,s}^{(\nu+1)})^\top$ and R_h is the amplification matrix given by

$$R_h = (I + hA \otimes M)^{-1} hA \otimes N. \tag{8.3.16}$$

Now $\rho(R_h)$ controls the rate of convergence of the discretized waveform iterates with the smaller the $\rho(R_h)$ the faster the waveforms converge. In the case that a Runge-Kutta method is semi-implicit then A is lower triangular with $\lambda_1, ..., \lambda_s$ on the diagonal and so R_h is a block lower triangular matrix whose diagonal blocks are $(I+h\lambda_j M)^{-1} h\lambda_j N, \quad j = 1, ..., s$. Consequently,

$$\rho(R_h) = \max_{j=1,\dots,s} \rho\left(\left(\frac{1}{\lambda_j} I + hM\right)^{-1} hN\right) \tag{8.3.17}$$

and this can be used to compare the rates of convergence of different discretized Runge-Kutta waveform algorithms. In particular, if $\mathcal{R}_1$ and $\mathcal{R}_2$ represent two different semi-implicit Runge-Kutta methods with elements $(\lambda_1, ..., \lambda_s)$ and $(\mu_1, ..., \mu_s)$ on the diagonal, respectively, and assuming that each of these diagonal elements is non-negative and M has eigenvalues with positive real part, then in a number of important cases

$$\rho(R_{1h}) < \rho(R_{2h})$$

if

$$\max_{j=1,\dots,s} \{\lambda_j\} < \max_{j=1,\dots,s} \{\mu_j\}. \tag{8.3.18}$$

This will happen, for example, in the case of Richardson iteration or if M is diagonalizable and Q is triangularizable under the same similarity transformation.

This condition gives a criterion for selecting between different semi-implicit Runge-Kutta methods in a waveform relaxation algorithm. For example, one consequence of (8.3.18) is that, in general, the trapezoidal rule discretized waveform algorithm will converge faster than the implicit Euler discretized waveform algorithm since the respective elements in (8.3.18) satisfy

$$\max\{0, \frac{1}{2}\} \leq \max\{1\}.$$

The trapezoidal method has other advantages in that it is order 2 while the implicit Euler method is order 1 and their implementation costs are essentially the same.

Similarly, it is known (see Burrage (1978a)) that a necessary condition for a two-stage SDIRK of order 2 with λ on the diagonal to be A-stable is

$$\lambda \geq 1/4.$$

Such an SDIRK with $\lambda = \frac{1}{4}$ in conjunction with waveform relaxation will converge faster than the trapezoidal approach for these important cases previously mentioned.

In fact for DIRKs there is a very natural relationship between $\rho(R_h)$ and $\rho(K)$, given in (7.4.12) for the continuous waveform case. Thus, for example, if M and N are such that the continuous waveform converges then clearly $\rho(R_h) < 1$ for all $h > 0$ for all DIRKs with nonnegative diagonal elements. This was first pointed out by Skålin (1988). In fact Skålin (1988) has exploited this natural relationship to generalize most of the convergence results of Miekkala and Nevanlinna (1987a) for the continuous waveform case to discretized waveform relaxation based on DIRKs. Thus there is a dynamic DIRK counterpart of Theorems 7.5.1, 7.5.3, 7.5.5 and 7.5.6 and Corollary 7.5.2 and 7.5.4. Skålin (1988) has also given a waveform implementation of SIMPLE on both a Cray and a shared-memory parallel machine (Alliant).

Bjørhus (1990) has considered an extension of this work to second-order problems (see section 7.12) when solved by discretized waveform methods based on diagonally implicit Runge-Kutta Nyström methods.

Bjørhus (1994) has also obtained error estimates for dynamic iteration when either the Euler or implicit Euler methods are used to compute the waveform iterates. The proofs are exact analogues of the proof for the continuous case. In particular, if the splitting F is globally Lipschitz continuous with Lipschitz constants K_1 and K_2 then an error estimate for the Euler method is given by

$$\begin{aligned} ||E^{(k)}||_N &\leq \left(\tfrac{hK_2}{1+hK_1}\right)^k \binom{N-1}{k} (1+hK_1)^{N-1}||E^{(0)}||_{N-k}, \quad k < N \\ ||E^{(k)}||_N &= 0, \quad k \geq N. \end{aligned}$$

The corresponding error estimate for the implicit Euler method for $h < \frac{1}{K_1+K_2}$ is given by

$$||E^{(k)}||_N \leq \left(\frac{hK_2}{1-hK_1}\right)^k \binom{N+k-1}{k} (1-hK_1)^{1-N}||E^{(0)}||_N.$$

Here

$$||Z||_N = \max_{0\leq n\leq N} ||Z_n||,$$

where $||\cdot||$ is some norm on $\mathbb{R}^m$.

8.4 Discretized multisplittings

The results presented in the previous two sections generalize naturally to the case of a multisplitting (M_l, N_l, E_l) of Q. For example, in the case of the application of a linear multistep with constant stepsize h to L different subsystems then, analogously to (8.2.4),

$$\varepsilon_l^{(\nu+1)} = K_{l,h}\varepsilon_l^{(\nu)}, \quad l = 1, ..., L,$$

where $K_{l,h}$ denotes the discrete convolution operator for the lth subsystem.

If K_h denotes the discrete convolution operator for the complete system and $K_h(w)$ its discrete Laplace transform then from (8.2.6) it is easily seen that

$$\begin{aligned} K_h(w) &= \sum_{l=1}^{L} E_l K_{l,h}(w) \\ &= \sum_{l=1}^{L} E_l \left(\frac{1}{h}\frac{\rho(w)}{\sigma(w)} I + M_l\right)^{-1} N_l \\ &= K\left(\frac{1}{h}\frac{\rho(w)}{\sigma(w)}\right) \end{aligned}$$

where $K(w)$ denotes the continuous Laplace transform for the multisplitting case.

Using the same argument as used in section 8.2 to show that $K_{l,h}(w)$ will be bounded for $| w |\geq 1$ if

$$\sigma(-hM_l) \subset \text{int } (S)$$

the multisplitting analogues of Theorems 8.2.1, 8.2.2, 8.2.3 and 8.2.4 can be proved. Similar remarks hold for algebraically stable Runge-Kutta methods.

Pohl (1993) and Frommer and Pohl (1993) have attempted to compare the convergence rates of discretized overlapping waveform algorithms and discretized nonoverlapping waveform algorithms. However, they have not been able to obtain a comparison result for infinitely long time windows because Theorem 7.7.6 is not necessarily valid when comparing two weak regular splittings.

However, by using r equidistant time points separated by distant h, it can be shown (see Pohl (1993), for example) that the iteration of the resulting multisplitting waveform algorithm is equivalent to linear fixed-point iteration on $\mathbb{R}^{rm}$. In the case of a linear multistep method, the iteration matrix is given by

$$H = \sum_{l=1}^{L} E_l \left(\frac{1}{\beta_0 h} I + M_l \right)^{-1} N_l.$$

If now an overlapping splitting $(\tilde{M}_l, \tilde{N}_l, E_l)$ is used with the same weighting matrices then

$$\tilde{H} = \sum_{l=1}^{L} E_l \left(\frac{1}{\beta_0 h} + \tilde{M}_l \right)^{-1} \tilde{N}_l.$$

Hence Frommer and Pohl (1993) have shown that if the E_l have the 0-1 weighting property then

$$\rho(\tilde{H}) \leq \rho(H),$$

if $\beta_0 > 0$, Q is an M-matrix, (M_l, N_l, E_l) and $(\tilde{M}_l, \tilde{N}_l, E_l)$ are regular and weak regular splittings of Q, respectively, and the (M_l, N_l, E_l) splittings represent a block Jacobi or block Gauss-Seidel multisplitting.

This result has been numerically verified in conjunction with a block Jacobi waveform relaxation algorithm and an equal-stepsize implementation of the implicit Euler method on the semi-discretized, two-dimensional heat equation defined on the unit square under the standard five-point central differencing operator.

8.5 Mesh refinement and the nonlinear problem

Although the results obtained in the previous section are pleasing from a theoretical viewpoint, they are deficient in that they only apply for linear problems and a constant-stepsize discretization implementation. Nevanlinna (1989) has considered a fixed-stepsize refining strategy for the standard linear problem.

The split linear system can in the continuous case be written in fixed-point form as

$$y = Ky + \phi, \tag{8.5.1}$$

while the discrete analogue is

$$y_h = K_h y_h + \phi_h. \tag{8.5.2}$$

Suppose now the discrete analogue is computed on a set of grids $\Delta = \{\Delta_0, \Delta_1, \ldots\}$, where each Δ_j is an equidistant grid given by

$$\Delta_j = \{0, h_j, 2h_j \ldots, n_j h_j = T\}.$$

Then Nevanlinna (1989) introduces a restriction operator, an interpolation operator and a prolongation operator in order to study the convergence of (8.5.2) on Δ. Given that the underlying linear multistep method is of order p, a similar convergence property on the restriction operator and a similar accuracy requirement on the interpolation operator is needed in order to get meaningful error bounds on the maximum error at the grid points.

Thus Nevanlinna (1989) defines the restriction operator R_h by

$$R_h y_j = y(jh) \tag{8.5.3}$$

and assumes

$$||R_h y - y_h||_T \leq c_1 h^p ||y^{(p+1)}||_T. \tag{8.5.4}$$

Nevanlinna (1989) also defines an interpolation operator Q_h mapping grid functions, on the space X_h, to continuous functions by

$$Q_h : X_h \rightarrow C[0, T]$$

and a prolongation operator $P_j : X_{\Delta_{j-1}} \rightarrow X_{\Delta_j}$ by

$$P_j = R_{\Delta_j} Q_{\Delta_{j-1}}, \tag{8.5.5}$$

satisfying

$$||y - Q_h R_h y||_T \leq c_2 h^p ||y^{(p)}||_T. \tag{8.5.6}$$

In addition to these assumptions, Nevanlinna (1989) also assumes

$$\| P_j y_{\Delta_{j-1}} \|_T \leq D \| y_{\Delta_{j-1}} \|_{T+Lh_{j-1}} \tag{8.5.7}$$

and that the kernel $\{k_{h,j}\}_{j=0}^{\infty}$ of K_h associated with the discretization method satisfies

$$||k_{h,j}|| \leq Ch, \quad j \geq 0, \quad h \leq h_0. \tag{8.5.8}$$

The iterations on Δ can be written as

$$y^{(\nu)} = K_{h_\nu} P_\nu y^{(\nu-1)} + \phi^{(\nu)}. \tag{8.5.9}$$

In the case of a one-step method $\phi^{(\nu)} = \phi_{h_\nu}$, while for a linear multistep method a starting process is needed which converges at least at the same rate as (8.5.9), with

$$||\phi_{h_\nu} - \phi^{(\nu)}||_T \leq c_4 h_\nu^p. \tag{8.5.10}$$

Now since (8.5.2) and (8.5.9) imply

$$y_{h_\nu} - y^{(\nu)} \quad = \quad K_{h_\nu}(y_{h_\nu} - P_\nu y^{(\nu-1)}) + \phi_{h_\nu} - \phi^{(\nu)}$$

$$= \quad K_{h_\nu}(y_{h_\nu} - y^{(\nu)}) + K_{h_\nu}(y_{h_\nu} - P_\nu y_{h_{\nu-1}}) + (\phi_{h_\nu} - \phi^{(\nu)}),$$

this can be written as

$$Y_\nu = K_{h_\nu} P_\nu Y_{\nu-1} + K_{h_\nu}\eta_\nu + \xi_\nu, \tag{8.5.11}$$

where

$$Y_\nu = y_{h_\nu} - y^{(\nu)}, \quad \eta_\nu = y_{h_\nu} - P_\nu y_{h_{\nu-1}}, \quad \xi_\nu = \phi_{h_\nu} - \phi^{(\nu)}.$$

But from (8.5.5)

$$y_{h_\nu} - P_\nu y_{h_{\nu-1}} = y_{h_\nu} - R_{h_\nu}y + R_{h_\nu}y - R_{h_\nu}Q_{h_{\nu-1}}R_{h_{\nu-1}}y + P_\nu(R_{h_{\nu-1}}y - y_{h_{\nu-1}})$$

so that

$$\begin{aligned} ||y_{h_\nu} - P_\nu y_{h_{\nu-1}}||_T \quad &\leq \quad ||y_{h_\nu} - R_{h_\nu}y||_T \\ +||R_{h_\nu}y - R_{h_\nu}Q_{h_{\nu-1}}R_{h_{\nu-1}}y||_T &+ ||P_\nu(R_{h_{\nu-1}}y - y_{h_{\nu-1}})||_T. \end{aligned}$$

Using bounds in (8.5.4), (8.5.6) and (8.5.7) and after applying K_{h_ν} then

$$||\eta_\nu|\,|_T \leq (CT)c_5 h^p_{\nu-1}, \tag{8.5.12}$$

where c_5 can be bounded in terms of the smoothness of y.

By introducing another operator $B_{\nu\mu}$ as

$$\begin{aligned} B_{\nu j} \quad &= \quad K_{h_\nu}P_\nu B_{\nu-1j} \\ B_{jj} \quad &= \quad \text{identity on grid functions on } \Delta_j, \end{aligned}$$

it can be shown from (8.5.11) (see Nevanlinna (1989)) that

$$Y_\nu = B_{\nu_0}Y_0 + \sum_{j=1}^{\nu} B_{\nu_j}(\eta_j + \xi_i)$$

and by bounding $||B_{\nu_j}||_T$, it can be shown that the maximum error at the grid points satisfies

$$\begin{aligned} ||R_{h_\nu}y - y^{(\nu)}||_T \quad \leq \quad & \tfrac{(CD\widetilde{T})^\nu}{\nu!}||y_{h_0} - y^{(0)}||_T \\ & + \left(c_1 + \left(c_4 + c_5\tfrac{CT}{\delta^p}\right)e^{CD\widetilde{T}/\delta^p}\right)h_0^p\delta^{p\nu}, \end{aligned} \tag{8.5.13}$$

with

$$h_\nu = \delta^\nu h_0, \quad \delta < 1, \quad \widetilde{T} = T + h_0\frac{L+1}{L-\delta}. \tag{8.5.14}$$

Equation (8.5.13) represents a generalization of Theorem 7.4.2 in a mesh refinement setting, but of course for a stiff problem the bound in (8.5.13) may not be particularly realistic. Nevanlinna (1989) has used the formulation in (8.5.13) to give an automatic mesh refinement and adaptive windowing scheme.

This analysis is of a constant-stepsize nature because a variable-stepsize analysis would add an extra degree of complexity of either a variable coefficient implementation or interpolation of past values in the linear multistep scheme. On the other hand, the fact that stage order and order of consistency are equivalent for linear multistep methods means that it is trivial to construct continuous approximations over the region of integration whose order is equal to the order of the underlying method. In the case of multistage methods, however, this is a non-trivial process.

Before considering the influence of multistage methods with continuous interpolants on the waveform convergence, a more general analysis than that given above for a general nonlinear splitting of (2.1.2) is presented, namely

$$y^{(k+1)'}(x) = F(x, y^{(k+1)}(x), y^{(k)}(x)), \quad y^{(k+1)}(0) = y_0. \qquad (8.5.15)$$

This is a generalization of a process described in Bellen and Tagliaferro (1991).

Thus assume that the discretization process has a continuous interpolant of order p and is applied with the same mesh refinement strategy described above on a window $[0, T]$.

Now if the discretized approximations $y_h^{(k+1)}$ for a fixed stepsize h satisfy

$$u^{(k+1)'}(x) = F(x, u^{(k+1)}(x), y_h^{(k)}(x)) \qquad (8.5.16)$$

and have an associated continuous interpolant of order p then

$$||y_h^{(k+1)} - u^{(k+1)}||_x \leq \theta x (h\delta^k)^p, \qquad (8.5.17)$$

where δ is the mesh refinement factor and where (8.5.17) is interpreted to hold for variable x.

Noting that

$$u^{(k+1)'} - y' = F(x, u^{(k+1)}, y_h^{(k)}) - F(x, u^{(k+1)}, y) + F(x, u^{(k+1)}, y) - F(x, y, y)$$

so that

$$\langle u^{(k+1)'} - y', u^{(k+1)} - y\rangle = \langle F(x, u^{(k+1)}, y_h^{(k)}) - F(x, u^{(k+1)}, y), u^{(k+1)} - y\rangle$$

$$+ \langle F(x, u^{(k+1)}, y) - F(x, y, y), u^{(k+1)} - y\rangle \qquad (8.5.18)$$

and letting

$$\varepsilon^{(k+1)} = u^{(k+1)} - y$$

(8.5.18) implies

$$\langle \varepsilon^{(k+1)'}, \varepsilon^{(k+1)} \rangle \leq L_2 \parallel \varepsilon^{(k+1)} \parallel \parallel y_h^{(k)} - y \parallel + l_1 \parallel \varepsilon^{(k+1)} \parallel^2 . \qquad (8.5.19)$$

Here l_1 is the one-sided Lipschitz constant associated with the second argument of the non-autonomous splitting function and L_2 is the two-sided Lipschitz constant associated with the third argument.

Under the assumption that $\parallel \varepsilon^{(k+1)} \parallel \neq 0$ and with $v^{(k)}(x) = \parallel \varepsilon^{(k)}(x) \parallel$, a positive function, and $w^{(k)}(x) = \parallel y_h^{(k)} - y \parallel$, (7.9.8) and (8.5.19) imply

$$v^{(k+1)'} \leq l_1 v^{(k+1)} + L_2 w^{(k)}. \qquad (8.5.20)$$

But since, from (8.5.17),

$$w^{(k)} \leq v^{(k)} + g_k, \quad g_k(x) = \theta x (h\delta^{k-1})^p, \quad k \geq 1, \quad g_0 = 0, \qquad (8.5.21)$$

(8.5.20) gives

$$v^{(k+1)'} \leq l_1 v^{(k+1)} + L_2 v^{(k)} + L_2 g_k, \quad v^{(k+1)}(0) = 0. \qquad (8.5.22)$$

Hence

$$v^{(k+1)}(x) \leq L_2 e^{l_1 x} \int_0^x e^{-l_1 s}(v^{(k)}(s) + g_k(s)) ds, \qquad (8.5.23)$$

and

$$v^{(1)}(x) \leq L_2 e^{l_1 x} \int_0^x e^{-l_1 s} ds ||v^{(0)}||_x$$

or

$$v^{(1)}(x) \leq \left(\frac{L_2}{-l_1}\right)(1 - e^{l_1 x}) ||v^{(0)}||_x . \qquad (8.5.24)$$

Substituting (8.5.24) into (8.5.23), with $k = 1$,

$$v^{(2)}(x) \leq \left(\frac{L_2}{-l_1}\right)^2 (1 - e^{l_1 x}(1 - l_1 x)) ||v^{(0)}||_x + L_2 \theta e^{l_1 x} h^p \int_0^x s e^{-l_1 s} ds$$

so that

$$\begin{aligned} v^{(2)}(x) \leq & \left(\frac{L_2}{-l_1}\right)^2 \left(1 - e^{l_1 x}(1 - l_1 x)\right) \|v^{(0)}\|_x \\ & + \left(\frac{L_2}{-l_1}\right)^2 \frac{\theta}{L_2} \left(e^{l_1 x} - (1 + l_1 x)\right) h^p. \end{aligned}$$

Thus by induction it is easy to verify that for $k = 1, 2, \ldots$

$$\begin{aligned} v^{(k)}(x) \leq & \left(\frac{L_2}{-l_1}\right)^k \left(1 - e^{l_1 x} \sum_{j=0}^{k-1} \frac{(-l_1 x)^j}{j!}\right) \|v^{(0)}\|_x \\ & + \sum_{j=2}^{k} \left(\frac{L_2}{-\delta^p l_1}\right)^j \frac{\theta}{L_2} \left(p_{j-2}(l_1 x) e^{l_1 x} - (l_1 x + j - 1)\right) (\delta^k h)^p. \end{aligned} \tag{8.5.25}$$

Here $p_k(x)$ is the unique polynomial of degree k such that

$$p_{k-1}(x) e^x - (x + k) = \frac{(-x)^{k+1}}{(k+1)!} + O(x^{k+2}). \tag{8.5.26}$$

The induction argument proceeds from the fact that upon substitution of (8.5.25) into (8.5.23), (8.5.25) will hold only if

$$k - \int_0^x p_{k-2}(s) ds = p_{k-1}(x), \quad p_0(x) = 1$$

or

$$p_k'(x) = -p_{k-1}(x), \quad p_k(0) = k + 1.$$

Consequently,

$$p_k(x) = \sum_{j=0}^{k} (k + 1 - j) \frac{(-x)^j}{j!}$$

and it is easily shown that this satisfies (8.5.26).

Finally, substituting (8.5.25) into (8.5.21) gives the discretized analogue of Theorem 7.9.3.

Theorem 8.5.1. *Consider the general splitting given in (8.5.15) defined on $[0, T]$ and let l_1 and L_2 be the one-sided and two-sided Lipschitz constants associated with the second and third arguments, respectively. Then if the discretized approximations have an associated continuous interpolant* *of order p which are applied with a constant mesh refinement strategy $\{h, \delta h, \delta^2 h, ...\}$ and denoting $\varepsilon_h^{(k)}(x)$ by $y_h^{(k)}(x) - y(x)$ then*

$$\begin{aligned} ||\varepsilon_h^{(k)}||_T \leq & \left(\frac{L_2}{-l_1}\right)^k \left(1 - e^{l_1 T}\sum_{j=0}^{k-1}\frac{(-l_1 T)^j}{j!}\right) ||\varepsilon_h^{(0)}||_T \\ & + \frac{\theta}{L_2}(h\delta^k)^p \left(\sum_{j=1}^{k} \psi^j (p_{j-2}(l_1 T)e^{l_1 T} - (l_1 T + j - 1))\right), \end{aligned} \quad (8.5.27)$$

where

$$\psi = \frac{L_2}{-l_1 \delta^p}$$

and

$$p_{-1}(x) = 0, \quad p_k(x) = \sum_{j=0}^{k}(k+1-j)\frac{(-x)^j}{j!}. \quad (8.5.28)$$

Remarks.

1. Theorem 8.5.1 generalizes the continuous result given in Theorem 7.9.3 to the discretized case. The error bound in (8.5.27) can be viewed as the sum of the continuous bound (the first term) plus the bound due to interpolation.
2. By writing the second term as

$$\frac{\theta}{L_2}(h\delta^k)^p \sum_{j=1}^{k} \psi^j p_{j-2}(l_1 T)(e^{l_1 T} - R_{j-2}(l_1 T))$$

with

$$R_{j-2}(x) = \frac{j-1+x}{p_{j-2}(x)},$$

R_{j-2} can be interpreted as a $(1, j-2)$ Padé approximation to the exponential function of order $j-1$ (that is it corresponds to entries in the second column of the Padé table).

3. (8.5.27) can be written as

$$||\varepsilon_h^{(k)}||_T \leq \left(\frac{L_2}{-l_1}\right)^k \alpha_k(T)||\varepsilon_h^{(0)}||_T + \frac{\theta}{L_2}(h\delta^k)^p \sum_{j=1}^{k}\psi^j \beta_j(T), \quad (8.5.29)$$

where for $k \geq 1$

$$\alpha_k(T) = 1 - e^{l_1 T}\sum_{j=0}^{k-1}\frac{(-l_1 T)^j}{j!}, \quad \beta_k(T) = p_{k-2}(l_1 T)e^{l_1 T} - (l_1 T + k - 1).$$

Now it is already known from Remark 3 of Theorem 7.9.3 that if $l_1 < 0$ then

$$0 < \alpha_{k+1}(T) < \alpha_k(T) < 1.$$

Furthermore, if $l_1 < 0$ then (8.5.28) implies

$$\beta_k(T) > 0, \qquad (8.5.30)$$

while (8.5.28) implies

$$\begin{aligned} \beta_{k+1}(T) - \beta_k(T) &= (p_{k-1}(l_1T) - p_{k-2}(l_1T))\, e^{l_1T} - 1 \\ &= \sum_{j=0}^{k-1} \tfrac{(-l_1T)^j}{j!} e^{l_1T} - 1 \qquad (8.5.31) \\ &< 0, \quad \text{if} \quad l_1 < 0. \end{aligned}$$

Thus if $R = \frac{L_2}{-l_1} < 1$ the continuous term is a contraction mapping on all bounded intervals $[0, T]$. However, this condition does not guarantee that the second term is a contraction mapping. Indeed if $\delta = 1$ (so that there is no mesh refinement), the second term can never be a contraction and the convergence is "carried" by the continuous case.

This last remark leads to the question whether there is a restriction on R which guarantees that for a dissipative system the discretized waveform iteration is a contraction mapping on all bounded intervals $[0, T]$. Clearly $R < 1$ is a necessary condition. However the second term will be a contraction mapping on all bounded intervals if with $r = \delta^p < 1$ and $\psi = R/r$

$$\psi^{k+1}\beta_{k+1}(x) - \sum_{j=1}^{k} \psi^j \beta_j(x) \left(\frac{1-r}{r}\right) < 0, \quad \forall k \in Z^+, \quad \forall x \leq 0. \quad (8.5.32)$$

Now from (8.5.30) and (8.5.31) if $x < 0$ (so that $l_1 < 0$ and $\psi > 0$)

$$\psi^k \left(B_{k+1}(x) - B_k(x)\right) < \sum_{j=1}^{k-1} \psi^j B_j(x)$$

so that

$$\psi^k B_{k+1}(x) < \sum_{j=1}^{k-1} \psi^j B_j(x).$$

Hence (8.5.32) will hold if

$$\psi < \frac{1-r}{r}.$$

But since $\psi = R/r$ this implies

$$R < 1 - \delta^p$$

and the following result holds:

Theorem 8.5.2. *Under the assumptions of Theorem 8.5.1 in which* $l_1 < 0$, *the discretized waveform algorithm will be a contraction mapping on all bounded intervals if*

$$R = \frac{L_2}{-l_1} < 1 - \delta^p. \tag{8.5.33}$$

Remark. The condition expressed by (8.5.33) is of course only a sufficient one and convergence can of course take place if $\delta = 1$ (so that there is no refinement) after a sufficient number of iterations have been performed. However, Theorem 8.5.2 does show the importance of grid refinement.

8.6 Multistage methods in waveform relaxation

Consider the application of an s-stage Runge-Kutta method of order p to the nonautonomous problem (2.1.2) on a mesh $\Delta = \{x_0, x_1, ..., x_N\}$ with stepsize $h_{n+1} = x_{n+1} - x_n$. It can be represented by

$$\begin{aligned} Y_{i,n+1} &= u(x_n) + h_{n+1} \sum_{j=1}^{s} a_{ij} f(x_{j,n+1}, Y_{j,n+1}), \quad i = 1, ..., s \\ u(x_{n+1}) &= u(x_n) + h_{n+1} \sum_{j=1}^{s} b_j f(x_{j,n+1}, Y_{j,n+1}), \end{aligned} \tag{8.6.1}$$

where $u(x)$ is a numerical approximation to $y(x)$ on Δ and $x_{j,n+1} = x_n + c_j h_{n+1}$.

There are a number of possible approaches to applying (8.6.1) in a waveform implementation. The first is to apply (8.6.1) directly to the splitting function F defined in (8.5.15) and this gives, for $i = 1, \ldots, s$,

$$\begin{aligned} Y_{i,n+1}^{(k+1)} &= u^{(k+1)}(x_n) \\ &\quad + h_{n+1} \sum_{j=1}^{s} a_{ij} F(x_{j,n+1}, Y_{j,n+1}^{(k+1)}, Y_{j,n+1}^{(k)}), \\ u^{(k+1)}(x_{n+1}) &= u^{(k+1)}(x_n) \\ &\quad + h_{n+1} \sum_{j=1}^{s} b_j F(x_{j,n+1}, Y_{j,n+1}^{(k+1)}, Y_{j,n+1}^{(k)}). \end{aligned} \tag{8.6.2}$$

Another approach involves generating a continuous approximation $u^{(k)}(x)$ by some interpolation procedure of the discrete approximations computed during the previous iteration. By denoting for the rest of this section

$$\alpha_{j,n+1}^{(k)} = u^{(k)}(x_{j,n+1})$$

this gives, for $i = 1, \ldots, s$,

$$\begin{aligned} Y_{i,n+1}^{(k+1)} &= u^{(k+1)}(x_n) \\ &\quad + h_{n+1} \sum_{j=1}^{s} a_{ij} F(x_{j,n+1}, Y_{j,n+1}^{(k+1)}, \alpha_{j,n+1}^{(k)}), \\ u_{n+1}^{(k+1)}(x) &= u^{(k+1)}(x_n) \\ &\quad + h_{n+1} \sum_{j=1}^{s} b_j F(x_{j,n+1}, Y_{j,n+1}^{(k+1)}, \alpha_{j,n+1}^{(k)}). \end{aligned} \tag{8.6.3}$$

The approach characterized by (8.6.2) has been considered by Lie and Skålin (1992) while (8.6.3) has been analysed by Bellen and Zennaro (1993). Bellen and Zennaro (1993) remark that the limit of (8.6.2), if it exists, is the underlying Runge-Kutta method and that (8.6.2) does not allow for the mesh to be changed while iterating which, as seen from the previous section, appears to be important. Moreover, this implementation does not easily allow for different meshes for different component blocks.

As a consequence, Bellen et al. (1990b, 1993, 1994) and Bellen and Zennaro (1993) have studied the convergence, contractivity and limit properties of (8.6.3) and a more general algorithm based on a splitting which allows a Gauss-Seidel approach between the subsystems.

Denoting this splitting by

$$F^* : [x_0, x_f] \times \mathbb{R}^m \times \mathbb{R}^m \times \mathbb{R}^m \to \mathbb{R}^m,$$

where

$$F^*(x, y, y, z) = F(x, y, z),$$

this leads to the following modification of (8.6.3), for $i = 1, \ldots, s$,

$$\begin{aligned} Y_{i,n+1}^{(k+1)} &= u^{(k+1)}(x_n) \\ &\quad + h_{n+1} \sum_{j=1}^{s} a_{ij} F^*(x_{j,n+1}, Y_{j,n+1}^{(k+1)}, \alpha_{j,n+1}^{(k+1)}, \alpha_{j,n+1}^{(k)}), \\ u^{(k+1)}(x_{n+1}) &= u^{(k+1)}(x_n) \\ &\quad + h_{n+1} \sum_{j=1}^{s} b_j F^*(x_{j,n+1}, Y_{j,n+1}^{(k+1)}, \alpha_{j,n+1}^{(k+1)}, \alpha_{j,n+1}^{(k)}). \end{aligned} \tag{8.6.4}$$

Bellen and Zennaro (1993) consider a very general interpolation scheme which includes Lagrangian or Hermitian interpolation of the nodal values, one-step interpolants (Enright et al. (1986)), and multistep interpolants. This leads to the numerical scheme, for $i = 1, \ldots, s$,

$$\begin{aligned} Y_{i,n+1}^{(k+1)} &= u^{(k+1)}(x_n) \\ &\quad + h_{n+1} \sum_{j=1}^{s} a_{ij} F^*(x_{j,n+1}, Y_{j,n+1}^{(k+1)}, \alpha_{j,n+1}^{(k+1)}, \alpha_{j,n+1}^{(k)}), \\ u^{(k+1)}(z_n) &= u(x_n) \end{aligned} \tag{8.6.5}$$

$$+h_{n+1} \sum_{i=-\alpha}^{\beta} \sum_{j=1}^{s} b_{ij}^{(n+1)}(\theta) F^*(x_{j,n+1}, Y_{j,n+i+1}^{(k+1)}, \alpha_{j,n+i+1}^{(k+1)}, \alpha_{j,n+i+1}^{(k)}).$$

Here $z_n = x_n + \theta h_{n+1}$, α and β are non-negative integers, the $b_{ij}^{(n+1)}(\theta)$ are certain polynomials in θ of degree less than or equal to p which depend on the ratios between h_{n+1} and each of the stepsizes in the $\alpha + \beta$ subintervals of Δ which influence the interpolation procedure.

Bellen and Zennaro (1993) are able to prove the uniform convergence of (8.6.5) to a limit method (described by suppressing the iteration index k) which is no longer a Runge-Kutta method. The proof is similar to those given in White and Sangiovanni-Vincentelli (1987) in that the difference of two different methods is considered and a contraction principle is established. The proof is straight-forward but technical and is not given here.

Theorem 8.6.1. *If F^* satisfies three two-sided Lipschitz conditions on each of its dependent arguments and*

$$Q := \max_{0 \le n \le N-1} \max_{0 \le \theta \le 1} \sum_{j=1}^{s} \left| b_{ij}^{(n+1)}(\theta) \right| \le Q^*, \tag{8.6.6}$$

where Q^* is independent of the mesh Δ, then $\exists h_0 > 0$ such that for all meshes which satisfy (8.6.6) with $h_{\max} = \max_{0 \le n \le N-1}\{h_{n+1}\} \le h_0$, the $\{u^{(k)}\}$ described by (8.6.5) converge uniformly on $[x_0, x_f]$ to a limit function $u(x)$, given that the mesh is the same for all components and all iterations.

Remarks.

1. The proof relies on bounding the maximum stepsize by the reciprocal of a two-sided Lipschitz constant. This means that for stiff problems the convergence will be very slow even on moderately sized windows.
2. The proof can be restructed in terms of one-sided and two-sided Lipschitz constants to give more realistic error bounds on the iterates but the proof will not be given here.

3. Bellen and Zennaro (1993) note that the restriction (8.6.6) is not severe as it is satisfied by all meshes with

$$\frac{h_{\max}}{h_{\min}} \leq K, \quad K \text{ some constant.}$$

They also note that Theorem 8.6.1 holds when different meshes are used for different subsystems.

Bellen and Zennaro (1993) also consider the more complicated case when the meshes are changed from iteration to iteration. Again a contraction principle is used but rather than the static iteration

$$u^{(k+1)} = \phi(u^{(k)})$$

a dynamic iteration of the form

$$u^{(k+1)} = \phi^{(k+1)}(u^{(k)})$$

is considered. The proof then requires the existence of a limit contractive mapping which implies that the varying meshes $\{\Delta_k\}$ must converge to a limit mesh, Δ, say, such that

$$\lim_{k\to\infty} \operatorname{dist}(\Delta_k, \Delta) = 0.$$

Here dist is some appropriate measure between grids.

Because the mesh in Theorem 8.6.1 defines an underlying norm, then rather than use a different norm for each iteration, Bellen and Zennaro (1993) assume that there exists a nonnegative integer r such that for any pair of meshes Δ_j, Δ_i and for all $x \in [x_0, x_f]$

$$x \in [x_{i,n}, x_{i,n+1}] \cap [x_{j,p}, x_{j,p+1}] \Rightarrow |n-p| \leq r.$$

Under these two assumptions, Bellen and Zennaro (1993) prove the analogue of Theorem 8.6.1 given that the meshes can vary from iteration to iteration.

Finally, Bellen and Zennaro note that the limit method of (8.6.5) has uniform order $w = \min\{p, q\}$ and nodal order p, where p is the order of the underlying Runge-Kutta method and $q \geq p-1$ is the uniform order of the interpolation procedure.

In a series of other papers Bellen et al. (1990b, 1993, 1994) study the stability and order properties of a number of waveform Runge-Kutta methods as well as the limit methods for a more restricted class of interpolation procedures. For example, in Bellen et al. (1990b) the waveform relaxation Heun method is studied in conjunction with a Jacobi waveform over a constant-stepsize mesh. This leads to the method

$$\begin{aligned} Y &= y_n + hf(x_n, y_n) \\ y_{n+1}^{(k)} &= u^{(k)}(x_{n+1}), \quad u^{(0)}(x_n + \theta h) = y_n, \end{aligned}$$

where, for $i = 1, \ldots, m$,

$$u_i^{(k+1)}(x_n + \theta h) = y_{in} + \frac{h\theta}{2}(f_i(x_n, y_{in}) + f_i(x_{n+1}, Z)), \tag{8.6.7}$$

and where

$$\begin{aligned} y_n &= (y_{1n}, ..., y_{mn})^{\top} \\ \alpha_i^{(k)} &= (y_{1,n+1}^{(k)}, ..., y_{i-1,n+1}^{(k)})^{\top} \\ \beta_i^{(k)} &= (y_{i+1,n+1}^{(k)}, \ldots, y_{m,n+1}^{(k)})^{\top} \\ Y &= (Y_1, \ldots, Y_m)^{\top} \\ Z &= (\alpha_i^{(k)}, Y_i, \beta_i^{(k)})^{\top}. \end{aligned}$$

A simple Taylor series expansion shows that the order of (8.6.7) is $\min\{k, 2\}$, where k is the number of iterations of (8.6.7).

Bellen et al. (1990b) also analyse the stability properties of (8.6.7) when applied to the linear test equation

$$y' = Qy, \quad Q = \begin{pmatrix} \lambda & -\mu \\ \mu & \lambda \end{pmatrix} \tag{8.6.8}$$

with the window chosen to be equal to the stepsize of integration and with a fixed number of iterations. If S_k represents the stability region of (8.6.7) after k iterations, Bellen et al. (1990b) show

$$S_H \subset S_k, \quad k \geq 3,$$

where S_H is the stability region of the underlying Heun method.

Bellen et al. (1994) consider a generalization of (8.6.7) in which the underlying method is an s-stage continuous Runge-Kutta method in which the interpolation is supplied by natural continuous extensions (NCE) (see Zennaro (1986)). These methods are derived from (2.7.2) and can be written as

$$Y_i = u(x_n) + h\sum_{j=1}^{s} a_{ij} f(x_n + c_j h, Y_j), \quad i = 1, ..., s$$

$$\begin{aligned} u(x_{n+1}) &= u(x_n) + h\sum_{j=1}^{s} b_j f(x_n + c_j h, Y_j) \\ u(x_n + \theta h) &= y(x_n) + h\sum_{j=1}^{s} b_j(\theta) f(x_n + c_j h, Y_j). \end{aligned} \tag{8.6.9}$$

The $b_j(\theta)$ satisfy

$$b_j(0) = 0, \quad b_j(1) = b_j, \quad j = 1, ..., s$$

and are polynomials of degree d where $[\frac{p+1}{2}] \leq d \leq p$ and p is the underlying order of (8.6.9). Thus $u(x)$ is a piecewise polynomial continuous on the whole interval of integration and is called an NCE of degree d. This formulation fits into the more general format described in (8.6.5) and such methods are called time-point relaxation Runge-Kutta methods. Bellen et al. (1994) are able to show that each iteration of a time-point relaxation method increases the order by one until the nodal order is reached if relaxation is based on Jacobi, Gauss-Seidel or SOR iteration and that the order of the limit method is also the order of the underlying method. They also study the stability regions of such methods when applied to (8.6.8) and they note that, in general, stability properties improve as the iteration index k increases. However, the increase is not monotonic and S_k does not always entirely contain the stability region of the underlying method. Bellen et al. (1994) also explore how the order of the interpolant affects S_k.

8.7 Contractivity of waveform RK methods

It was shown in Theorem 8.3.1 that if an algebraically stable Runge-Kutta method is used to solve a linear problem based on an appropriate waveform splitting then the spectral radius of the discretization operator is less than or equal to the spectral radius of the continuous waveform operator. Bellen et al. (1994) have investigated how to choose (8.6.9) in such a way that waveform RK methods based on Jacobi, Gauss-Seidel or SOR relaxation have contractivity properties in the maximum norm for dissipative systems. For sufficiently smooth problems this is equivalent to requiring

$$u[J] \leq 0, \quad \forall y \in \mathbb{R}^m \tag{8.7.1}$$

(see Dekker and Verwer (1984), for example), where $u[J]$ represents the logarithmic norm of the associated Jacobian matrix. Note that an equivalent property to (8.7.1) is that J is weakly diagonally dominant.

Spijker (1983) originally studied the contractivity properties of many classes of numerical methods in the maximum norm, while Kraaijevanger

(1991) has shown that if a Runge-Kutta method of order p is to be unconditionally contractive in the maximum norm then necessarily $p \leq 1$. However, Bellen et al. (1994) are able to show that waveform Runge-Kutta methods break this order barrier, and that they will be contractive in the maximum norm if the underlying Runge-Kutta method and its interpolant (see (8.6.9)) satisfy a generalization of AN-stability introduced by Bellen and Zennaro (1992) known as $AN_f(0)$-stability.

Definition 8.7.1. *The $AN_f(0)$-stability region of a Runge-Kutta method given by (2.7.2) is the maximum segment $[-r, 0)$ such that $I - AZ$ is non-singular and*

$$\left|1 + b^{\top} Z(I - AZ)^{-1} e\right| + \left\|b^{\top} Z(I - AZ)^{-1}\right\|_1 = 1$$

for $-rI \leq Z \leq 0$, where $Z = \operatorname{diag}(z_1, ..., z_s)$ and $z_i = z_j$ if $c_i = c_j$. The method is $AN_f(0)$-stable if $r = \infty$.

Remark.
This definition is easily modified to $A_f(0)$-stability and extended to continuous Runge-Kutta methods by replacing b_i by $b_i(\theta)$ and r by $r(\theta)$ in the above definition. The segment $[-R^*, 0]$, where $R^* = \min\{r(c_1), ..., r(c_s), 1\}$ is the region of semi-$AN_f(0)$-stability of the **continuous Runge-Kutta** (CRK) method which is said to be semi-$AN_f(0)$-stable if $R^* = \infty$.

Let $y(x)$ and $z(x)$ represent two different solutions to (2.1.2) with different initial conditions $y(x_0) = y_0$, $z(x_0) = z_0$, respectively, and let $u^{(0)}$ and $v^{(0)}$ be initial approximations to y and z such that

$$\| u^{(0)}(x) - v^{(0)}(x) \|_\infty \leq \| y_0 - z_0 \|_\infty, \quad x \in [x_0, x_f].$$

Then Bellen et al. (1994) have shown that the iterates $u^{(k)}$ and $v^{(k)}$ generated from (8.6.9) in conjunction with Jacobi, Gauss-Seidel or SOR relaxation (with $0 < w \leq 1$) satisfy, for $n = 0, \ldots, N-1$,

$$\max_{\theta \in \{c_1, \ldots, c_s, 1\}} \| u^{(k)}(x_n + \theta h) - v^{(k)}(x_n + \theta h) \|_\infty \leq \| y_0 - z_0 \|_\infty . \quad (8.7.2)$$

Furthermore, if the underlying CRK method is semi $AN_f(0)$-stable, (8.7.2) holds for any dissipative problem in the maximum norm with negative diagonal elements for the Jacobian, and for any mesh. Examples of $AN_f(0)$-stable methods are given in Bellen and Zennaro (1992) and Bellen et al. (1994) and include the two-stage method of order 2

$$\begin{array}{c|cc} c & a & c-a \\ 1 & b & 1-b \\ \hline & b & 1-b \end{array} \qquad (8.7.3)$$

where

$$b = \frac{1}{2(1-c)}, \quad a \geq b, \quad c \in [0, \frac{1}{2}],$$

and the three-stage method of order 3

$$\begin{array}{c|ccc} 0 & \frac{5}{2} & -2 & -\frac{1}{2} \\ \frac{1}{2} & -1 & 2 & -\frac{1}{2} \\ 1 & \frac{1}{6} & \frac{2}{3} & \frac{1}{6} \\ \hline & \frac{1}{6} & \frac{2}{3} & \frac{1}{6} \end{array} \tag{8.7.4}$$

If in (8.7.3) a linear interpolation given by

$$b_j(\theta) = b_j\theta, \quad j = 1, 2$$

and in (8.7.4) a quadratic interpolation given by

$$b_1(\theta) = -\frac{5}{6}\theta^2 + \theta, \quad b_2(\theta) = \frac{2}{3}\theta^2, \quad b_3(\theta) = \frac{1}{6}\theta^2$$

are used then these methods are both semi-$AN_f(0)$-stable.

Bellen and Zennaro (1992) have also analysed the regions of $AN_f(0)$-stability for explicit CRKs and have shown that there is an upper bound of four on the existence of methods with nonempty $AN_f(0)$-stability regions. In 't Hout (1994) has generalized this work and shown that a waveform relaxation Runge-Kutta method will be unconditionally contractive in the maximum norm if the underlying Runge-Kutta method satisfies

$$|1 + b^\top Z(I - AZ)^{-1}e| + |b^\top Z(I - AZ)^{-1}d| \leq 1 \tag{8.7.5}$$

whenever $Z = \mathrm{diag}(z_1, ..., z_s) \leq 0$ and $\| d \|_\infty \leq 1$. Note that a method is called $AN_f(0)$-stable if this condition is satisfied. Using this condition, In 't Hout (1994) is able to show that if a waveform relaxation Runge-Kutta method of order p is unconditionally attractive in the maximum norm, then necessarily $p \leq 4$.

In fact the preceding theory developed by Bellen et al. is unsatisfactory in many ways in that it suggests that there is an order barrier to the convergence of Runge-Kutta waveform methods where, in fact, none may exist. This is confirmed in a private communication by In t' Hout in which convergence is studied in a different norm and is related to the algebraic stability of the underlying Runge-Kutta method. In this case there is no order barrier. The analysis of In t' Hout is presented below. A crucial point here is that the complicated interpolation schemes described in (8.6.5) are replaced by schemes of the form, for $i = 1, \ldots, s$,

$$\begin{aligned} u_i^{(k+1)} &= y_n^{(k+1)} + h\sum_{j=1}^{s} a_{ij} F(x_n + c_j h, u_j^{(k+1)}, u_j^{(k)}), \\ y_{n+1}^{(k+1)} &= y_n^{(k+1)} + h\sum_{j=1}^{s} b_j F(x_n + c_j h, u_j^{(k+1)}, u_j^{(k)}), \end{aligned} \tag{8.7.6}$$

where $y_0^{(k+1)} = y_0$ when applied to the splitting (8.5.15).

If this scheme is now applied with constant stepsize h over N steps then a similar scheme to (8.7.6) is obtained except a_{ij}, b_j and c_j are replaced by $\hat{a}_{1j}, \hat{b}_j$ and $\hat{c}_j$ where

$$\begin{aligned} \hat{b}^\top &= (b^\top, b^\top, \ldots, b^\top) \\ \hat{c}^\top &= (c^\top, (e+c)^\top, \ldots, ((N-1)e + c)^\top)^\top \\ \hat{A} &= \begin{pmatrix} A & & & \\ eb^\top & A & & \\ \vdots & & \ddots & \\ eb^\top & \ldots & eb^\top & A \end{pmatrix} \end{aligned}$$

and

$$Y^{(k)} = (u_{11}^{(k)}, \ldots, u_{1s}^{(k)}, \ldots, u_{N1}^{(k)}, \ldots, u_{Ns}^{(k)})^\top.$$

This describes a composition scheme with $\bar{s} = sN$ stages, and the limit of this waveform is the Runge-Kutta method

$$\begin{aligned} Y_i &= y_n + h\sum_{j=1}^{\bar{s}} \hat{a}_{ij} f(x_n + \hat{c}_j h, Y_j), \quad i = 1, \ldots, \bar{s} \\ y_{n+1} &= y_n + h\sum_{j=1}^{\bar{s}} \hat{b}_j f(x_n + \hat{c}_j h, Y_j). \end{aligned}$$

Applying the Mean Value theorem now to the composition variant of (8.7.6) and its limit method gives, for $j = 1, \ldots, \bar{s}$,

$$\begin{aligned} F(x_n + \hat{c}_j h, Y_j, Y_j) &- F(x_n + \hat{c}_j h, Y_j^{(k+1)}, Y_j^{(k)}) \\ &= L_j(Y_j - Y_j^{(k+1)}) + M_j(Y_j - Y_j^{(k+1)}), \end{aligned}$$

where

$$L_j = \int_0^1 \frac{\partial}{\partial x} F(x_n + \hat{c}_j h, uY_j + (1-u)Y_j^{(k+1)}, uY_j + (1-u)Y_j^{(k)})du$$

$$M_j = \int_0^1 \frac{\partial}{\partial y} F(x_n + \hat{c}_j h, uY_j + (1-u)Y_j^{(k+1)}, uY_j + (1-u)Y_j^{(k)})du.$$

By writing

$$L = \text{diag}(L_1, \ldots, L_{\bar{s}}), \quad M = \text{diag}(M_1, \ldots, M_{\bar{s}})$$

and letting

$$\Delta Y^{(k+1)} = (Y_1^{(k+1)} - Y_1, \ldots, Y_{\bar{s}}^{(k+1)} - Y_{\bar{s}})^\top$$

then

$$\Delta Y^{(k+1)} = (\hat{A} \otimes I)hL\Delta Y^{(k+1)} + (\hat{A} \otimes I)hM\Delta Y^{(k)}$$

so that

$$\Delta Y^{(k+1)} = ((\hat{A} \otimes I)^{-1} - hL)^{-1} hM\Delta Y^{(k)}. \qquad (8.7.7)$$

Now let $\langle \cdot, \cdot \rangle$ and $\| \cdot \|$ be the standard inner product and its associated norm on $\mathbb{R}^m$ and define on $\mathbb{R}^{\bar{s}}$

$$\langle \xi, \eta \rangle_{\hat{D}} = \sum_{j=1}^{\bar{s}} \hat{d}_j \xi_j \eta_j \qquad (8.7.8)$$

and on $\mathbb{R}^{m\bar{s}}$

$$[u, v]_{\hat{D}} = \sum_{j=1}^{\bar{s}} \hat{d}_j \langle u_j, v_j \rangle, \qquad (8.7.9)$$

where

$$\hat{D} = \text{diag}(D, D, \ldots, D), \quad \text{and} \quad D = \text{diag}(d_1, \ldots, d_s),$$

with $d_j > 0$, $j = 1, \ldots, s$. Let $|||\cdot|||_{\hat{D}}$ and $\|\cdot\|_{\hat{D}}$ denote the corresponding norms to (8.7.8) and (8.7.9), respectively.

Writing (8.7.7) in the form

$$(\hat{A} \otimes I)^{-1} Y = hLY + hMX$$

then

$$[(\hat{A} \otimes I)^{-1} Y, Y]_{\hat{D}} = [hLY + hMX, Y]_{\hat{D}}.$$

From Dekker and Verwer (1984) this implies

$$\psi_{\hat{D}}(\hat{A}^{-1})[Y, Y]_{\hat{D}} \le [(\hat{A} \otimes I)^{-1} Y, Y]_{\hat{D}} = [hLY + hMX, Y]_{\hat{D}},$$

where

$$\psi_D(\hat{A}^{-1}) := \min_{v \neq 0} \frac{\langle A^{-1}v, v\rangle_D}{\langle v, v\rangle_D}.$$

Thus

$$\psi_{\hat{D}}(\hat{A}^{-1})[Y,Y]_D \leq \sum_{j=1}^{\bar{s}} \hat{d}_j(\langle hL_jY_j, Y_j\rangle + \langle hM_jX_j, Y_j\rangle)$$

so that

$$0 \leq \sum_{j=1}^{\bar{s}} \hat{d}_j(h\mu[L_j] - \psi_{\hat{D}}(\hat{A}^{-1}))\langle Y_j, Y_j\rangle + \langle hM_jX_j, Y_j\rangle), \tag{8.7.10}$$

where $\mu[L]$ denotes the logarithmic norm of L.

Assume now that

$$h\mu[L_j] < \psi_{\hat{D}}(\hat{A}^{-1}), \quad \forall j \tag{8.7.11}$$

and let

$$\begin{aligned} \bar{d}_j &= \hat{d}_j(\psi_{\hat{D}}(\hat{A}^{-1}) - h\mu[L_j]) \\ \bar{M}_j &= M_j/(\psi_{\hat{D}}(\hat{A}^{-1}) - h\mu[L_j]). \end{aligned}$$

Then (8.7.10) implies

$$0 \leq \sum_{j=1}^{\bar{s}} \hat{d}_j(-\langle Y_j, Y_j\rangle + \langle h\bar{M}_jX_j, Y_j\rangle).$$

Thus

$$|||Y|||_{\bar{D}}^2 \leq [h\bar{M}X, Y]_{\bar{D}}$$

which by the Cauchy-Schwarz inequality gives

$$|||Y|||_{\bar{D}} \leq h|||\bar{M}|||_{\bar{D}}\ |||X|||_{\bar{D}}. \tag{8.7.12}$$

Since

$$|||\bar{M}|||_{\bar{D}} = \max_j ||\bar{M}_j||, \quad j = 1, \ldots, \bar{s}$$

(8.7.12) gives

Theorem 8.7.2. *Given* $h\mu[L_j] < \psi_{\hat{D}}(\hat{A}^{-1})$, $j = 1, \ldots, \bar{s}$, *then*

$$|||\Delta Y^{(k+1)}|||_{\hat{D}} \leq \frac{h \max_{j=1,\ldots,\bar{s}} ||M_j||}{\psi_{\hat{D}}(\hat{A}^{-1}) - h \max_{j=1,\ldots,\bar{s}} \mu[L_j]} |||\Delta Y^{(k)}|||_{\hat{D}}. \quad (8.7.13)$$

Remarks.
1. For algebraically stable Runge-Kutta methods with A nonsingular, then D can be chosen by

$$D = \text{diag}(b_1, \ldots, b_s) > 0$$

and in this case $\psi_{\hat{D}}(\hat{A}^{-1}) \geq 0$ (Dekker and Verwer (1984)).
2. The result (8.7.13) is a general convergence result for Runge-Kutta waveform methods and imposes no order barriers and additional conditions on the underlying Runge-Kutta method (apart from algebraic stability). The result was proved by using a different norm to the maximum norm approach of Bellen et al. which imposes too stringent conditions for convergence analysis.

9

IMPLEMENTATION OF WAVEFORM ALGORITHMS

The previous two chapters have focused on both waveform and discretized waveform algorithms. It has been shown that convergence results can be obtained for general splittings based on one-sided (l_1) and two-sided (L_2) Lipschitz constants, and that there is a remarkable similarity between these results for both the continuous and the discretized case.

However, for many problems, especially nonlinear ones, it is difficult to get a priori estimates for the coupling ratio $R = -L_2/l_1$ and so more empirical approaches may be necessary to determine the efficacy of the waveform approach. Thus this final chapter will be devoted to parallel implementational issues of waveform relaxation. In particular, a parallel block Jacobi waveform relaxation code will be described and implementational details and numerical results on a 96-processor Paragon will be described in some detail. The chapter will conclude with some general comments on the parallel algorithms described in this monograph and outline possible future directions. Thus this chapter will cover the following material:

- section 9.1: the effect of coupling and reordering on convergence behaviour;
- section 9.2: adaptive waveform algorithms based, for example, on variable meshes, variable code tolerance and adaptive windows;
- section 9.3: a discussion of VODE and VODEPK which will be the core differential equations solver in the waveform implementation. The waveform relaxation algorithm is based on block Jacobi relaxation;
- section 9.4: a case study based on the two-dimensional chemical Brusselator problem;
- section 9.5: a parallel implementation of a waveform code, numerical results and critical analysis;
- section 9.6: conclusions and future directions.

9.1 Coupling and reordering

It has already been noted that the ordering of the components and the coupling between the components can have enormous impact on the convergence properties of WR algorithms. The effect of the ordering can be dramatically seen by applying a waveform Gauss-Seidel (WGS) algorithm to the problem $y' = Qy$ where Q is an upper triangular matrix. By writing

$$Q = D + U$$

where D denotes the diagonal part of Q and U the strictly upper triangular part, then the WGS algorithm is given by

$$y^{(k+1)'} - Dy^{(k+1)} = Uy^{(k)}.$$

In this case convergence will be slow if $T\rho(D^{-1}U)$ is large, where T denotes the length of the window in which the iterations take place.

However, if the components are reordered by writing

$$z = Py,$$

where P is the permutation matrix which reverses the order of the components, then z satisfies the equation

$$z' = Q_1 z, \quad Q_1 = PQP.$$

Here Q_1 is now lower triangular and the WGS algorithm applied to z will solve this problem exactly in one iteration.

Of course, not all reordering strategies have such impressive effects as the one just described. Gear and Juang (1991) and Juang (1989) have examined the rate of increase of the order of accuracy of the iterates and a speed of convergence can be defined in terms of the average number of additional terms in the power series expansion that are correct in each iteration. In the case of waveform Jacobi the accuracy increase is one, while for waveform Newton the accuracy increase doubles per iteration (because of the quadratic convergence of Newton iteration). For waveform Gauss-Seidel the increase is greater than one and depends on the ordering of the system components, and the cycles in the directed dependency graph of the differential system.

For the general problem given by (2.1.2.) a **dependency graph** can be constructed in which if the ith component of f depends on the jth component of y then a directed edge in the dependency graph is drawn from node j to node i. Thus the dependency graph shows the sparsity structure of the Jacobian.

Consider the problems whose Jacobians have the structures given by

$$\begin{pmatrix} x & & x \\ x & x & \\ & x & x \end{pmatrix}, \begin{pmatrix} x & x & x \\ & x & x \\ & & x \end{pmatrix}, \begin{pmatrix} x & & & & x \\ x & x & & & \\ & x & x & & \\ x & & x & x & \\ & & & x & x \end{pmatrix}.$$

Then the respective dependency graphs are shown in Figure 9.1.

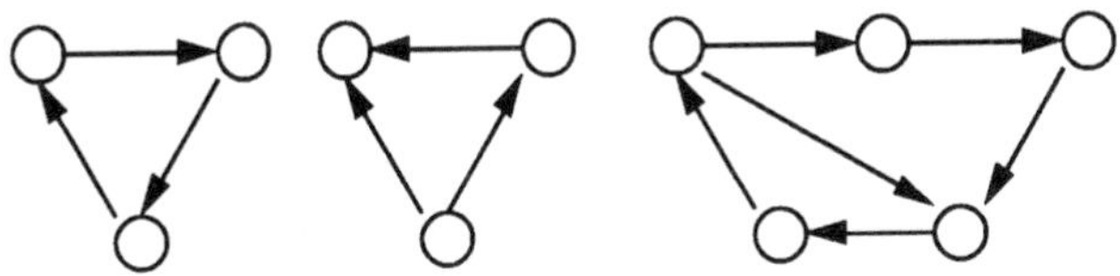

FIG. 9.1. dependency graphs

Here the nodes of the graph are numbered in ascending fashion in clockwise order. The cycles in these graphs are, respectively,

$$\{1,2,3\}$$

$$\{\}$$

$$\{1,2,3,4,5\} \quad \text{and} \quad \{1,4,5\}.$$

Juang and Gear (1989) have shown that the accuracy increase per iteration of the WGS algorithm is governed by the length of the shortest cycle in the dependency graph. In the case of the first problem, the shortest cycle is the complete system itself and so WGS will have an accuracy increase of three iterations per cycle. For the second problem, convergence will be slow because there are no cycles, while for the third problem the accuracy increase per iteration will be three which is the length of the shortest cycle $\{1,4,5\}$.

In addition to the dependency graph having an effect on accuracy so too can the numbering of the nodes affect the average accuracy increase. Thus for the problem whose Jacobian has the structure

$$\begin{pmatrix} x & x & \\ & x & x \\ x & & x \end{pmatrix},$$

the dependency graph is the same as the first dependency graph in Figure 9.1 but now the nodes are numbered 1, 3, 2 in clockwise order. Hence two iterations are needed to get an accuracy increase of three (the cycle length). The number two here represents the number of times that the numbering of the nodes decreases as the cycle is traversed once in the direction specified by the edges. This quantity is denoted in Juang and Gear (1989) by the number of **ascending chains** in the cycle.

Thus Juang and Gear (1989) have proved the general result that the accuracy increase of WGS is bounded by the length of the shortest cycle of dependency graph every d iterations, where d is the number of ascending chains of that cycle. In particular if $d = 1$ so that all the nodes of a cycle are solved in cyclic order then the accuracy increase in one WGS iteration is bounded by the length of the smallest cycle.

This approach can, of course, be extended to block WGS algorithms. In this case the dependency graph has a vertex for each subsystem and an edge from vertex B to vertex A if a variable in subsystem A appears in subsystem B.

Of course the convergence of the WGS algorithm for the second problem can be improved by reordering the components in reverse order. Thus the basis of an effective WGS algorithm is the reordering of components to produce long cycles.

These concepts can be generalized to the nonlinear problem. In this case a permutation matrix $P(x)$, which may vary as the independent variable varies, must be found such that WGS is applied to the reordered system

$$z'(x) = P(x)f(x, P^{\top}(x), z(x)).$$

The difficulty here is that the reordering may have to change as the integration proceeds.

Consider now the differential system $y' = Qy$ where

$$Q = \begin{pmatrix} -\lambda_1 & a & 0 \\ a & -\lambda_2 & 0 \\ 0 & b & -\lambda_3 \end{pmatrix}, \quad a, b, \lambda_1, \lambda_2, \lambda_3 > 0.$$

Then a Gauss-Seidel splitting would result in an amplification matrix whose spectral radius is given by

$$\max_{\mathrm{Re}(z)\geq 0} \frac{a^2}{(z+\lambda_1)(z+\lambda_2)} = \frac{a^2}{\lambda_1\lambda_2}. \tag{9.1.1}$$

Consequently, a Gauss-Seidel waveform will converge if and only if

$$\lambda_1\lambda_2 > a^2. \tag{9.1.2}$$

Note that if $a = 0$, so that Q is lower triangular, then Gauss-Seidel relaxation will converge in one iteration, while the condition given in (9.1.1) reflects the existence of a diagonally dominant loop (see below). Clearly if the coupling between the first and second components in the differential system is too strong then Gauss-Seidel relaxation will not converge, since Gauss-Seidel relaxation splits these two components.

Consider, now a relaxation process in which the first and second components are coupled together. Such a relaxation scheme can be written as

$$\begin{aligned} y_1^{(k+1)'} &= -\lambda_1 y_1^{(k+1)} + a y_2^{(k+1)} \\ y_2^{(k+1)'} &= a y_1^{(k+1)} - \lambda_2 y_2^{(k+1)} \\ y_3^{(k+1)'} &= b y_2^{(k)} - \lambda_3 y_3^{(k+1)}. \end{aligned}$$

In this case the splitting (M, N) of Q is given by

$$M = \begin{pmatrix} -\lambda_1 & a & 0 \\ a & -\lambda_2 & 0 \\ 0 & 0 & -\lambda_3 \end{pmatrix}, \quad N = \begin{pmatrix} 0 & 0 & 0 \\ 0 & 0 & 0 \\ 0 & b & 0 \end{pmatrix}$$

and the spectral radius of the amplification matrix is 0 and the waveform algorithm will converge in one iteration. Here y_3 depends only on the block of y_1 and y_2 which itself is independent of y_3.

This simple example suggests that it is not always appropriate to apply a relaxation algorithm such as Jacobi or Gauss-Seidel which splits components that are strongly coupled to one another. This effect has been extensively commented on in the case of integrated circuit models by, for example, Carlin and Vachoux (1984), White et al. (1985) and White and Sangiovanni–Vincentelli (1987). These authors attempt to construct methodologies which enable the automatic lumping of tightly coupled components.

In particular, Carlin and Vachoux (1984) propose a functional extraction method in which an integrated circuit system is examined in an attempt to find functional blocks (subcircuits) which group all the components together in one subsystem. A difficulty here is the automation of the approach.

Another approach is the so-called **dominant loop method** (Carlin and Vachoux (1984)) in which a sequence of 2×2 subdiagonal blocks of the Jacobian matrix are analysed. If for any i and j

$$q_{ij} q_{ji} > \alpha q_{ii} q_{jj},$$

where α is problem dependent, then components i and j are lumped together. Although this is a simple approach it can be inappropriate as it only tends to cluster components together in pairs. White and Sangiovanni-Vincentelli (1987) have suggested a combination of both functional extraction and the dominant loop method. Adaptive schemes and various robustness aspects associated with these are considered in Odeh et al. (1983).

White et al. (1985) also note that in the case of Gauss-Seidel relaxation, waveform relaxation will converge in one iteration if the Jacobian matrix is lower triangular. Thus the reordering of components can be very important if the reordering can produce a system or subsystems which are lower triangular or nearly lower triangular. In some MOS digital circuit problems such a structure often arises because of the nature of transistors which are highly directional. (However the presence of certain physical components which lead to loops will complicate this nearly lower triangular structure.)

Indeed, White and Sangiovanni-Vincentelli (1987) present a waveform algorithm based on a trapezoidal discretization and Gauss-Seidel relaxation designed for digital circuit problems. This code attempts to find subcircuits (whose components are then coupled together into one subsystem) and order the components automatically. More recently, Ruehli and Zukowski (1992) note that as the interconnect models increase in complexity when modelling large digital VLSI circuits, it becomes more and more important to partition accurately the system. Ruehli and Zukowski (1992) thus consider the special case of a low-pass RC circuit which cannot be partitioned effectively with the diagonally dominant method and analyse the convergence properties and optimal window size for the waveform relaxation algorithm. It is claimed that this simple example captures some of the properties of large interconnect subcircuits.

Zukowski et al. (1990) note that when gallium arsenide circuits are modelled, the mostly unidirectional coupling of MOS integrated circuits may reverse and areas of weak coupling are less readily identified. Thus Zukowski et al. (1990) derive an analysis tool that measures subcircuit coupling and error attenuation in the waveform space, which can be used to estimate convergence and coupling properties possibly in an adaptive manner.

By noting that any waveform relaxation algorithm can be written in the form

$$\varepsilon^{(k+1)}(x) = F(\varepsilon^{(k)}(x)), \quad \varepsilon^{(k)}(x) = y^{(k)}(x) - y(x),$$

Zukowski et al. (1990) produce a linearization about the solution trajectory that gives a linear time-varying mapping of the form

$$\varepsilon^{(k+1)}(x) = H\varepsilon^{(k)}(x),$$

where H represents the error propagation matrix. In the case of integrated circuit modelling, H can be calculated by a set of impulse responses to a set of finite triangular impulses. If the inputs are piecewise linear, the output is calculated using convolution. In particular, if the simulation interval is divided into N windows, the error mapping between variable i and variable j is described as a matrix H_{ij} of size N. Column k in H_{ij} indicates how the error at node i at window k affects errors at all subsequent points in node j. Because the system is causal H_{ij} is lower triangular. The total error mapping between all components is described by a matrix of order N^2 with blocks H_{ij}.

Some of these techniques have been adapted to general systems of equations, but with less success because the automatic reordering and efficient groupings of the subsystems is often performed on the basis of physicality which is not always available (see Juang and Gear (1989), for example) for general systems.

The waveform code that will be presented in this chapter, however, is intended to be a general one, and so very little can be assumed about general structures. On the other hand a very important class of problems that have been considered in this monograph arise from the semi-discretization of parabolic partial differential equations. Such problems often have a symmetric Jacobian structure and are block banded. It will make sense, where appropriate, to choose an ordering of components which makes the bandwidth of each subsystem as small as possible. This will be discussed in greater detail later in section 9.4. However, for such problems it is possible to estimate the coupling ratio for various splittings. This will be illustrated for the problem (2.2.4), where Q is the block tridiagonal matrix given by $Q = (m+1)^2(I_m, T, I_m)$, $T = (1, -4, 1)$.

For a (M, N)-splitting of Q, the coupling factor is given by (7.9.16) and is

$$R_p = \frac{\|N\|_p}{-\mu_p[M]}.$$

Clearly the value of R_p depends on both M and the norm $\|\cdot\|_p$. For a Jacobi splitting, (4.5.9) and Example 2.3.1 give

$$R_\infty = 1, \quad R_2 = \cos h\pi, \quad h = 1/(m+1),$$

whilst for a block Jacobi splitting with $M_l = T$, $l = 1, ..., m$, (4.5.9) implies

$$R_\infty = 1, \quad R_2 = \frac{\cos h\pi}{2 - \cos h\pi}, \quad h = 1/(m+1).$$

The expressions for the 2-norm are just the values given in (4.5.9) for $\rho(J)$ and $\rho(J_B)$ which suggested that block Jacobi will converge twice as fast as Jacobi. Hence a block Jacobi waveform implementation will prove to be an appropriate one for a coarse-grained parallel environment.

9.2 Adaptive waveform algorithms

In the application of waveform relaxation techniques, Gear (1989, 1991a) distinguishes between two types of problems: heterogeneous and homogeneous. In the former case, different subsystems may exhibit very different convergence properties as in the case of VLSI circuit modelling, while in the latter case subsystems have a uniform structure, as for many classes of partial differential equations. Thus for heterogeneous problems partitioning can exploit weaknesses in some portions of the overall system. As mentioned in Leimkuhler and Ruehli (1992), very good performances have been obtained by using waveform relaxation techniques on both homogeneous (see, for example, Vandewalle and Piessens (1991)) and heterogeneous (Johnson and Zukowski (1991)) problems.

Gear (1991a) has discussed a number of implementational details associated with an efficient parallel implementation of waveform relaxation. The organizational problems that must be addressed depend on whether the problem is homogeneous or heterogeneous. For homogeneous problems a static load balancing approach can be used along with a synchronization of all processors at the end of each subsystem computation of a waveform on the same window. However, for heterogeneous problems this approach may not be appropriate. In this case, each processor may compute waveforms for different window lengths and then a number of issues must be addressed. These include

- dealing with load balancing due to different subsystems taking different processing times;
- minimizing synchronization overheads due to irregular communication between subsystems;
- deciding when and how the waveform process should finish;
- the scheduling and descheduling of tasks depending on changes in the coupling of the components.

It was seen in Chapter 7 that mesh refinement is an important strategy in any waveform implementation. Nevanlinna (1989) discusses an adaptive windowing implementation in which the refinement only ever doubles the number of interpolation points. Other strategies have been considered by Dyke (1995) in which an averaged convergence factor over a few initial iterations is used to determine whether the convergence on a particular window is satisfactory. If it is not, a smaller window is chosen and the initial waveform iterate for this window is the appropriate restriction on the larger window of the last waveform iterate on that window. A new window is then chosen based on available convergence rates. The initial waveform iterate is again the appropriate restriction on any larger window (which contains the present window) of the last iterate on that larger window.

As Vandewalle and Piessens (1990) note, any stiff integrator can be used as the core solver. Vandewalle and Roose (1989) use a variable-step, variable-order integrator from ODEPACK but Vandewalle and Piessens (1990) claim that this is not an efficient approach due to overheads of stepsize and order selection recurring at each grid point. Instead Vandewalle and Piessens (1990) advocate the use of a simple stiff integrator such as one based on the trapezoidal rule. However, in the code presented in this monograph, a more flexible approach has been adopted.

Vandewalle and Piessens (1990) report on fixed-stepsize implementation of the trapezoidal rule in conjunction with a multigrid waveform relaxation based on a four-colour, nine-point Gauss-Seidel smoother on a 16-processor iPSC/2 hypercube on various parabolic partial differential equations which have two spatial dimensions. Various multigrid options are considered based on V cycles, W cycles and full multigrid. On a problem of dimension 65×65, multigrid waveform relaxation is faster than a parallel implementation of the trapezoidal rule by about a factor of 7 on 16 processors, with waveform relaxation being about 1.3 times as fast on a single processor. The reason for this latter speed-up is due to vectorization of the multigrid in the temporal direction and the efficacy of this depends on the number of (equidistant) time steps.

As Worley (1992) notes, the waveform relaxation multigrid algorithm is usually implemented sequentially in the temporal direction. But as described in Chapter 4, if the ODEs are solved with a fixed stepsize then in the case of linear problems a linear recurrence is generated. Thus for a one-step method implementation, cyclic reduction of a block bidiagonal system could be combined with a multigrid waveform relaxation. In the case of a k-step linear multistep implementation the linear recurrence has a bandwidth of k. The parallel complexity of such an approach has been considered by Worley (1992). This same approach to parallelizing in time has also been discussed for constant-coefficient hyperbolic partial differential equations by solving ordinary differential systems along the associated characteristics.

Vandewalle and Van de Velde (1993) and Vandewalle (1993) compare a space-concurrent multigrid waveform relaxation implementation with a space-time concurrent multigrid waveform relaxation implementation using a constant and global stepsize approach which is determined a priori. The parallelism across the steps is, in the case of linear problems, performed using the parallel solution of linear recurrences as described in Chapter 4.

More recent work on VLSI circuit simulation has resulted in the necessity of solving nonlinear equations with several hundred thousand components in which the effects of parasitics lead to problems of mixed heterogeneous-homogeneous type. Such problems have been studied by Leimkuhler and Ruehli (1993), who note that the standard spectral approach does not take into account all of the aspects of the problem – this is especially so in the case of large VLSI circuit problems with regular and symmetric structures. Indeed this spectral radius approach may not be an accurate approximation to the actual rate of convergence, which is very much window dependent, in early sweeps of the iteration.

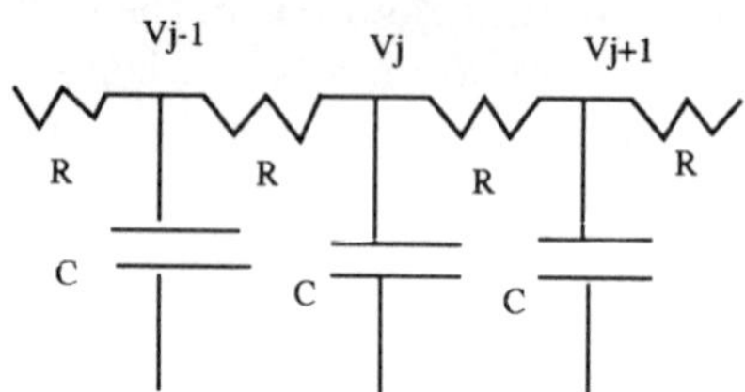

FIG. 9.2. one-dimensional RC circuit problem

Thus Leimkuhler and Ruehli (1993) consider a number of cases of a model one-dimensional RC circuit (described in Figure 9.2) in which the chain is considered to be either infinite in length, finite in length with imposed boundary conditions, or finite in length but not terminated. This approach can of course be extended to higher-dimensional lattices. Because of this regular structure, there is an intimate relationship with the heat equation of corresponding dimension when solved by a symmetric, second-order spatial differencing technique. Thus there is a natural relationship between regular circuit problems and semi-discretized PDEs. For example, the jth differential equation associated with Figure 9.2 can be written as

$$v_j' = \theta(v_{j-1} - 2v_j + v_{j+1}), \quad \theta = \frac{1}{RC}. \tag{9.2.1}$$

Leimkuhler and Ruehli (1993) argue that for chains of infinite length the spectral radius technique does not provide useful information because of the restrictions on the window size. Furthermore, for the unterminated case Leimkuhler et al. (1991) have shown unit spectral radius.

For this reason, Leimkuhler and Ruehli (1993) attempt to relate the convergence of the Jacobi waveform for (9.2.1) in terms of global and local convergence effects. From (7.4.3)

$$\varepsilon^{(k)}(x) = \theta^k K^k \varepsilon^{(0)}(x),$$

where K is the convolution operator defined as in (7.4.2) whose kernel function r is given by

$$r(x) = \theta e^{-2\theta x} \begin{pmatrix} 0 & 1 & & \\ 1 & 0 & \ddots & \\ & \ddots & \ddots & 1 \\ & & 1 & 0 \end{pmatrix}.$$

Leimkuhler and Ruehli, by a simple induction process, relate the nonzero elements of K^k to r^k and show

$$||\varepsilon^{(k)}||_T \leq (2\theta)^k T \, |h(x)|_T \, ||\varepsilon^{(0)}||_T, \tag{9.2.2}$$

where

$$h(x) = e^{-2\theta x} \frac{x^k}{k!}.$$

Since, on $[0, \infty)$, $h(x)$ is maximized when $x = k/2\theta$, (9.2.2) implies

$$||\varepsilon^{(k)}||_T \leq \frac{T}{k!} \left(\frac{k}{e}\right)^k ||\varepsilon^{(0)}||_T, \quad k \leq 2\theta T. \tag{9.2.3}$$

On the other hand, when $k > 2\theta T$, $|h(x)|_T < h(T)$ and this along with Stirling's approximation implies

$$||\varepsilon^{(k)}||_T \leq \frac{Te^{-2\theta T}}{\sqrt{2\pi k}} \left(\frac{2\theta e T}{k}\right)^k ||\varepsilon^{(0)}||_T. \tag{9.2.4}$$

Thus when $\frac{2e\theta T}{k} < r < 1$, rapid convergence with geometric decay rate r will take place. For the equivalent two-dimensional lattice the appropriate condition for a Jacobi waveform is $\frac{4e\theta T}{k} < r < 1$. These rapid-convergence conditions can be used to estimate appropriate windows, possibly in an adaptive manner. Furthermore, by noting the relationship between the RC-line and the spatially discretized form of the heat equation, Leimkuhler and Ruehli (1993) try to use information regarding the decay of solutions in space to obtain more reasonable estimates of convergence for localized regions of the network, which can be used to adaptively evaluate the influence of a subsystem.

In order to obtain some understanding of how waveform convergence behaves as a function of the window size for nonlinear problems, a block step Jacobi waveform relaxation is applied to a two-dimensional reaction-diffusion equation based on the interaction between two chemical species, described in (9.4.1).

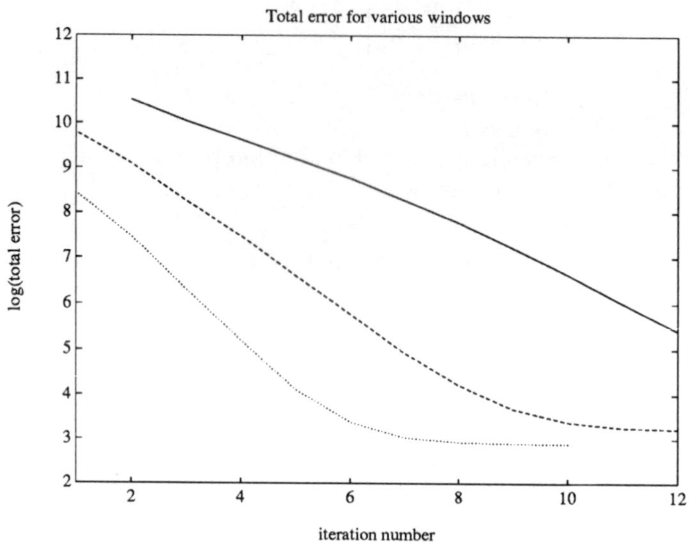

FIG. 9.3. $\log_e$(total error) for various windows

After a suitable semi-discretization the dimension of this problem is 1152 and it is solved using 40 step points with 15 subsytems on a 16-processor subset of the 96-processor Paragon sited at ETH Zürich over three windows: $[0, 6]$, $[0, 3]$ and $[0, 1.5]$. The results are presented in Figure 9.3. The total error here represents the L1 norm of the global error computed over all step points and all components. As can be seen, for each of the three windows convergence is initially approximately constant and then tails away. One reason for this tailing away is that the exact solution is approximated by a numerically computed solution computed using a stringent tolerance and so the errors are only approximations to the true errors. These initial convergence factors are given in Table 9.1.

Table 9.1 *convergence factors*

window	ratio
6	0.62
3	0.43
1.5	0.32

It can be seen that as the window length is halved in each case the convergence factor is reduced by approximately 0.7. This suggests that the error bounds given in Theorem 7.4.2 are too pessimistic and that the error at the kth sweep does not depend on T^k. Rather the bounds given in (9.2.3) which are linear in the window size seem much more realistic, even though they were derived for a linear problem.

Peterson (1992) has developed a dynamic partitioning method for merging and dividing blocks based on the theory of iterative methods and an electrical definition of coupling, which has been implemented in a MIMD environment using a block Jacobi waveform approach. A scheduler program synchronizes a circuit simulation program for each window iteration and decides whether to iterate the same window again or to move to the next window or to split the window. Decisions to relocate circuit nodes to processors only takes place at the end of each window iteration.

The division criterion allows for both bidirectional coupling and loops of unidirectional coupling. Thus for block j a return voltage transfer ratio between circuit nodes r and s belonging to the block is computed to see if they are still to belong to the same block, namely

$$\left| \frac{(J_{jj}^{-1})_{rr}(J_{jj}^{-1})_{ss}}{(J_{jj}^{-1})_{rs}(J_{jj}^{-1})_{sr}} \right| > \gamma,$$

where γ is some division limit. If this condition is satisfied then the two circuit nodes are not to be divided. Here J_{jj} represents the diagonal block of the Jacobian matrix in the (j, j) position. Thus the inversion of the block diagonal matrices of the Jacobian matrix associated with the nonlinear difference equation mapping is the core computational procedure and this can be done efficiently using the already LU factorized Jacobian matrices from the last Newton iteration in the current step point.

The merging criterion is based on computing the iteration matrix H for the block Jacobi method and then applying the diagonal dominant loop criterion discussed previously to the system matrix $I - H$. As Peterson (1992) notes this approach is not just limited to MOS circuit problems because it concentrates on the underlying Jacobian rather than the circuit structures. Some comparisons with static partitioning methods show that this dynamic partitioning approach can gain almost an order of magnitude improvement in a parallel implementation over the static approach if the static heuristic partitioning yields large block sizes, but very little, if any, improvement if the static heuristic yields small bock sizes.

9.3 VODE, VODEPK, DASPK – ODE solvers

The waveform algorithm that will be described in this chapter is based on block Jacobi relaxation and is consequently coarse grained. Thus the appropriate parallel target architecture is a MIMD one. Assuming that a problem has been decomposed into a number of subproblems by some form of waveform relaxation, the question that remains is how best to solve these individual subproblems. It is at this stage that existing sequential packages that are known to be efficient and robust can be exploited. Two such packages are VODE (Brown et al. (1989)) and VODPK (Brown and Hindmarsh (1989)) which are available from NETLIB.

Because an existing sequential package can be used in SPMD mode, with each processor running the same form of VODE or VODPK, most of the parallel implementation of the waveform algorithm involves communication issues. Here standard message-passing environments such as *p4* and *PVM* can be used. An important advantage in using either of these environments is that they can be used to combine a network of workstations as one single computational resource. This makes code generation and debugging a much more efficient task, and indeed the waveform code presented in this chapter was developed and tested in a distributed environment of Sparc 2 workstations at the Department of Mathematics in the University of Queensland using both *PVM* and *p4*. However, in order to gain some uniformity and portability across architectures the code is currently being rewritten using the MPI interface.

One of the first differential equation packages (DIFSUB) developed by Gear (1971a) was (in the stiff case) based on fixed-coefficient Backward Differentiation Formulae. Although codes based on these methods can be very efficient they can have difficulties coping with stiff problems whose eigenvalues have a large imaginary component. There are also additional difficulties associated with the fixed-coefficient approach. A later code (EPISODE), developed by Byrne and Hindmarsh (1975)), implemented the nonstiff Adams methods and stiff BDF methods in a variable-coefficient code with improved local error control.

In EPISODE the order of the Adams methods and the BDF methods can vary between 1 and 12 and 1 and 5, respectively. The coefficients are computed as functions of current and past stepsizes with the past history being stored in a Nordsieck array of scaled derivatives.

EPISODE was further refined (by Hindmarsh) within LSODE (Hindmarsh (1980)) and ODEPACK (Hindmarsh and Byrne (1983)) which cater for a variety of different problems (full, banded, linearly implicit) and which also allows much greater flexibility with respect to user controls and options. The latest variant, VODE, has a user interface which is almost identical to LSODE but improves on the efficiency of problems which require frequent and large changes in stepsize.

In the solution of the nonlinear systems for the update point, VODE offers a choice of either functional iteration or modified Newton iteration in which the Jacobian is either user-supplied or computed internally. VODE also caters for full or banded problems. Since most of the computational work in any differential equation solver involves the solution of the nonlinear equations defining the update, VODE uses a number of techniques to reduce this work. These include reusing the LU factors of the amplification matrix until convergence properties deteriorate, as well as accelerating the convergence of the iterations within the Newton step by relaxing the amplification matrix based on estimates of the extreme eigenvalues of the problem. (These techniques are also used in other ODE solvers such as STRIDE (Butcher et al. (1979)), which is based on an implementation of singly implicit Runge-Kutta methods.)

At the end of each step within VODE the local error is estimated and a stepsize change is considered for the current step or subsequent steps depending on the magnitude of the error. Periodically, order changes of one lower or one higher are considered based on estimation of the local errors. The error control takes the form

$$rtol * abs(y(i)) + atol(i),$$

where $rtol$ and $atol(i)$ are the relative tolerance and absolute tolerance parameters, respectively.

All of the codes mentioned above use direct linear algebra techniques for solving the linear systems associated with implicit methods. These are of the form (4.4.1) where $Q = I - h\beta_0 J$. Recent advances in iterative techniques for linear systems based on preconditioned Krylov GMRES methods, have meant that these approaches can now be incorporated into ODE solvers and this has been done within VODPK and DASPK (Brown et al. (1993)). VODPK has a very similar structure to VODE except that a preconditioned GMRES technique based on a scaled preconditioned incomplete version of GMRES (SPIGMR) is used for the linear solver, while DASPK is based on DASSL (a code for solving differential-algebraic equations). A version of SIMPLE has also been implemented in a matrix-free from (see Grevskott (1992) and Hindmarsh and Nørsett (1987), for example).

VODPK allows for both left and right preconditioned matrices P_1 and P_2 in which the iterative method is applied to $P_1^{-1}QP_2^{-1}$, where P_1P_2 is an approximation to Q. In the stiff mode, the user must supply two routines: JAC, which evaluates and processes any part of the Jacobian involving P_1 and P_2, and PSOL, which solves the linear systems involving P_1 or P_2 as the coefficient matrix. In the case of DASPK only a left preconditioner is used.

As Brown and Hindmarsh (1989) note, these preconditioners have to be chosen very carefully, but if they are chosen well then VODPK can be very effective. One of the advantages of iterative solvers is that the factored preconditioners can often be re-used over many more integration steps than is the case with the iteration matrix in a direct solve. Another advantage (not exploited in VODPK but which is used in DASPK) is that the Jacobian matrix need never be computed explicitly, since only Qv need ever be computed, where v is some vector in the GMRES implementation. This matrix-vector quantity can be computed by a differencing of the form

$$(f(x, y + \sigma v) - f(x, y))/\sigma.$$

The waveform implementation described in section 9.4 has been used with both VODE and VODPK as the core solver. Since linear iterative techniques are, in general, easier to parallelize than direct techniques some attempts were made to parallelize the linear algebra within VODPK and some performance improvements of a modest nature were observed. However, the authors of VODPK are currently developing a full parallel version of this code and this should be available in 1995. The availability of such parallel code introduces, however, additional complexities in the parallel implementation of waveform relaxation algorithms. These will be discussed in section 9.6 under future directions.

Maier et al. (1994) have used preconditioned GMRES iteration for the associated linear systems within DASSL when solving large-scale DAEs. Two parallel versions have been implemented: DASPKF90 (a Fortran90 implementation) and DASPKMP (a message-passing implementation written in Fortran77). Both versions have been implemented on a 512-processor CM5, but the message-passing version did not support the vector units and so performance is very poor.

Maier et al. (1994) solve a number of large-scale DAEs including a three-dimensional heat equation, a multispecies reaction diffusion problem and a Cahn-Hilliard equation (used in modelling phase separation of alloys). For each problem a number of numerical experiments were made with various types of suitable preconditioners for the GMRES iterations, including preconditioners based on least square polynomials, operator splitting and fast fourier transforms. They showed that the quality of the preconditioner can significantly influence the stepsize, and that no one preconditioner performed uniformly well on each of the three test problems. Thus it would appear likely that the flexibly preconditioned restarted GMRES algorithm introduced in Chapter 4 may be of some benefit in this respect since it does produce a very accurate preconditioner.

Two versions of DASPKF90 were developed in Maier et al. using either data shaping or no shaping. In the reshaped version it is necessary to reshape from one to three dimensions and vice-versa when solving three-dimensional problems. This introduces very considerable overheads but gives a consistent and portable code. A non-reshaped version, using array aliasing, proved to be much more efficient but at the cost of losing robustness, elegance and portability.

Results on the CM5 show that DASPKF90 performs very much better on large-scale problems than on small or moderately sized problems. However, even with 512 processors on a problem size of 2^{24} ($\approx$ 16,000,000), only a sustained rate of 3.2 Gflops is produced, which is only 5% of peak performance. Typical performance of DASPKF90 on 32 processors of the CM5 was approximately that of the sequential implementation running on a single-processor Cray Y-MP. In general it was found that performance did not scale linearly with problem size, when the mesh is refined. This is because mesh refinement increases the stiffness of the problem and this leads to an increase in the number of GMRES iterations and integration steps.

9.4 A case study

In order to test the performance of the waveform code (henceforth called BJWR), only large structured problems have been considered. This will allow the code to exploit the parallelism associated with the problem. The main test equation is a two-dimensional reaction-diffusion equation based on the interaction between two chemical species. This equation is known as the diffusion Brusselator equation (see Hairer et al. (1987)) and is defined on the unit square and takes the form

$$\begin{aligned} \frac{\partial u}{\partial t} &= B + u^2 v - (A+1)u + \alpha\left(\frac{\partial^2 u}{\partial x^2} + \frac{\partial^2 u}{\partial y^2}\right) \\ \frac{\partial v}{\partial t} &= Au - u^2 v + \alpha\left(\frac{\partial^2 v}{\partial x^2} + \frac{\partial^2 v}{\partial y^2}\right), \end{aligned} \tag{9.4.1}$$

with initial conditions

$$u(0,x,y) = 0.5 + \theta_1 y, \quad v(0,x,y) = 1 + \theta_2 x,$$

$$A = 3.4, \quad B = 1, \quad \alpha = 0.002,$$

and Neumann boundary conditions

$$\frac{\partial u}{\partial n} = 0, \quad \frac{\partial v}{\partial n} = 0.$$

Here u and v denote chemical concentrations of reaction products, A and B are concentrations of input reagents which are taken to be constant and $\alpha = \frac{d}{L^2}$ where d is a diffusion coefficient and L a reactor length.

Central differencing of (9.4.1) leads to a system of coupled nonlinear equation of order $2N^2$ (with $\hat{\alpha} = \alpha(N-1)^2$) of the form

$$\begin{aligned} u'_{ij} &= B + u_{ij}^2 v_{ij} - (A+1)u_{ij} \\ &\quad + \hat{\alpha}(u_{i+1,j} + u_{i-1,j} + u_{i,j+1} + u_{i,j-1} - 4u_{ij}) \\ v'_{ij} &= Au_{ij} - u_{ij}^2 v_{ij} \\ &\quad + \hat{\alpha}(v_{i+1,j} + v_{i-1,j} + v_{i,j+1} + v_{i,j-1} - 4v_{ij}). \end{aligned} \tag{9.4.2}$$

The constants θ_1 and θ_2 are chosen to take on two different sets of values

$$\theta_1 = 0.25, \quad \theta_2 = 0.8 \tag{9.4.3}$$

and

$$\theta_1 = 1, \quad \theta_2 = 5. \tag{9.4.4}$$

The effect of these values on the performance of the code is discussed in section 9.5.

This appears to be an ideal test problem for a number of reasons as will be seen later. In particular, the problem is large and structured, the stiffness can be varied by the parameter α, it allows the analysis of different parallel communication protocols and allows the effects of different orderings and clusterings on the waveform convergence to be thoroughly investigated. In this section a number of implementational issues which can affect the parallel performance of the block Jacobi waveform relaxation code will now be discussed.

Ordering

A perusal of (9.4.2) suggests that there are two natural ways of ordering the components of the system. It is of course well known that the nature of the ordering of the components can have enormous influence on the convergence of waveform algorithms. For problems with no particular observable structure and for which the physicality of the underlying model cannot be exploited, a waveform implementation can be very much a hit-or-miss affair as there are no simple techniques for knowing how to group the components. However for problems such as (9.4.2) which arise from the method of lines applied to some parabolic partial differential equation there is usually a very natural ordering.

In the case of (9.4.2) the two natural orderings, respectively (A) and (B), are:

$$u_{11}, \ldots, u_{1N}, u_{21}, \ldots, u_{2N}, \ldots, u_{NN}, v_{11}, \ldots, v_{1N}, v_{21}, \ldots, v_{2N}, \ldots, v_{NN}$$

and

$$u_{11}, v_{11}, u_{12}, v_{12} \ldots, u_{NN}, v_{NN}.$$

For the ordering (A), the Jacobian of the problem has the structure

$$\left(\begin{array}{c|c} T_1 & D_1 \\ \hline D_2 & T_2 \end{array}\right).$$

Here each T_i is a block-tridiagonal matrix consisting of N blocks of blocksize N. The diagonal blocks of the T_i are tridiagonal matrices, while the off-diagonal blocks are diagonal. D_1 and D_2 are also diagonal matrices. The (B) ordering leads to a block-tridiagonal Jacobian, consisting of N blocks of dimension $2N$. The diagonal blocks are themselves pentadiagonal matrices, and the off-diagonal blocks are diagonal matrices.

Ignoring for the moment the effect of the (A) or (B) ordering on the convergence of the waveforms it is seen that the (B) ordering is more suitable than the ordering (A) in that the bandwidth is a factor $N/2$ smaller with the concomitant substantial reduction in linear algebra costs and memory storage which can become significant when N is large.

Of course these comments only apply to a sequential implementation. In the case of a parallel implementation, the situation is more complicated as will now be explained.

Consider first the (A) ordering. It can be seen from the above comments that if L, the number of subsystems, satisfies

$$\frac{2N^2}{3} \geq L \geq 2N,$$

then a block Jacobi approach leads to the solution of a tridiagonal problem on each processor. If $L < 2N$, then portions of the upper diagonal and lower diagonal blocks of T_1 or T_2 must be included into each subsystem.

On the other hand if for the (B) ordering L satisfies

$$\frac{2N^2}{5} \geq L \geq N,$$

then a block Jacobi approach leads to the solution of a banded problem with bandwidth 5 on each system. Thus whereas in a sequential environment a (B) ordering is to be preferred in terms of reduced bandwidth, in a parallel environment an (A) ordering appears to be more appropriate if $L \geq 2N$, while a (B) ordering is preferable if $L < 2N$.

Communication

This analysis, however, is still incomplete because it does not cover overlapping or communication issues. In the case of a dense problem, a waveform algorithm requires communication between all subsystems to update the waveform at the end of each iterate and to compute the input for the next step. Rather than perform multibroadcasts (with possible risk of deadlock) a master-slave model can be used. In this model, the master collects and sends at the end of each waveform iteration all the necessary information in order to proceed with the next iteration. This involves each node sending the computed waveform for its subsystem as well as error diagnostics to the host at the end of each iteration. This option is used as the default option if the problem is dense or if the user does not wish to exploit any structure within the problem. In this model it does not matter whether there is overlap or no overlap. Nor does it particularly matter what subset of processors are allocated to a task or what the connection topology is.

However, in the case of block banded systems this master-slave model may be an inappropriate model for efficiency reasons. If an (A) ordering is used it is not possible to write the problem in a block-tridiagonal form to allow for local communication in an efficient manner whereas in the case of the (B) ordering this is possible and hence local communication can be exploited. This means that when the waveform is to be communicated to the relevant processors for the next sweep only the processors to the left and right of a processor need communicate their results to that processor. Of course the overlap must be small enough to allow this to be the case, since the overlap can be chosen to include the complete problem. This local communication effect associated with the (B) ordering thus has to be taken into account when considering linear algebra costs.

The numerical effect of using local communication as opposed to master-slave communication will be explored in section 9.5.

Finally, it should be noted that in spite of the previous comments on (A) and (B) orderings, the (B) ordering is much more efficient, in that the number of iterations needed to achieve satisfactory convergence is considerably less for the (B) ordering. The reason for this is that there is a natural coupling in (9.4.2) between the u_{ij} and v_{ij} which the (B) ordering exploits, while the (A) ordering breaks this coupling by putting corresponding elements u_{ij} and v_{ij} in separate blocks far removed from one another. However, the partitioning of the system for the (B) ordering must always be such that there is no splitting between a u_{ij} and a v_{ij} component.

Load balancing

Some load balancing issues have already been discussed in the previous section, including a dynamic scheduling algorithm of Petersen (1992) which can move components backwards and forwards between subsystems as appropriate. Given the fact that communication performance of most extant parallel machines still lags behind computation performance, this approach will entail considerable overheads for a general-purpose parallel implementation. Consequently, an initial implementation uses a static load balancing heuristic (see Burrage and Pohl (1993)) in which it is assumed that each subsystem needs more or less the same amount of processor time. Thus given p processors, $L = p - 1$ subsystems, a problem of dimension m and an overlap of θ, the dimensions of the L subsystems are for $m = qL + r$ given by:

$$\begin{aligned} d(\theta) &= q + \theta && \text{for } l = 1 \\ d(\theta) &= q + \gamma * \theta && \text{for } l = 2, \ldots, L - r \\ d(\theta) &= q + \gamma * \theta + 1 && \text{for } l = L - r + 1, \ldots, L - 1 \\ d(\theta) &= q + \theta + 1 && \text{for } l = L. \end{aligned} \tag{9.4.5}$$

If overlap in both directions is used, $\gamma = 2$, while in the case of one-sided overlap, $\gamma = 1$. The underlying assumption here is that each subsystem needs more or less the same amount of processor time. This of course may not be true because of varying stiffness and coupling properties between the subsystems and these properties may also vary across the steps. However for this problem this assumption is not perhaps unreasonable.

In spite of the apparent uniformity of the subsystems in (9.4.2), it will be seen in the next section, through a study of the space-time diagram created by ParaGraph on a run of the BJWR code, that there are still considerable load balancing difficulties. Thus a dynamic load balancing approach has been implemented by exploiting the use of overlap. This is now explained.

Initially the work load is distributed as in (9.4.5) with a small initial overlap, θ_0, of say two or three. After a few waveform sweeps, say four, on a particular window, each processor, i, computes the average time, T_i, taken for one waveform sweep for its own subsystem. A native global routine is then used to compute the maximum (T_{max}), minimum (T_{min}) and average (T_{av}) of these times over all the processors. These numbers now reside on all the processors.

Suppose now the dimension of the ith subsystem is $d_i(\theta)$, given as in (9.4.5), and let the dimension of the ith subsystem be $d_i(0)$ when there is no overlap. Each processor now computes a load balancing factor

$$B_i = C \frac{T_{\text{av}} - T_i}{T_{\text{max}} - T_{\text{min}}}, \tag{9.4.6}$$

where C is a safety factor chosen to be 0.5 which mitigates against changing the dimensions of the subsystems by too much.

If $T_i > T_{\text{av}}$ then (9.4.6) suggests that the dimension of the subsystem should be reduced. However, $d_i(\theta)$ cannot be reduced below the dimension of the original subsystem with overlap 0, $d_i(0)$. The reason for this is that the code BJWR is structured to cope automatically with overlapping components by just changing the dimensions of the vectors that are communicated between the processors. However, the dimensions of these vectors cannot be less than the original dimensioning due to the multisplitting as this will cause inconsistencies and erroneous results.

Similarly, if $T_i < T_{\text{av}}$, then (9.4.6) implies that the dimension of the ith subsystem should be increased. But in the case that local communication is used, these dimensions cannot be increased beyond a limit which would allow more than the original components of the neighbouring processors to appear on the ith processor.

These constraints lead to the following changes:

$$B_i < 0, \quad d_i(\theta) = \min\{d_i(\theta_0)(1 + B_i), d_i(0)\}$$
$$B_i > 0, \quad d_i(\theta) = \min\{d_i(\theta_0)(1 + B_i), l_i(0)\},$$

where

$$l_i(0) = d_i(0) + d_{i+1}(0), \quad \gamma = 1$$
$$l_i(0) = d_i(0) + d_{i-1}(0) + d_{i+1}(0), \quad \gamma = 2.$$

Periodically these quantities are updated after a suitable number of sweeps. This approach is a new one and allows for efficient and effective dynamic load balancing by adapting the overlap. Previously, there have been difficulties in knowing, with a static implementation, how to choose the fixed overlap. However, this new approach avoids these difficulties and provides good load balancing by adapting the overlap to suit the needs of the integration process. The efficacy of this approach is shown in section 9.5.

Using VODE

As already remarked, VODE is perhaps the most well-known package for solving ordinary differential equations in a sequential environment. The reasons for using this as the core solver in a parallel waveform implementation have already been addressed; however, the main reason can be emphasized again. It is important to make use of, where possible, sophisticated and user-friendly packages that have been developed with care over a long period of time. These production packages provide a robustness, portability and ease of use that would otherwise not be available in a parallel implementation. However, in using such sequential packages, some changes have to be made either to the package or the interface to allow for incorporation into a parallel code. Some of these changes are now briefly mentioned.

- The VODE interface has to be modified to add the processor number to the error messages.

- The error messages that VODE produces have to be modified and shortened, otherwise 96 processors, for example, each printing the same one or two-line error message will completely swamp the screen.
- Some care has to be taken in the dimensioning of the arrays for each subsystem and the complete system, as there are three dimensions to the waveform iterates (subsystem, component and step point). In addition, function evaluations have to be modified in order to take into account the interpolation procedure and the building of the vectors appropriate to each subsystem.

Two versions of the code BJWR have been developed: an adaptive window and a fixed window implementation. The adaptive window implementation has already been described in the previous section through the work of Dyke (1995). In the case of the fixed window implementation, the user selects a window size and BJWR attempts the integration on that window. If this is successful the same window size is used for the next window. In some cases, VODE may experience severe computation difficulties due to the unbounded growth of the waveform iterations. This is described in more detail in the next section. If this occurs, and it is observed by placing a bound on the number of steps that can be taken on a given window, then the window is shortened to the point at which the difficulties occur and the calculations are repeated on that smaller window. After a successful convergence of the waveforms on this smaller window, the originally user-selected window size is used for integration over the next window.

In order to compute the waveform, VODE is required to provide continuous output at N equidistant step points over the interval of integration. The value N can be specified by the user, but the default value is set to 80. These values are then interpolated by a piecewise linear polynomial to provide the continuous waveform at each iteration.

If a number of integration points are specified at which the solution values are required (as is the case above), then VODE provides an option that allows the stepsize control to compute the solution values at precisely those integration points, rather than interpolating the solution values near these points. Initially this option was not used within BJWR but when it was used, an improvement of approximately 10% in timings was noted. This is due to the fact that smoother waveforms are computed through this option and this improves the convergence rates.

Adapting tolerances

Some care must be taken in selecting the convergence criteria. There are in fact two criteria: the tolerance that VODE uses (the VODE tolerance ϵ_{vo}) in controlling the local error and choosing its step size and the waveform tolerance (ϵ_{wr}) which determines when successive waveform iterates are sufficiently close. Since the initial sweeps are often quite inaccurate, VODE does not have to solve these initial equations very accurately. Thus an adaptive tolerance is used in which the VODE tolerance is gradually tightened as the waveform iterations proceed. In the case of ϵ_{wr}, $\sum_{n=1}^{N}\sum_{i=1}^{m}|y_{i,n}^{(k+1)} - y_{i,n}^{(k)}|$ is computed for successive waveform iterates, and when this difference is less than ϵ_{wr}, the iterates are deemed to have converged. ϵ_{vo} is chosen as a mixture of absolute and relative errors.

9.5 Numerical results

Before presenting numerical results a brief description of the machine on which the code BJWR was run is presented. The machine is an Intel Paragon which is a distributed-memory MIMD computer, sited at ETH Zürich. This Paragon has 96 compute nodes, four service nodes and eight I/O nodes arranged in a rectangular grid. Each compute node consists of two, 50 MHz, i860XP processors with 32 Mbytes of memory and a 16 Kbyte cache. Only one processor per node is accessible and this processor has a peak rating of 75 Mflops. The network can be partitioned into subnets and the operating system is OSF version 1.2. Wormhole routing gives a peak communication performance of 200 Mbytes/s between adjacent nodes.

The first set of results explores the effect of fixed overlap and local versus master-slave communication. The problem is described by (9.4.2) and (9.4.4) with $m = 5000$, the interval of integration is [0,6] and the (B) ordering is used. BJWR uses 40 windows and 64 processors.

Table 9.2 *overlap and communication times (secs)*

overlap	0	1	2	3	4	5
local	211.8	207.7	207.7	208.7	211.7	260.7
master-slave	470.1	463.9	458.7	462.9	462.7	508.7

These results show that there is a slight improvement in time if a small (nonzero) fixed overlap is used. More importantly, Table 9.2 shows that a master-slave implementation is at least twice as slow as a local communication implementation which exploits the tridiagonal structure of the problem. Although it is claimed that long messages can be sent by wormhole routing efficiently in a global manner, local communication is clearly important if locality of data can be guaranteed.

In other numerical tests (Burrage and Pohl (1993)) the difference between using message passing based on *p4* or the native Intel communication routines, NX, has been explored. Although there is some improvement in using NX (at most 10% or so), it is by no means clear the effort in modifying the *p4* commands is worthwhile since the use of *p4* allows a code to be developed, debugged and tested on a cluster of workstations before migration to a parallel computer.

As well as investigating the effect of overlap and communication protocols on timings, the effect of the ordering of the components and the initial conditions has also been explored. Denoting the two different sets of initial conditions given in (9.4.3) and (9.4.4) by the terms "easy" and "difficult" the following results were obtained for the BJWR code running on one processor of the Paragon using both the (A) ordering and the (B) ordering.

Table 9.3 *(A) ordering timings (secs)*

m	200	392	512	800	968	1152
easy	12.64	75.29	182.95	561.18	973.44	1361.45
difficult	38.83	190.88	398.22	1319.32	2263.49	3666.29

Table 9.4 *(B) ordering timings (secs)*

m	800	1800	3200	5000
easy	23.44	85.79	239.81	566.35
difficult	73.22	267.83	694.32	1348.66

The following conclusions can be drawn from these results.

- Much larger problems can be solved with the (B) ordering than with the corresponding (A) ordering. This is due to the smaller bandwidth associated with the (B) ordering.
- Even for moderately sized problems, the (B) ordering is at least an order of magnitude faster than the (A) ordering. This is again due to reduced linear algebra costs, but more importantly due to much faster convergence rates arising from the preservation of the coupling between tightly coupled components within the (B) ordering.
- There is a considerable disparity in timings between the "easy" and "difficult" initial values. Furthermore, if the values for θ_1 and θ_2 in (9.4.4) were to further increase in size then this disparity would become even more pronounced. This effect is initially puzzling, as it does not

appear in the sequential case. The reason for its appearance in the parallel waveform case is due to the fact that the initial waveform iterate is always chosen as $y^{(0)}(x) = y_0$. In the case of (9.4.4) this gives an initial value for the v components of 6 and, as can be seen from the function in (9.4.2), this will cause the $u_{ij}^2 v_{ij}$ terms to grow dramatically. This in turn limits the window size and affects the speed of convergence. Thus, although this effect has nothing to do with coupling or stiffness it can still severely compromise the efficacy of waveform relaxation techniques.

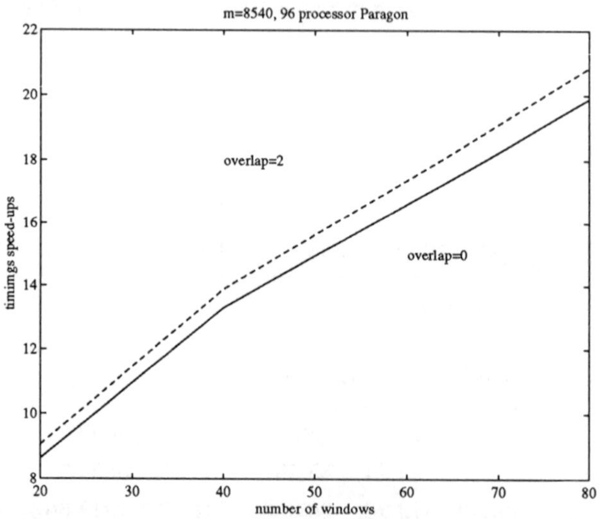

FIG. 9.4. effect of window size on speed-up

The size of the window can also have a dramatic effect on timings. In Figure 9.4 a problem of dimension 8540 is solved on 96 processors using the (B) ordering and "difficult" initial values. The results show that as the (equidistant) window size is reduced there is a factor of 20 improvement in timings when using a window size of $\frac{3}{40}$ compared with a window size of 6 (the complete region of integration).

Table 9.5 shows the timings obtained for the (B) ordering compared with VODE running on a single processor on a problem of dimension 5000. In the parallel execution, L denotes the number of subsystems (with $L-1$ processors) and the results are only given for the optimal window size, and by adapting the tolerance.

Table 9.5 *timings (secs), $m = 5000$*

VODE	L	Time
1348.66	3	13555.1
	7	6493.2
	15	4956.6
	31	991.3
	63	64.8

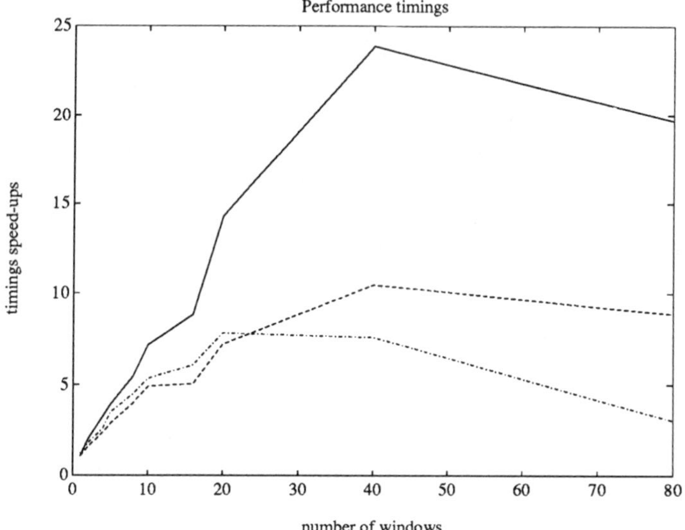

FIG. 9.5. effect of adapting and local communication

It can be seen here that only when the number of processors exceeds 30 is there any improvement over VODE running on one processor. However, if 64 processors are used then the speed-up over VODE is more than 20 which is very respectable. One of the reasons for this sudden improvement is due to the fact that only when $L \geq N$ (which in this case is equivalent to $L \geq 50$) does the (B) ordering give a small bandwidth (5) on each subsystem. Otherwise the bandwidth is $2N$.

The final graph of results (Figure 9.5) gives a summary of speed-ups obtained for differing window sizes. The three curves show, respectively, the effects of both adapting the tolerance and using local communication (the solid line), or just using local communication (the dashed line), or just adapting the tolerance (the dot-dashed line). As can be seen the improvements are quite significant. For this problem, clearly a small window size is an efficient choice.

Finally in this section, some results are presented which indicate the overall efficacy of the BJWR code. These were obtained from ParaGraph, which is an extremely useful software tool for analysing the parallel performance of a code. The first results give a concurrency profile for a 16 processor implementation of BJWR using, respectively, static load balancing or dynamic load balancing by controlling the overlap.

For the static load balancing it appears that the full 16 processors are busy for approximately only 30% of the execution time and that, for example, for 20% of the execution time only seven processors are fully busy. This suggests a poor load balancing. In the case of dynamic load balancing, however, the profile indicates that all 16 processors are busy for 60% of the time and there is an even distribution of busy time in general. This shows the success of dynamic load balancing by controlling the overlap. These results have been confirmed by analysing a space-time diagram created by ParaGraph over the entire execution of the program. Here dynamic load balancing was enabled for the first two iteration sweeps and then turned off. For the first two iterations the space-time diagram shows that the processors are much busier than in later iterations.

9.6 Conclusions

This monograph has attempted to give a thorough and up-to-date account of "the state of the art" of parallel methods for ordinary differential equations (mainly of initial value type). Most direct methods, discussed in Chapters 5 and 6, offer only small-scale parallelism while some direct methods presented there are computationally useless in a parallel setting. However, this latter class of methods has been included in this monograph because they are illustrative of what can be expected from more general approaches. This is especially the case for certain predictor-corrector methods claimed to be suitable for massive parallelism.

In fact direct methods such as those based on extrapolation or on prediction-correction are all examples of small-scale parallelism. While the efficacy of prediction-correction can be improved by careful analysis of error coefficients or by preconditioning the residual or by splitting the coefficient matrix, such an approach offers at best moderate parallelism. This is hardly surprising since a prediction step is for the most part an extrapolatory step so that if the blocksize is too large it takes a large number of corrections to produce an acceptable error. These direct methods seem best suited to shared-memory machines with a small number of very powerful processors.

On the other hand it is not always necessary to try to obtain large speed-ups over sequential implementations. In some cases it may be appropriate to exploit parallelism in order to enhance robustness or the accuracy of an algorithm. Parallel implementations of extrapolation or deferred correction techniques can be interpreted in this way.

For certain classes of problems, such as large linear problems, effective massive parallelism is sustainable by exploiting parallelism across the steps. If a one-step method is applied to such a problem with a constant stepsize over a large number of steps, the ensuing block bi-diagonal system can be solved efficiently by parallel linear algebra techniques such as cyclic reduction or multigrid. However, this approach is not readily extendable to nonlinear problems. In this case a fixed-point formulation is obtained and convergence can be very slow for stiff problems or on long windows of integration.

One very simple way of attempting to obtain good parallel performance is through the solution of large-scale stiff problems. In this case, parallelism can be exploited through the linear systems of equations that have to be solved at each integration step, and this is one reason why a significant amount of material on numerical linear algebra has been presented in this monograph. However, the fact that most parallel machines are still ill balanced with communication performance being inferior to individual processor performance means that speed-ups can be much less than peak performance figures may imply. Nevertheless, efficient parallel linear solver algorithms are now becoming available and these are being imported into existing ODE packages (for example, DASPK (Maier et al. (1994)). Furthermore, at the time of writing a parallel version of VODE is under development.

In a coarse-grained MIMD environment, it has been stated that, where possible, parallel algorithmic development should make use of existing sequential packages which have been fine-tuned over a number of years and which have proven to be robust and efficient. Not only does this provide some robustness to the parallel algorithms but means that the programmer only has to focus on the interprocessor communications. Waveform relaxation algorithms are very appropriate algorithms for MIMD architectures and offer the hope of good parallel performance on large-scale, well-structured problems. However, a significant number of problems still need to be addressed.

In Chapter 9 it was shown that a parallel waveform code (BJWR) can give good parallel performance and run in a efficient and robust manner by incorporating existing sequential software packages such as VODE and through the use of some simple dynamic techniques such as refining of tolerances and allowing dynamic overlap. In spite of these developments, this code merely represents an initial attempt at creating robust and efficient parallel routines for solving ordinary differential equations. Further enhancements could include.

- The use of asynchronous iteration to reduce the large synchronization overheads. This would include allowing the subsystems to be solved completely independently with differing stepsize strategies.
- Better automation strategies for window selection and more sophisticated dynamic load balancing strategies.
- The incorporation of parallel linear solve routines and parallel ODE packages. This raises a number of issues including processor allocation for the different parallel components.
- Combining waveform parallelism with parallelism across the method and across the steps. However, communication characteristics of most present-day parallel machines suggest that it would be very difficult to get enhanced performance from this approach at present.
- Accelerating the convergence of the waveform algorithms, through the use of multigrid techniques or by applying the deflation techniques of Chapter 4 to the waveform process or by the use of other preconditioning techniques.

This restriction of needing short window lengths occurs time and time again when solving ordinary differential equations of initial value type in a parallel environment. The restriction arises because of the inherent seriality of numerical methods applied to IVPs and because many parallel methods are based on some fixed-point iteration process (predictor-corrector methods, waveform relaxation methods, etc.). The convergence of this fixed-point process is governed by the stiffness of the underlying problem and for many stiff problems convergence can be very slow. Thus much of the research into developing efficient parallel methods concerns the acceleration of this convergence process. Some of this has been done in a very ad hoc manner with apparently different techniques being used for different classes of methods. However, it is possible that the deflation approach introduced in Chapter 4 for linear and nonlinear systems of equations can also be applied to ordinary differential equations in a very general way, with just as dramatic an improvement. In a similar way the preconditioning techniques described in section 7.13 may also be of some significance.

It is clearly true that not all algorithms will fare equally well on a wide range of architectures. It seems likely that in the immediate future there will have to be a greater variety, than in the sequential case, of codes which are both problem dependent and architecture dependent. However, with an apparent vendor trend towards massively parallel MIMD machines it may well be that this situation is temporary and that there will be a uniformity and portability of codes across a large set of parallel machines not just in the area of differential equations but in all areas of scientific computation. Clearly, there are still many significant advancements that can be made in the field of parallel methods for differential equations.

REFERENCES

Abbas, S. and Delves, L.M. (1989). *Parallel solution of ODEs by one-step block methods.* Report CSMR, University of Liverpool, U.K.

Abou-Rabia, O. and Boglaev, A.Y. (1994). *Parallel solution of ODEs in symmetrical form.* Report, Dept. Math. and Comp. Sci., Laurentian University, Sudbury, Ontario, Canada.

Adams, L. and Jordan, H. (1985). Is SOR color-blind? *SIAM J. Sci. Stat. Comput.*, **7**, 490–506.

Addison, C.A. (1984). A comparison of several formulas on lightly damped, oscillatory problems. *SIAM J. Sci. Stat. Comp.*, **5**, 920–936.

Aiken, R.C. (1985). *Stiff Computation.* Oxford University Press, New York.

Alekseev, V.M. (1961). An estimate for the perturbations of the solution of ordinary differential equations. *Vestn. Mosk. Univ. Ser. I, Math. Mech.*, **2**, 28–36.

Alexander R. (1977). Diagonally implicit Runge-Kutta methods for stiff ODES. *SIAM J. Numer. Anal.*, **14**, 1006–1021.

Anderson, E., Bai, C., Bischof, C., Demmel, J., Dongarra, J.J., Du Croz, J., Greenbaum, A., Hammarling, S. and Sorensen, D. (1990).*LAPACK: a portable linear algebra library for high performance computers.* Comp. Sci. Tech. Rep. CS-90-105, Univ. of Tennessee, U.S.A.

Ascher, U.M. and Chan, S.Y.P. (1991). On parallel methods for Boundary Value ODEs. *Computing*, **46**, 1–17.

Ascher, U.M., Christiansen, J. and Russell, R.D. (1981). Collocation software for boundary value ODEs. *ACM Trans. Math. Soft.*, **7**, 209–229.

Ascher, U., Mattheij, R.M.M. and Russell, R.D. (1988). Numerical solution of boundary value problems for ordinary differential equations. Prentice Hall.

Ashby, S.F. (1987). Polynomial preconditioning for conjugate gradient methods. Ph.D. Thesis, Univ. of Illinois at Urbana-Champaign, U.S.A.

Ashby, S.F. (1989). Adaptive polynomial preconditioning for Hermitian indefinite linear systems. *BIT*, **29**, 583–609.

Ashby, S.F., Manteuffel, T.A. and Saylor, P.E. (1989). Adaptive polynomial preconditioning for Hermitian indefinite linear systems. *BIT*, **29**, 583–609.

Ashby, S.F., Manteuffel, T.A. and Otto, J.S. (1992). A comparison of adaptive Chebyshev and least squares polynomial preconditioning for Hermitian positive definite linear systems. *SIAM J. Sci. Stat. Comput.*, **13, 1**, 1–29.

Aslam, S. (1990). Asynchronous waveform relaxation methods for ordinary differential equations on multiprocessors. Ph.D. Thesis, Dept. Comp. Sci., Univ. of Illinois at Urbana-Champaign, U.S.A.

Aslam, S. and Gear, C.W. (1989). *Asynchronous integration of ordinary differential equations on multiprocessors.* Report UIUCDCS-R-89-1525, Univ. of Illinois at Urbana-Champaign, U.S.A.

Athas, W. and Seitz, C.L. (1988). Multicomputers: message passing concurrent computers. *IEEE Comp.*, **21, 8**, 9–24.

Augustyn, R. and Ueberhuber, C.W. (1992). *Parallel defect correction algorithms for ordinary differential equations.* ACPC/TR 92-22, Austrian Centre for Parallel Computation, Wien, Austria.

Auzinger, W. and Frank, R. (1988). *Asymptotic error expansions for stiff equations – an analysis for the implicit midpoint and trapezoidal rules in the strongly stiff case.* Rep. 77/88., Inst. für Angewandte und Numerische Mathematik, TU, Wien, Austria.

Auzinger, W. and Frank, R. (1989). *Asymptotic error expansions for stiff equations: the implicit mid point rule.* Rep. 73/89, Inst. für Angewandte und Numerische Mathematik, TU, Wien, Austria.

Bader, G. and Deuflhard, P. (1983). A semi-implicit mid-point rule for stiff systems of ordinary differential equations. *Numer. Math.*, **41**, 373–398.

Bai, Z., Hu, D. and Reichel, L. (1992). An implementation of the GMRES method using QR factorization. In *Fifth SIAM Conf. on Parallel Processing for Scientific Computing* (ed. J. Dongarra et al.). Philadelphia.

Baker, C.T.H. and Paul, C.A.H. (1993). Parallel continuous Runge-Kutta methods and vanishing delay differential equations. *Advances in Comp. Math.*, **1**, 367–394.

Baker, C.T.H. and Paul, C.A.H. (1994). A global convergence theorm for a class of parallel explicit Runge-Kutta methods and vanishing lag delay differential eqautions. To appear *SIAM J. NUmer. Anal.*

Baker, C.T.H., Paul, C.A.H. and Will/e, D.R. (1994). Issues in the numerical solution of evolutionary delay differential equations. To appear *Advances in Comp. Math.*

Bales, L., Karakashian, O. and Serbin, S. (1988). On the A_0-acceptability of rational approximations to the exponential function with only real poles. *BIT*, **28**, 70–79.

Barret, R., Berry, M., Chan, T., Demmel, J., Donato, J., Dongarra, J., Eijkhout, V., Pozo, R., Romine, C. and van der Vorst, H. (1993). Templates for the solution of linear systems: building blocks for iterative methods. *SIAM / NETLIB.*

Barton, D., Willers, I.M. and Zahar, R.V.M (1971). Taylor series methods for ordinary differential equations – an evaluation. In *Mathematical Software,* (ed. I.R. Rice), pp 369–390. Academic Press, New York.

Bellen, A. and Tagliaferro, F. (1991). *A combined WR-parallel steps method for ODEs.* Rep. Progetto Finalizzato Sistemi Informatici e Calcolo Parallelo, 1/49, Italy.

Bellen, A. and Zennaro, M. (1989). Parallel algorithms for initial value problems for difference and differential equations. *J. Comp. Appl. Math.*, **25**, 341–350.

Bellen, A. and Zennaro, M. (1992). Strong contractivity properties of numerical methods for ordinary and delay differential equations. *Appl. Numer. Math.*, **9**, 321–346.

Bellen, A. and Zennaro, M. (1993). The use of Runge-Kutta formulae in waveform relaxation methods. *Appl. Numer. Math.*, **11**, 95–114.

Bellen, A., Vermiglio, R. and Zennaro, M. (1990a). Parallel ODE solvers with step–size control. *J. Comp. Appl. Math.*, **31**, 277–293.

Bellen, A., Jackiewicz, Z. and Zennaro, M. (1990b). *Stability analysis of time-point relaxation Heun method.* Rep. Progetto Finalizzato Sistemi Informatici e Calcolo Parallelo 1/30, Italy.

Bellen, A., Jackiewicz, Z. and Zennaro, M. (1993). Time point relaxation Runge–Kutta methods for ordinary differential equations. *J. Comp. Apppl.*, **45**, 121–137.

Bellen, A., Jackiewicz, Z. and Zennaro, M. (1994). Contractivity of waveform relaxation Runge–Kutta methods for dissipative systems in the maximum norm. *SIAM J. Numer. Anal.*, **31, 4**, 499–523.

Bellman, R.E., Buell, J.D. and Kalaba, R.E. (1965). Numerical intergation of a differential-difference equation with a decreasing time-lag. *Comms. ACM*, **8**, 227–228.

Bemmerl, T. and Bode, A. (1991). An integrated environment for programming distributed memory multiprocessors. In *Distributed Memory*

Computing (ed. A. Bode), Lecture Notes in Computer Science, 487, pp 130–142. Springer, Berlin.

Berman, A. and Plemmons, R.J. (1979). *Nonnegative matrices in the mathematical sciences.* Academic Press, New York.

Bertsekas, D.P. and Tsitsiklis, J.N. (1989). *Parallel and distributed computation: numerical methods.* Prentice Hall, Englewood Cliffs, New Jersey.

Bickart, T.A. and Rubin, W.B. (1974). Composite multistep methods and stiff stability. In *Stiff Differential Systems* (ed. R. A. Willoughby). Plenum Press, New York.

Birta, L.G. and Abou–Rabia, O. (1987). Parallel block predictor methods for ODEs. *IEEE Trans. Comput.*, **C–36**, 299–311.

Bischoff, C. (1991). *Issues in parallel automatic differentiation.* Preprint MCS-P235-0491, Maths. and Comp. Sci. Division, Argonne Nat. Labs., U.S.A.

Bisseling, R.H. and van de Vorst, G.G. (1989). *Parallel LU decomposition on a transputer network.* Report Koninklijke/Shell - Laboratorium, Amsterdam, The Netherlands.

Björck, A. (1994). Numerics of Gram-Schmidt orthogonalization. *Lin. Alg. and its Applications*, **197**, 297–316.

Bjørhus, M. (1990). *Dynamic iteration for second order linear initial value problems.* Student report, Mathematical Sciences Div., Norwegian Inst. of Tech., Trondheim, Norway.

Bjørhus, M. (1992). *Dynamic iteration for time-dependent partial differential equations: a first approach.* Rep. Numerics 1/92, Mathematical Sciences Div., Norwegian Inst. of Tech., Trondheim, Norway.

Bjørhus, M. (1993). *On dynamic iteration for delay differential equations.* Rep. Numerics 2/1993, Mathematical Sciences Div., Norwegian Inst. of Tech., Trondheim, Norway.

Bjørhus, M. (1994). *A note on the convergence of discretized dynamic iteration.* Rep. Numerics 3/1994, Mathematical Sciences Div., Norwegian Inst. of Tech., Trondheim, Norway.

Boglaev, Y.P. (1993). Parallel integration of ODEs based on convolution algorithms. *Computing*, **51**, 185–207.

Bohte, Z. (1975). Bounds for rounding errors in the Gaussian elimination for band systems. *J. Inst. Math. Appls.*, **16**, 133–142.

Boillat, J.E. and Kropf, P.G. (1990). A fast distributed mapping algorithm. In *CONPAR 90 - VAPP IV. Joint Int. Conf. on Vector and Parallel Processing* (ed. H. Buckhart), pp 405–416. Springer-Verlag.

Bondeli, S. (1989). *Divide and Conquer: a new parallel algorithm for the solution of a tridiagonal linear system of equations.* Inst. für Wissenschafliches Rechnen, ETH, Zürich, Switzerland.

Brandt, A. (1977). Multigrid adaptive solutions to Boundary Value Problems. *Math. Comp.,* **31,** 333–390.

Brankin, R.W., Gladwell, I. and Shampine, L.F. (1992). *RKSUITE: a suite of Runge-Kutta codes for the initial value problem for ODEs.* Rep. 92-1, Math. Dept., Southern Methodist University, Dallas, Texas 75275, U.S.A.

Brenan, K.E., Campbell, S.L. and Petzold, L.R. (1989). *Numerical solution of initial-value problems in differential algebraic equations.* North-Holland, New York.

Briggs, W.L. (1987). *A multigrid tutorial.* SIAM, Philadelphia.

Brown, P.N. and Hindmarsh, A.C. (1986). Matrix-free methods for stiff systems of ODEs. *SIAM J. Numer. Anal.*, **23**, 610–638.

Brown, P.N. and Hindmarsh, A.C. (1989). Reduced storage matrix methods in stiff ODE systems. *J. Appl. Math. and Comp.*, **31**, 40–91.

Brown, P.N., Byrne, G.D. and Hindmarsh, A.C. (1989). VODE: A variable-coefficient ODE solver. *SIAM J. Sci. Stat. Comp.*, **20**, 1038–1051.

Brown, P.N., Hindmarsh, A.C. and Petzold, L.R. (1993). *Using Krylov methods in the solution of large-scale differential-algebraic systems.* Rep. TR 93–37, Dept. Comp. Sci., University of Minnesota, U.S.A.

Brunner, H. and van der Houwen, P.J. (1986). *The numerical solution of Volterra equations.* CWI Monograph No. 3, North-Holland, Amsterdam.

Bulirsch, R. and Stoer, J. (1966). Numerical treatment of ordinary differential equations by extrapolation methods. *Numer. Math.,* **8**, 1–13.

Burrage, K. (1978a). Stability and efficiency properties of implicit Runge-Kutta methods. Ph.D. Thesis, Dept. of Math., Univ. of Auckland, New Zealand.

Burrage, K. (1978b). A special family of Runge-Kutta methods for solving stiff differential equations. *BIT*, **18**, 22-41.

Burrage, K. (1978c). High order algebraically stable Runge-Kutta methods. *BIT*, **18**, 373–383.

Burrage, K. (1982). Efficiently implementable algebraically stable Runge-Kutta methods. *SIAM J. Numer. Anal.*, **19**, 245–258.

Burrage, K. (1987). High order algebraically stable multistep Runge-Kutta methods. *SIAM J. Num. Anal.*, **16**, 46–57.

Burrage, K. (1988). Order properties of implicit multivalue methods for ordinary differential equations. *IMA J. of Num. Anal.*, **8**, 43–69.

Burrage, K. (1989). *Solving nonstiff IVPs in a transputer environment.* Report, CSMR, Liverpool University, U.K.

Burrage, K. (1990a). The hunting of the SNARK, or why the trapezoidal rule is algebraically stable. In *Numerical Analysis* (ed. D.F. Griffiths and G.A. Watson), pp 40–59. Longman Scientific and Technical.

Burrage, K. (1990b). *Parallel block predictor-corrector methods with an Adams type predictor.* Report, CSMR, Liverpool University, U.K.

Burrage, K. (1991). The error behaviour of a general class of predictor–corrector methods. *App. Num. Math.*, **8**, 201–216.

Burrage, K. (1993a), Efficient block predictor-corrector methods with a small number of corrections. *J. of Comp. and App. Math.*, **45**, 139–150.

Burrage, K. (1993b). Parallel methods for initial value problems, *App. Num. Math.*, **11**, 5–25.

Burrage, K. (1993c). The search for the Holy-Grail or predictor–corrector methods for solving IVPs. *App. Num. Math.*, **11**, 125–141.

Burrage, K. (1993d). Parallel methods for systems of ordinary differential equations. *SIAM News*, 12–22.

Burrage, K. and Butcher, J.C. (1979). Stability criteria for implicit Runge-Kutta methods. *SIAM J. Numer. Anal.*, **16**, 46–57.

Burrage, K. and Butcher, J.C. (1980). Non-linear stability of a general class of differential equation methods. *BIT*, **20**, 185–203.

Burrage, K. and Chan, R.P.K. (1993). On smoothing and order reduction effects for implicit Runge-Kutta formulae. *J. of Comp. and App. Maths.*, **45,** 17–27.

Burrage, K. and Chipman, F.H. (1985). The stability properties of singly implicit general linear methods. *IMA J. Num. Anal.*, **5**, 287–295.

Burrage, K. and Chipman, F.H. (1989). Construction of A-stable diagonally implicit multivalue methods. *SIAM J. Numer. Anal.*, **26**, 391–413.

Burrage, K. and Hundsdorfer, W.H. (1987). The order of B-convergence of algebraically stable Runge-Kutta methods. *BIT*, **27**, 62–71.

Burrage, K. and Plowman, S. (1990). The numerical solution of ODEIVPs in a transputer environment. In *Applications of Transputers 2*, (eds. D.J. Pritchard, C.J. Scott), pp 495–505. IOS Press.

Burrage, K. and Pohl, B. (1994). Implementing an ODE code on distributed-memory computers. *Special Issue of Computers and Mathematics with Applications*, **28, 10–12**, 235–252.

Burrage, K. and Sharp, P. (1994). A family of explicit Nordsieck methods. *SIAM J. Num. Anal.*, **31, 5**, 1434–1451.

Burrage, K. and Jackiewicz, Z. (1995). *Parallel iterated methods based on multistep Runge-Kutta methods of radau type.* In preparation.

Burrage, K., Hundsdorfer, W.H. and Verwer, J.G. (1986). A study of B-convergence of Runge-Kutta methods. *Computing*, **36**, 17–34.

Burrage, K., Erhel, J. and Pohl, B. (1994a). *A deflation technique for linear systems of equations.* Res. Rep. 94-02, Seminar für Angewandte Mathematik, ETH Zürich, Switzerland.

Burrage, K., Williams, A., Erhel, J. and Pohl, B. (1994b). *The implementation of a generalized cross validation algorithm using deflation techniques for linear systems.* Res. Rep. 94-05, Seminar für Angewandte Mathematik, ETH Zürich, Switzerland.

Burrage, K., Jackiewicz, Z., Nørsett, S. P. and Renaut, R. (1995). *Preconditioned waveform relaxation iterations.* Submitted to *BIT*.

Burrage, P., Oliver, T., Phillips, C. and Watson, D. (1990). *The performance of LAPACK on a transputer array.* Supernode II Working paper, Liverpool University, U.K.

Butcher, J.C. (1963). Coefficients for the study of Runge-Kutta Integration Processes. *J. of the Australian Math. Soc.*, **3**, 185-201.

Butcher, J.C. (1964a). Implicit Runge-Kutta processes. *Math. Comp.*, **18**, 50–64.

Butcher, J.C. (1964b). Integration processes based on Radau quadrature formulas. *Math. Comp.*, **18**, 233–244.

Butcher, J.C. (1964c). On Runge-Kutta processes of high order. *J. Austral. Math. Soc.*, **4**, 179–194.

Butcher, J.C. (1965a). A modified multistep method for the numerical integration of ordinary differential equations. *J. Assoc. Comput. Mach.*, **12**, 124–135.

Butcher, J.C. (1965b). On the attainable order of Runge-Kutta methods. *Math. Comp.*, **19**, 408–417.

Butcher, J.C. (1966). On the convergence of numerical solutions to ordinary differential equations. *Math. Comp.*, **20**, 1–10.

Butcher, J.C. (1972). An algebraic theory of integration methods. *Math. Comp.*, **26**, 79–106.

Butcher, J.C. (1973). The order of numerical methods for ordinary differential equations. *Math. Comp.*, **27**, 793-806.

Butcher, J.C. (1974). Order conditions for general linear methods for ordinary differential equations. *ISNM*, **19**, 77-81.

Butcher, J.C. (1975). A stability property of implicit Runge-Kutta methods. *BIT*, **15**, 358–361.

Butcher, J.C. (1976). On the implementation of implicit Runge-Kutta methods. *BIT*, **6**, 237–240.

Butcher, J.C. (1979), A transformed implicit Runge-Kutta method. *J. ACM*, **26**, 731-738.

Butcher, J.C. (1981). A generalization of singly implicit methods. *BIT*, **21**, 175–189.

Butcher, J.C. (1985). The non-existence of ten stage eighth order explicit Runge-Kutta methods. *BIT*, **25**, 521–540.

Butcher, J.C. (1987a). Linear and nonlinear stability for general linear methods. *BIT*, **27**, 182–189.

Butcher, J.C. (1987b). *The numerical analysis of ordinary differential equations.* Wiley, Chichester.

Butcher, J.C. (1993a). Diagonally implicit multi-stage integration methods. *App. Num. Math.*, **11**, 347–363.

Butcher, J.C. (1993b). General linear methods for the parallel solution of ordinary differential equations. *World Scientific series in Applicable Analysis*, **2**, 99–111.

Butcher, J.C. (1994). A transformation for the analysis of DISIMs. *BIT*, 34, 25–32.

Butcher, J.C. and Chan, R.P.K. (1990). *On symmetrizers for Gauss methods.* Report No. 249, Dept. of Math. and Stat., Univ. of Auckland, New Zealand.

Butcher, J.C. and Jackiewicz, Z. (1993). Diagonally implicit general linear methods for ordinary differential equations. *BIT*, 33, 452–472.

Butcher, J.C., Burrage, K. and Chipman, F.H. (1979). *STRIDE: Stable Runge-Kutta integrator for differential equations.* Report No. 150, Math. Dept., Univ. of Auckland, New Zealand.

Butler, R. and Lusk, E. (1992). *User's Guide to the p4 Programming System.* Argonne Nat. Lab., Rep. ANL-92/17, U.S.A.

Byrne, G.D. (1992). Pragmatic experiments with Krylov methods in the stiff ODE setting. In *Computational Ordinary Differential Equations* (ed. J.R. Cash). Oxford University Press.

Byrne, G.D. and Hindmarsh, A.C. (1975). A polyalgorithm for the numerical solution of ordinary differential equations. *ACM Trans. Math. Software,* **1**, 71–96.

Byrne, G.D. and Hindmarsh, A.C. (1986). *Stiff ODE solvers: a review of current and coming attractions.* Rep., Lawrence Livermore Nat. Lab., U.S.A.

Calvetti, D., Golub, G.H. and Reichel, L. (1994). An adaptive Chebyshev iterative method for nonsymmetric linear systems based on modified moments. *Numer. Math.*, **67**, 21–40.

Cameron, I.T. (1983). Solution of differential-algebraic systems using diagonally implicit Runge-Kutta methods. *IMA J. Num. Anal.*, **3**, 273–289.

Carlin, C.H. and Vachoux, C. (1984). On partitioning for waveform relaxation time-domain analysis of VLSI circuits. In *Proc. of Int. Conf. on Circ. and Syst.*, Montreal.

Cash, J.R. (1975). A class of implicit Runge-Kutta methods for the numerical integration of stiff ordinary differential equations. *J. Assoc. Comput. Mach.*, **22**, 504–511.

Cash, J.R. (1979). Diagonally implicit Runge-Kutta formulae with error estimates. *J. Inst. Math. Appl.*, **24**, 293–301.

Cash, J.R. (1980). On the integration of stiff systems of ODEs using extended backward differential formulae. *Numer. Math.*, **34**, 235–246.

Cash, J.R. (1983). The integration of stiff initial value problems in ODEs using modified extended backward differentiation formulae. *Comput. Math. Appl.*, **9**, 645–657.

Cash, J.R. and Liem, C.B. (1980). On the design of a variable order, variable step diagonally implicit Runge-Kutta algorithm. *J. Inst. Maths. Applics.*, **26**, 87–91.

Ceschino, F. (1961). Modification de la longeur du pas dans l'intégration numérique par les méthodes à pas lies. *Chiffres*, **2**, 101–106.

Chan, R.P.K. (1989). Extrapolation of Runge-Kutta methods for stiff initial value problems. Ph.D. Thesis, Dept. of Math. and Stat., Univ. of Auckland, New Zealand.

Chan, T.F. and Jackson, K.R. (1986). The use of iterative linear equation solvers in codes for large systems of stiff IVPs for ODEs. *SIAM J. Sci. Statist. Comput.*, **7**, 378–417.

Chandy, K.M. and Misra, J. (1988). *Parallel program design: a foundation.* Addison-Wesley.

Chartier, P. (1993). Parallélisme dans la résolution numerique des problèmes de valeur initiale pour les équations différentielles ordinaires et algébriques. Ph.D Thesis, Univ. of Rennes, France.

Chen, S.C., Kuck, P.J. and Sameh, A.H. (1978). Practical parallel band triangular systems. *ACM Trans. Math. Software*, **4**, 270–277.

Cheng, D.Y. (1993). *A survey of parallel programming languages and tools.* Tech. Rep. RND-93-005, NASA Ames, Moffet Field, CA, U.S.A.

Chipman, F.H. (1971). A-stable Runge-Kutta processes. *BIT*, **11**, 384–388.

Choi, J., Dongarra, J., Pozo, R. and Walker, D.W. (1992). SCALAPACK: a scalable linear algebra library for distributed memory concurrent computers. In *Proc. Fourth symposium on frontiers of massively parallel computation.* IEEE Computer Society Press.

Chu, M.T. and Hamilton, H. (1987). Parallel solution of ODEs by multi–block methods. *SIAM J. Sci. Stat. Comput.*, **3**, 342–353.

Clémencon, C., Decker, K.M., Endo, A., Fritscher, J. Jost, G., Masuda, N., Müller, A. Rühl, R., Sawyer, W., de Sturler, E. and Wylie, B.J.N. (1994). *Application-driven development of an integrated tool environment for distributed memory parallel processors.* Rep. CSCS-TR-94-01, Swiss Scientific Computing Center, Manno, Switzerland.

Concus, P., Golub, G. and Meurant, G. (1985). Block preconditioning for the conjugate gradient method. *SIAM J. Sci. and Stat. Comp.,* **6**, 220–252.

Cong, N.H. (1993a). Note on the performance of direct and indirect Runge-Kutta-Nyström methods. *J. Comp. Appl. Math.*, **45**, 347–355.

Cong, N.H. (1993b). A-stable diagonally implicit Runge-Kutta-Nyström methods for parallel computers. *Numerical Algorithms*, **4**, 263–281.

Conrad, V. and Wallach, Y. (1979). Alternating methods for sets of linear equations. *Numer. Math.,* **27,** 371–372.

Cook, S.A. (1985). A taxonomy of problems with fast parallel algorithms. *Information and Control*, **64**, 2–22.

Cooper, G.J. (1978). The order of convergence of general linear methods for ordinary differential equations. *SIAM J. Numer. Anal.*, **15**, 643–661.

Cooper, G.J. (1986). An algebraic condition for A-stable Runge-Kutta methods. In *Pitman Research notes in Mathematics* **140** (ed. D.F. Griffiths and G.A. Watson), pp 32–46.

Cooper, G.J. (1991). On the implementation of singly implicit Runge-Kutta methods. *Math. Comp.*, **57**, 663–672.

Cooper, G.J. and Sayfy, A. (1979). Semi-explicit A-stable Runge-Kutta methods. *Math. Comp.*, **33**, 541–556.

Cooper, G.J. and Vignesvaran, R. (1993). On the use of parallel processors for implicit Runge-Kutta methods. To appear, *Computing*.

Crisci, M.R., Russo, E. and Vecchio, A. (1991a). *Stability plots of parallel Volterra Runge-Kutta methods for Volterra integral equations.* Rep., Tech. Inst. per Applicazoni della Matematica, Napoli, Italy.

Crisci, M.R., van der Houwen, P.J., Russo, E. and Vecchio, A. (1991b). *Parallel Volterra Pouzet Runge-Kutta methods for Volterra integral equations.* Rep. Tech. 87/91, Inst. per Applicazoni della Matematica, Napoli, Italy.

Crouch, P.E. and Grossman, R. (1993). Numerical integration of ordinary differential equations on manifolds. *Nonlinear Science*, **3**, 1–33.

Crouzeix, M. (1979). Sur la B-stabilité des méthodes de Runge-Kutta. *Numer. Math.*, **32**, 75–82.

Dahlquist, G. (1956). Convergence and stability in the numerical integration of ordinary differential equations. *Math. Scand.*, **4**, 33–53.

Dahlquist, G. (1958). Stability and error bounds in the numerical integration of ordinary differential equations. In *Trans. Royal Inst. of Technology*, **130**, Stockholm.

Dahlquist, G. (1975). Error analysis for a class of methods for stiff nonlinear initial value problems. In *Lecture Notes in Math.* **506** (ed. G.A.Watson), pp 1-41. Springer-Verlag, Berlin.

Dahlquist, G. (1978). G-stability is equivalent to A-stability. *BIT*, **18**, 384–401.

Dahlquist, G. and Lindberg, B. (1973). *On some implicit one-step methods for stiff differential equations.* Report TRITA-NA-7302, Dept. of Computer Sciences, Royal Inst. Tech., Stockholm, Sweden.

Decker, K. M. (1993). Methods and tools for programming massively parallel distributed systems. *SPEEDUP Journal* **7(2)**.

Decker, K., Dvorak, J. and Rehmann, R. (1993). *A knowledge-based scientific parallel programming environment.* Rep. CSCS-TR-93-07, Swiss Scientific Computing Center, Manno, Switzerland.

Dekker, K. (1981). *Algebraic stability of general linear methods.* Comp. Sci. Rep. No. 25, Univ. of Auckland, New Zealand.

Dekker, K. and Verwer, J.G. (1984). *Stability of Runge-Kutta methods for stiff nonlinear differential equations.* North Holland, Amsterdam.

Dekker, K., Kraaijevanger, J.F.B.M. and Schneid, J. (1990). On the relation between algebraic stability and B-convergence for Runge-Kutta methods. *Numer. Math*, **57**, 249–262.

Delves, L.M. and Brown, N.G. (1988). Numerical libraries for transputer arrays. In *Transputer Applications*, (ed. G. Harp), pp 114–141. G. Pitman Publishing.

Demmel, J., Dongarra, J.J., Du Croz, J., Greenbaum, A., Hammarling, S. and Sorenson, D. (1987). *Prospectus for the development of a linear algebra library for high-performance computers.* ANL-MCS-TM-97, Argonne National Lab. Rept., U.S.A.

Deuflhard, P. (1983). Order and step-size control in extrapolation methods. *Numer. Math.*, **41**, 399–422.

Deuflhard, P., Bader, G. and Nowak, V. (1981). LARKIN – A software package for the numerical solution of large systems arising in chemical reaction kinetics. In *Modelling of Chemical Reaction Systems* (ed. K.H. Ebert, P. Deuflhard and W. Jaegar).

Dickinson, R.P. and Gelinas, R.J. (1976). Sensitivity analysis of ordinary differential equation systems - a direct method (atmospheric chemical kinetics application). *J. Comput. Phys.*, **21**, 123.

Dimitriadis, S. and Karplus, W.J. (1988a). Automatic generation of ordinary differential equations application software for multiprocessor computers. In *MAPCON IV: Special Processing – Proceedings of the Fourth Society for Computer Simulation Multiconference on Multiprocessors and Array Processors*, (ed. H.L. Johnson), pp 52–57. Society for Computer Simulation, San Diego.

Dimitriadis, S. and Karplus, W.J. (1988b). Scheduling the solution of ordinary differential equations on multiprocessor computers. In *MAPCON IV: Special Processing - Proceedings of the Fourth Society for Computer*

Simulation Multiconference on Multiprocessors and Array Processors, (ed. H.L. Johnson), pp 58–66. Society for Computer Simulation, San Diego.

DiNucci, D.C. and Babb, R.G. (1990). Development of portable parallel programs with Large-Grain Data Flow 2. In *CONPAR 90 - VAPP IV. Joint. Int. Conf. on Vector and Parallel Processing* (ed. H. Buckhart), pp 253–264.

Dongarra, J.J. and Johnsson, L. (1987). Solving banded systems on parallel processor. *Par. Comp.*, **5**, 219–246.

Dongarra, J.J. and Sameh, A.H. (1984). On some parallel banded system solvers. *Par. Comp.*, **1**, 223–235.

Dongarra, J.J., Bunch, J. Moler, C. and Stewart, G. (1979). *LINPACK User's Guide.* SIAM, Philadelphia.

Dongarra, J.J., Duff, I.S., Sorensen, D.C. and van der Vorst, H. A. (1991). *Solving linear systems on vector and shared memory computers.* SIAM, Philadelphia.

Dormand, J.R. and Prince, P.J. (1980). A family of embedded Runge-Kutta formulae. *J. Comput. Appl. Math.*, **6**, 19–26.

Duff, I.S. (1986). Parallel implementation of multifrontal schemes. *Parallel Computing*, **3**, 193–204.

Duff, I.S., Erisman, A.M. and Reid, J.K. (1986). *Direct methods for sparse matrices.* Oxford University Press, London.

Dyke, C.T. (1995). The solution of differential equations in a parallel environment. Ph.D. Thesis, Dept. of Math., Univ. of Queensland, Australia.

Edelson, D. (1976). A simulation language and compiler to aid computer solution of chemical kinetic problems. *Comp. and Chem.*, **1**, 29.

Ehle, B.L. (1969). *On Padé approximations to the exponential function and A-stable methods for the numerical solution of initial value problems.* Research Rep. CSRR2010, Dept. AACS, Univ. of Waterloo, Canada.

Elsner, L. (1989). Comparisons of weak regular splittings and multisplitting methods. *Numer. Math.*, **56**, 283–289.

Enenkel, R.F. (1988). Implementation of parallel predictor-corrector Runge-Kutta methods. M.Sc. Thesis, Dept. Comp. Sci., Univ. of Toronto, Canada.

England, R. (1969). Error estimates for Runge-Kutta type solutions to systems of ordinary differential equations. *Computing J.*, **12**, 166–170.

England, R. (1982). Some hybrid implicit stiffly stable methods for ordinary differential equations. In *Lecture Notes in Maths.*, **909**, pp 147–158. Springer-Verlag.

Enright, W.H. (1978). Improving the efficiency of matrix operations in the numerical solution of stiff ordinary differential equations. *ACM Trans. on Math. Software*, **4**, 127–136.

Enright, W.H. (1989). A new error-control for initial value solvers. *App. Maths. and Computation,* **31**, 288–301.

Enright, W.H. and Higham, D.J. (1991). Parallel defect control. *BIT*, **31**, 647–663.

Enright, W.H., Hull, T.E. and Lindberg, B. (1975). Comparing numerical methods for stiff systems of ODEs. *BIT*, **15**, 10–48.

Enright, W.H., Jackson, K.R., Nørsett, S.P. and Thomsen, P.G. (1986). Interpolants for Runge-Kutta formulae. *ACM Trans. Math. Softw.*, **12**, 193–218.

Erhel, J., Burrage, K. and Pohl, B. (1994). *Restarted GMRES preconditioned by deflation.* Res. Rep. 94-04, Seminar für Angewandte Mathematik, ETH Zürich, Switzerland.

Esser, R. and Knecht, R. (1993). *Intel Paragon XP/S - Architecture and software environment.* ZAM, Forschungszentrum Jülich, Germany.

Evans, D.J. and Megson, G.M. (1987). Construction of extrapolation tables by systolic arrays for solving ordinary differential equations. *Parallel Computing*, **4**, 33–48.

Fang, M. (1991). A parallel multicoloured Gauss–Seidel solver. Contributed paper in *The International Conference on parallel methods for ordinary differential equations, Grado, Italy.*

Fehlberg, E. (1968). *Classical fifth-, sixth-, seventh- and eighth order Runge-Kutta formulas with step size control.* NASA Technical Report, 287, U.S.A.

Fei, J.G. (1986). Parallel algorithms of linear multistep formulas for the numerical integration of ordinary differential equations. *Mathematica Numerica Sinica*, **8**, 113–120.

Fei J.G. (1994). A class of parallel explicit Runge-Kutta formulas. *Chinese Journal of Num. Maths. and Apps.,* **16, 1**, 23–36.

Field, R.J. and Noyes, R.M. (1974). Oscillations in chemical systems. IV: limit cycle behaviour in a model of a real chemical reaction. *J. Chem. Phys.*, **60**, 1877–1884.

Flynn, M.F. (1972). Some computer organizations and their effectiveness. *IEEE Trans. Comput.*, **C-21**, 948–960.

Fox, G. and Otto, S. (1986). Concurrent computation and the theory of complex systems. In *Hypercube multiprocessors* (ed. M.J. Heath), 244. SIAM.

Fox, G.C., Johnson, M.A., Lyzenga, G.A., Otto, S.W., Salmon, J.K. and Walker, D.W. (1988). *Solving problems on concurrent processors: volume 1, general techniques and regular problems.* Prentice Hall, Englewood Cliffs, NJ.

Frank, R. and Ueberhuber, C.W. (1978). Iterated defect correction for differential equations, Part 1: theoretical results. *Computing*, **20**, 207–228.

Frank, R., Schneid, J. and Ueberhuber, C.W. (1981). The concept of B-convergence. *SIAM J. Numer. Anal.*, **18**, 753–780.

Frank, R., Schneid, J. and Ueberhuber, C.W. (1985). Order results for implicit Runge-Kutta methods applied to stiff systems. *SIAM J. Numer. Anal.*, **22**, 515–534.

Franklin, M.A. (1978). Parallel solution of ordinary differential equations. *IEEE Trans. Comp.*, **C-27**, 413–420.

Freeman, T.L. and Phillips, C. (1992). *Parallel numerical algorithms.* Prentice Hall.

Frommer, A. (1990). *Lösung linearer Gleichungssysteme auf Parallelrechnern.* Vieweg Verlag, Braunschweig.

Frommer, A and Mayer, G. (1990). Theoretische und praktische Ergebnisse zu Multisplitting-Verfahren auf Parallelrechnern. *ZAMM*, **70, 6**, 600–602.

Frommer, A. and Pohl, B. (1993). *A comparison result for multisplittings based on overlapping blocks and its application to waveform relaxation methods.* Research Report No. 93–05, Seminar für Angewandte Mathematik, ETH Zürich, Switzerland.

Gaffney, P.W. (1981). Some experiments in solving oscillatory differential equations. In *Elliptic Problem Solvers* (ed. M.H. Schultz), pp 301–305. Academic Press, New York.

Gallopoulos, E. and Saad, Y. (1989). On the parallel solution of parabolic equations. In *Proc. ACM SIGARCH-89*, 17–28.

Gallopoulos, E. and Saad, Y. (1990). *Efficient solution of parabolic equations by polynomial approximation methods.* Rep. 969, CSRD, Univ. of Illinois at Urbana-Champaign, U.S.A.

Gates, K. E. and Petersen, W. P. (1993). *A technical description of some parallel computers.* Rep. series, Interdisziplinäres Projekt für Supercomputing, ETH, Zürich, Switzerland.

Gear, C.W. (1965). Hybrid methods for initial value problems in ordinary differential equations. *SIAM J. Numer. Anal.*, **2**, 69–86.

Gear, C.W. (1971a). Algorithm 407 – DIFSUB for solution of ordinary differential equations. *Commun. ACM*, **14, 3**, 185–190.

Gear, C.W. (1971b). The simultaneous numerical solution of differential-algebraic equations. *IEEE Trans. Circuit Theory*, **CT-18**, 89–95.

Gear, C.W. (1986). *The potential of parallelism in ordinary differential equations.* Tech. Rep. UIUC–DCS–R–86– 1246, Comp. Sci. Dept., Univ. of Illinois at Urbana-Champaign, U.S.A.

Gear, C.W. (1988). Parallel methods for ordinary differential equations. *Calcolo*, **25**, 1–20.

Gear, C.W. (1989). Parallel solutions of ODEs. In *Real–time integration methods for mechanical system simulation* (ed. E.J. Haug and R.C. Deyo), **69**, pp 233–248. Springer-Verlag, Berlin.

Gear, C.W. (1991a). Massive parallelism across space in ODEs. *Conferentie Van Numerik Wiskundigen.*

Gear, C.W. (1991b). Parallelism across time in ODEs. *Proceedings of the International Conference on Parallel Methods for Ordinary Differential Equations, Grado, Italy.*

Gear, C.W. and Juang, F.L. (1991). The speed of waveform methods for ODEs. In *Applied and Industrial Mathematics* (ed. R. Spigler), pp 37–48. Kluwer, Netherlands.

Gear, C.W. and Xuhai, X. (1991). Massive parallelism across time in ODEs. *Proceedings of the International Conference on Parallel Methods for Ordinary Differential Equations, Grado, Italy.*

Geist, A., Beguelin, A., Dongarra, J., Jiang, W., Manchek, R. and Sunderan, V. (1993). *PVM 3.0 User's Guide and Reference Manual.* Report ORNL/TM–12187, Mathematical Sciences Section, Oak Ridge Nat. Lab. U.S.A.

Gelernter, D. (1985). Generative communication in LINDA. *ACM Trans. Prog. Lang. and systems*, **7**, 80–112.

Gill, S. (1951), A process for the step-by-step integration of differential equations in an automatic digital computing machine, *Proc. Cambridge Phil. Soc.*, **47**, 95–108.

Golub, G. and van Loan, C. (1989). *Matrix Computations.* Johns Hopkins University Press, Baltimore.

Gottwald, B.A. (1981). KISS – A digital simulation system for coupled chemical reactions. *Simulation*, **37**, 169.

Gragg, W.B. (1965). On extrapolation algorithms for ordinary initial value problems. *SIAM J. Numer. Anal.*, **2**, 384–403.

Gragg, W.B. and Stetter, H.J. (1964). Generalized multistep predictor-corrector methods. *J. Assoc. Comput. Mach.*, **11**, 188–209.

Grevskott, S. (1992). *Tidsintegratorene SIMPLE, med θ-metoden og KRYSI.* Student report, Mathematical Sciences Div., Norwegian Inst. of Tech., Trondheim, Norway.

Griewank, A. (1991). The chain rule revisited in scientific computing, Part II. *SIAM News*, **July**, 8–24.

Grigorieff, R.P. (1983). Stability of multistep methods on variable grids. *Numer. Math.*, **42**, 359–377.

Gröbner, W. (1960). Die Liereihen und ihre Anwendungen. In *D. verl. d. Wiss.* Berlin, 2nd ed. 1967.

Gustaffson, K., Lundh, M. and Söderlind, G. (1988). A PI stepsize control for the numerical solution of ordinary differential equations. *BIT*, **28**, 270–287.

Hackbusch, W. (1984). Parabolic multigrid methods. In *Computing methods in applied sciences and engineering, VI* (ed. R. Glowinski and J.L. Lions). North-Holland, Amsterdam.

Hackbusch, W. (1985). *Multigrid methods and applications.* Springer-Verlag, Berlin.

Hackbusch, W. and Trottenberg, U. (1982). *Multigrid methods.* Springer-Verlag, Berlin.

Hageman, L.A. and Young, D.M. (1981). *Applied iterative methods.* Academic Press.

Hairer, E. (1980). Highest possible order of algebraically stable diagonally implicit Runge-Kutta methods. *BIT*, **20**, 254–256.

Hairer, E. (1981). Order conditions for numerical methods for partitioned ordinary differential equations. *Numer. Math.*, **36**, 431–445.

Hairer, E. and Wanner, G. (1973). Multistep-multistage-multiderivative methods for ordinary differential equations. *Computing*, **11**, 287–303.

Hairer, E. and Wanner, G. (1974). On the Butcher group and general multivalue methods. *Computing*, **13**, 1–15.

Hairer, E. and Wanner, G. (1981). Algebraically stable and implementable Runge-Kutta methods of high order. *SIAM J. Numer. Anal.*, **18**, 1098–1108.

Hairer, E. and Wanner, G. (1991). *Solving ordinary differential equations II, stiff and differential-algebraic problems.* Springer-Verlag, Berlin.

Hairer, E., Bader, G. and Lubich, C. (1982). On the stability of semi-implicit methods for ordinary differential equations. *BIT*, **22**, 211–232.

Hairer, E., Nørsett, S.P. and Wanner, G. (1987). *Solving ordinary differential equations I, nonstiff problems.* Springer-Verlag, Berlin.

Hairer, E., Lubich, C. and Roche, M. (1988). Error of Runge-Kutta methods for stiff problems studied via differential algebraic equations. *BIT*, **28**, 678–700.

Hairer, E., Lubich, C. and Roche, M. (1989). The numerical solution of differential-algebraic systems by Runge-Kutta methods. *Lecture Notes in Math., 1409.* Springer-Verlag, Berlin.

Hall, G., Enright, W.H., Hull, T.E. and Sedgwick, A.E. (1973). *DETEST: a program for comparing numerical methods for ordinary differential equations.* TR 60, Dep. Comp. Sci., Univ.of Toronto, Canada.

He, Y.P. and Wang, N.C. (1988). The parallel Newton-Raphson iterative method for linear multistep formulas. *Mathematica Numerica Sinica*, **10**, 181–193.

Heller, D. (1978). Some aspects of the cyclic reduction algorithm for block tridiagonal systems. *SIAM J. Num. Anal.*, **13**, 484–496.

Hestenes, M. and Stiefel, E. (1952). Methods of conjugate gradients for solving linear systems. *J. Res. Natl. Bur. Stand. Sect. B*, **49,** 409–436.

Hey, T. (1988). Scientific applications. In *Transputer Applications* (ed. G. Harp), pp 170–203. G. Pitman Publishing.

Higham, D. (1989). Robust defect control with Runge-Kutta schemes. *SIAM J. Numer. Anal.*, **26**, 1175–1183.

Higham, D. and Hall, G. (1990). Embedded Runge-Kutta formulae with equilibrium states. *IMA J. Numer. Anal.*, **9**, 1–14.

Hindmarsh, A.C. (1974). *GEAR: Ordinary differential equation solver.* Rpt. UCID-30001, Rev. 3, Lawrence Livermore, U.S.A.

Hindmarsh, A. (1980). LSODE and LSODI, two new initial value ordinary differential equation solvers. *ACM SIGNUM News L.*, **15**, 10–11.

Hindmarsh, A.C. and Byrne, G.D. (1976). Applications of EPISODE: an experimental package for the integration of systems of ordinary differential equations. In *Numerical methods for differential systems* (ed. L. Lapidus and W.E. Schiesser), pp 147–166. Academic Press.

Hindmarsh, A.C. and Byrne, G.D. (1983). ODEPACK, A systematized collection of ODE solvers. In *Scientific Computing* (ed. R.S. Stepleman), pp 55–64. North-Holland, Amsterdam.

Hindmarsh, A. and Nørsett, S.P. (1987). *KRYSI, an ode solver combining a semi-implicit Runge-Kutta method and a preconditioned Krylov method.* Rep. 8/87, Mathematical Sciences Div., Norwegian Inst. of Tech., Trondheim, Norway.

Hoare, C.A.R. (1985). *Communicating sequential processes.* Prentice Hall, Englewood Cliffs, NJ.

Horn, M.K. (1983). Fourth- and fifth-order scaled Runge-Kutta algorithms for treating dense output. *SIAM J. Numer. Anal.*, **20**, 558–568.

Houbak, N. and Thomsen, P.G. (1979). *SPARKS, A Fortran subroutine for the solution of large systems of stiff ODES with sparse Jacobians.* NI-79-02, Inst. for N.A., Tech. Univ. of Denmark, Lyngby, Denmark.

Huang, Y. and van der Vorst, H.A. (1989). *Some observations on the convergence behavior of GMRES.* Rep. 89-09, Faculty of Technical Mathematics and Informatics, Delft, The Netherlands.

Hull, T.E., Enright, W.H. and Jackson, K.R. (1976). *User's Guide for DVERK- a subroutine for solving non-stiff ODEs.* Rep. TR100, Dept. Comp. Sci., Univ. of Toronto, Canada.

Hundsdorfer, W.H. (1981). *Nonlinear stability analysis for a simple Rosenbrock method.* Report No. 81/31, Inst. of Appl. Math. and Comp. Sci., Univ. of Leiden, The Netherlands.

Hundsdorfer, W.H. (1984). One-step methods for the numerical solution of stiff initial value problems. Ph.D. Thesis, Inst. of Appl. Math. and Comp. Sci., Univ. of Leiden, The Netherlands.

Hundsdorfer, W.H. (1991). *On the error of general linear methods for stiff dissipative differential equations.* Tech. Rep., Dept. of Math., Univ. of Genève, Switzerland.

Hundsdorfer, W.H. and Spijker, M.N. (1981a). A note on B-stability of Runge-Kutta methods. *Numer. Math.*, **36**, 319–333.

Hundsdorfer, W.H. and Spijker, M.N. (1981b). *On the existence of solutions to the algebraic equations in implicit Runge-Kutta methods.* Report No. 81149. Inst. of Appl. Math. and Comp. Sci., Univ. of Leiden, The Netherlands.

Huťa, A. (1956), Une amélioration de la méthode de Runge-Kutta-Nyström pour la résolution numérique des équations différentielles du premier ordre. *Acta Math. Univ. Comenian*, **1**, 201–224.

In 't Hout, K.J. (1994). *A note on unconditional maximum norm contractivity of waveform relaxation methods.* Preprint, CWI, Amsterdam, The Netherlands.

Iserles, A. and Nørsett, S.P. (1985). Bi-orthogonal polynomials. In *Polynomes Orthogonaux et Applications* (ed. A. Dold and B. Eckman), Lecture Notes in Mathematics, 1171, pp 92–100. Springer-Verlag, Berlin.

Iserles, A. and Nørsett, S.P. (1988). On the theory of biorthogonal polynomials. *Trans. Amer. Math. Soc.*, **306**, 455–474.

Iserles, A. and Nørsett, S.P. (1990). On the theory of parallel Runge-Kutta methods. *IMAJNA*, **10**, 463–488.

Jackiewicz, Z. and Zennaro, M. (1990). *Variable stepsize explicit two-step Runge-Kutta methods.* Tech. Rep. 125, Dept. of Math., Arizona State Univ., Tempe, U.S.A.

Jackiewicz, Z. and Kwapisz, M. (1994). *Convergence of waveform relaxation methods for differential-algebraic systems.* Rep., Dept. of Math., Arizona State Univ., Tempe, U.S.A.

Jackiewicz, Z., Kwapisz, M. and Lo, E. (1994). *Waveform relaxation methods for functional differential systems of neutral type.* Rep., Dept. of Math., Arizona State Univ., Tempe, U.S.A.

Jackson, K.R. and Nørsett, S.P. (1990). *The potential for parallelism in Runge–Kutta methods. Part 1: RK Formulas in Standard Form.* Tech. Rep. 239/90, Comp. Sci. Dept., Univ. of Toronto, Canada.

Jackson, K.R. and Nørsett S.P. (1991). *The potential for parallelism in Runge–Kutta methods. Part 2: RK predictor–corrector formulas.* Tech. Rep., Comp. Sci. Dept., Univ. of Toronto, Canada.

Jackson, K.R., Kvaernø, A. and Nørsett S.P. (1992). *The order of Runge-Kutta methods when an iterative method is used to compute the internal stage values.* Numerical Math. Div., Norwegian Inst. of Tech., Trondheim, Norway.

Jarausch, H. (1993). *Analyzing stationary and periodic solutions of systems of parabolic partial differential equations by using singular subspaces as reduced basis.* Rep. 92, Inst. für Geometrie und Praktische Mathematik, Aachen, Germany.

Jarausch, H. and Mackens, W. (1987). Numerical treatment of bifurcation problems by adaptive condensation. In *Numerical Methods for Bifurcation Problems* (ed. T. Küpper, H.D. Mittelmann and H. Weber). Birkhäuser Basel.

Jeltsch, R. and Nevanlinna, O. (1981). Stability of explicit time discretizations for solving initial value problems. *Numer. Math.*, **37**, 61–91.

Jeltsch, R. and Nevanlinna, O. (1982). Stability and accuracy of time discretizations for initial value problems. *Numer. Math.*, **40**, 245–256.

Jeltsch, R. and Pohl, B. (1991). *Waveform relaxation with overlapping systems.* Rep 91–02, Seminar für Angewandte Mathematik, ETH Zürich, Switzerland.

Johnson, O.G., Michelli, C.A. and Paul, G. (1983). Polynomial preconditioners for conjugate gradient calculations. *SIAM J. Numer. Anal.*, **20, 2**, 362–376.

Johnson, T.A. and Zukowski, D.J. (1991). Waveform relaxation based circuit simulation on the VICTOR (V256) parallel processor. *IBM J. Res. Develop.*

Jones, M.T. and Szyld, D.B. (1994). *Two stage multi-splitting methods with overlapping blocks.* Report 94-31, Dept. of Math., Temple University, U.S.A.

Joubert, J. (1994). On the convergence behaviour of the restarted GMRES algorithm for solving nonsymmetric linear systems. To appear, *Journal of Numerical Linear Algebra with Applications.*

Joubert, G.R. and Cloeth, E. (1984). The solution of tridiagonal linear systems with an MIMD parallel computer. In *Proc. 1984 GAMM Conference*, Z. Angew. Math. Mech.

Juang, F.L. (1989). Waveform methods for ordinary differential equations. Ph.D. Thesis, Dept. Comp. Sci., Univ. of Illinois at Urbana-Champaign, U.S.A.

Juang, F. and Gear, C.W. (1989). *Accuracy increase in waveform Gauss Seidel.* Report 1518, Comp. Sci. Dept., Univ. of Illinois at Urbana–Champaign, U.S.A.

Kalvenes, J. (1989). *Experimentation with parallel ODE-solvers.* Student report, Mathematical Sciences Div., Norwegian Inst. of Tech., Trondheim, Norway.

Kaps, P. and Rentrop, P. (1979). Generalized Runge-Kutta methods of order four with stepsize control for stiff ordinary differential equations. *Numer. Math.*, **33**, 55–68.

Kaps, M. and Schlegl, M. (1987). A short proof for the existence of the WZ-factorization. *Par. Comp.*, **4**, 229–232.

Kaps, P. and Wanner, G. (1981). A study of Rosenbrock-type methods of high order. *Numer. Math.*, **38**, 279–298.

Karakashian, O.A. and Rust, W. (1988). On the parallel implementation of implicit Runge-Kutta methods. *SIAM J. Sci. Stat. Comput.*, **9**, 1085–1090.

Kastlunger, K.H. and Wanner, G. (1972). On Turan type implicit Runge-Kutta methods. *Computing*, **9**, 317–325.

Katz, I.N., Franklin, M.A. and Sen, A. (1977). Optimally stable parallel predictors for Adams–Moulton correctors. *Comp. and Maths. with Appls.*, **3**, 217–233.

Keeling, S.L. (1989). On implicit Runge–Kutta methods with a stability function having distinct real poles. *BIT*, **29**, 91–109.

Khalaf, B.M.S. and Hutchinson, D. (1992). Parallel algorithms for initial value problems: parallel shooting. *Parallel Computing*, **18**, 661–673.

Kiehl, M. (1992). *Parallel one-step methods with minimal parallel stages.* TUM M-9210 Math. Inst., Tech. Univ. of München, Germany.

Kiehl, M. (1993). *Parallel multiple shooting for the solution of initial value problems.* TUM M-9309, Math. Inst., Tech. Univ. of München, Germany.

Kirkpatrick, S., Gelatt, C.D. and Vecchi, M.P. (1983). Optimization by simulated annealing. *Science*, **220**, 671.

Kohler, W. H. (1975). Preliminary evaluation of the critical path method for scheduling tasks on a multiprocessor system. *IEEE Trans. Comp.*, **C24**, 1235–1238.

Kraaijevanger, J.F.B.M. (1985). B-convergence of the implicit midpoint rule and the trapezoidal rule. *BIT*, **25**, 652–666.

Kraaijevanger, J.F.B.M. (1991). Contractivity of Runge-Kutta methods. *BIT*, **31**, 482–528.

Kramarz, L. (1980). Stability of collocation methods for the numerical solution of $y'' = f(x, y)$. *BIT*, **20**, 215–222.

Kreiss, H.O. (1978). Difference methods for stiff ordinary differential equations. *SIAM J. Numer. Anal.*, **15**, 21–58.

Krogh, F.T. (1969). A variable step variable order multistep method for the numerical solution of ordinary differential equations. *Information Processing*, **68**, pp 194–199. North-Holland, Amsterdam.

Krogh, F.T. (1973). Algorithms for changing the stepsize. *SIAM J. Numer. Anal.*, **10**, 949–965.

Krogh, F.T. (1974). Changing stepsize in the integration of differential equations using modified divided differences. In *Proc. Conf. on the numerical solution of ordinary differential equations, Univ. of Texas at Austin*, (ed. D. G. Bettis), Lecture Notes in Mathematics, No. 362, pp 22–71. Springer-Verlag, Berlin.

Kuck, P.J. (1977). A survey of parallel machine organization and programming. *Computing Surveys*, **9**, 29–59.

Kutta, W. (1901), Beitrag zur näherungsweisen Integration totaler Differentialgleichungen, *Zeitschr. für Math. u. Phys.*, **46**, 435–453.

Kuznetsov, Y.A. (1988). New algorithms for approximate realization of implicit difference schemes. *Sov. J. Numer. Anal. Math. Modeling*, **3, 2**, 99–114.

Kvaernø, A. (1992). *The order of Runge–Kutta methods applied to semi-explicit DAEs of index 1, using Newton-type iterations to compute the internal stage values.* Rep. Numerics 2/1992, Numerical Math. Div., Norwegian Inst. of Tech., Trondheim, Norway.

Lambert, J. D. (1991). *Numerical methods for ordinary differential equations.* Wiley, Chichester.

Lambiotte, J. and Voigt, R. (1974). The solution of tridiagonal linear systems on the CDC Star - 100 Computer. *ACM Trans. Math. Soft.*, **1**, 308–329.

Lanzkron, P.J., Rose, D.J. and Szyld, D.B. (1991). Convergence of nested classic iterative methods for linear systems. *Numer. Math.*, **58**, 685–702.

Leimkuhler, B., Miekkala, U. and Nevanlinna, O. (1991). Waveform relaxation for linear RC-circuits. *IMPACT*, July.

Leimkuhler, B. and Ruehli, A. (1993). Rapid convergence of waveform relaxation. *Appl. Numer. Maths.*, **11, 3**, 211–224.

Lelarasmee, E. (1982). The waveform relaxation method for the time domain analysis of large scale nonlinear dynamical systems. Ph.D. Thesis, Univ. of California, Berkeley, U.S.A.

Lelarasmee, E., Ruehli, A. and Sangiovanni-Vincentelli, A. (1982). The waveform relaxation method for time domain analysis of large scaled integrated circuits. *IEEE Trans. on CAD of IC and Syst.*, **1**, 131–145.

Leyk, T. and Stewart, D. (1993). *Solving linear parabolic differential equations by Krylov approximation methods.* Report, SMS Division, ANU, Canberra, Australia.

Lie, I (1987). *Some aspects of parallel Runge-Kutta methods.* Report 3/87, Numerical Math. Div., Norwegian Inst. of Tech., Trondheim, Norway.

Lie, I. and Nørsett, S. P. (1989). Superconvergence for multistep collocation. *Math. of Comp.*, **52**, 65–79.

Lie, I. and Skålin, R. (1992). *Relaxation based integration by Runge-Kutta methods and its application to the moving finite element method.* Preprint

Loveman, D.B. (1993). High Performance Fortran. In *Parallel and distributed technology,* (ed. M.J. Quinn), **1, 1**, pp 43-51. IEEE.

Lubich, C. (1992). Chebyshev acceleration of Picard-Lindelöf iteration. *BIT*, **32**, 535–538.

Lubich, C. and Ostermann, A. (1987). Multi–grid dynamic iteration for parabolic equations. *BIT*, **27**, 216–234.

Lustman, L., Neta, B. and Katti, C.P. (1991). Solution of linear systems of ordinary differential equations on an INTEL Hypercube. *SIAM Sci. Stat. Comput.*, **12, 6**, 1480–1485.

Lustman, L., Neta, B. and Gragg, W. (1992). Solution of ordinary differential initial value problems on an INTEL Hypercube. *Computers and Math. with Applications*, **23, 10**, 65–72.

Maier, R.S., Petzold, L.R. and Rath, W. (1994). *Parallel solution of large-scale differential-algebraic systems.* TR 94–10, Dept. Comp. Sci., Univ. of Minnesota, U.S.A.

Markowitz, H.M. (1957). The elimination form of the inverse and its application to linear programming. *Management Sci.*, **3**, 255–269.

May, D. (1989). *Toward general purpose parallel computers.* Working paper, Southampton University, U.K.

McColl, W.F. (1989). *Parallel algorithms and architectures.* Report, Programming Research Group, Oxford University, U.K.

Merson, R.H. (1957), An operational method for the study of integration processes. *Proc. Symp. Data Processing, Weapons Research Establishment,* Salisbury, Australia, 110–1 to 110–25.

Miekkala, U. and Nevanlinna, O. (1987a). Convergence of dynamic iterations for initial value problems. *SIAM J. Sci. Stat. Comp.*, **8**, 459–482.

Miekkala, U. and Nevanlinna, O. (1987b). Sets of convergence and stability regions. *BIT*, **27**, 557–584.

Miranker, W.L. and Liniger, W. (1967). Parallel methods for the numerical integration of ordinary differential equations. *Math. Comp.*, **21**, 303–320.

Missirlis, N.M. (1987). Scheduling parallel iterative methods on multiprocessor systems. *Par. Comp.*, **5**, 295–302.

Mitra, D. (1985). The numerical solution of linear differential equations on parallel processors by asynchronous relaxations. In *Proceedings of the Twenty-third Annual Allerton Conference on Communication, Control, and Computing*, Univ. of Illinois, Urbana-Champaign, pp 710–719.

Mitra, D. (1987). Asynchronous relaxations for the numerical solution of differential equations by parallel processors. *SIAM Journal on Scientific and Statistical Computing*, **8**, 43–58.

MPIF (1993). *Message Passing Interface Forum. Document for a standard Message-Passing Interface.* Tech. Rep., Oak Ridge Nat. Lab., TN, U.S.A.

Moler, C.B. and van Loan, C.F. (1978). Nineteen dubious ways to compute the exponential of a matrix. *SIAM Review*, **20, 4**, 801–836.

Moulton, F.R. (1926). *New methods in exterior ballistics.* University of Chicago Press, Chicago.

Mouney, G., Authie, G. and Gayraud, T. (1991). Parallel solution of ODEs. Implementation of block methods on transputer networks. *Contributed paper at ICIAM91, Washington D.C.*

Mühlenbein, H., Gorges-Schleuter, M. and Krämer, O. (1987). New solutions to the mapping problem of parallel systems: the evolution approach. *Parallel Computing*, **4**, 269–279.

Nachtigal, N.N., Reichel, L. and Trefethen, L.N. (1992). A hybrid GMRES algorithm for nonsymmetric linear systems. *SIAM J. Matrix Anal. Appl.*, **13, 3**, 796–825.

Nevanlinna, O. (1989). Remarks on Picard–Lindelöf iteration. *BIT*, **29**, 328–346 and 535–562.

Nevanlinna, O. (1990). Linear acceleration of Picard–Lindelöf iteration. *Numer. Math.*, **57**, 147– 156.

Neves, K.W. and Feldstein, M.A. (1976). Characterization of jump discontinuities for state-dependent delay differential equations. *J. Math. Anal. Applic.*, **56**, 689–707.

Neves, K.W. and Thompson, S. (1992). Software for the numerical solution of systems of functional differential systems with state-dependent delay. *Appl. Numer. Math.*, **9**, 385–401.

Nievergelt, J. (1964). Parallel methods for integrating ordinary differential equations. *Comm. ACM*, **7**, 731–733.

Nørsett, S.P. (1974). *Semi explicit Runge-Kutta methods.* Mathematics and Computing Rpt 6/74, Norwegian Inst. of Tech., Trondheim, Norway.

Nørsett, S.P. and Simonsen, H.H. (1987). Aspects of parallel Runge-Kutta methods. In *Numerical methods for ordinary differential equations* (ed. A. Bellen et al.), Lecture Notes in Mathematics, pp 103–117. Springer-Verlag, Berlin.

Nørsett, S.P. and Thomsen, P.G. (1986). Local error control in SDIRK methods. *BIT*, **26**, 100–113.

Nørsett, S.P. and Wanner, G. (1979). The real-pole sandwich for rational approximations and oscillation equations. *BIT*, **19**, 79–94.

Nørsett, S.P. and Wolfbrandt, A. (1977). Attainable order of rational approximations to the exponential function with only real poles. *BIT*, **17**, 200–208.

Nordsieck, A. (1962). On numerical integration of ordinary differential equations. *Math. Comp.*, **16**, 22–49.

Obreshkov, N. (1942). On mechanical quadrature. *Spisanie Bulgar. Akad. Nauk.*, **65**, 191–289.

Odeh, F. and Zein, D. (1983). A semidirect method for modular circuits. *Proc. Int. Symp. on circuits and systems.* Newport Beach, California.

Odeh, F., Ruehli, A.F. and Carlin, C.H. (1983). Robustness aspects of an adaptive waveform relaxation scheme. In *Proc. of Int. Conf. on Computer Design.* Port Chester, New York.

O'Leary, D.P. and White, R.E. (1985). Multi-splittings of matrices and parallel solution of linear systems. *SIAM J. Alg. Disc. Meth.*, **6**, **4**, 630–640.

Orel, B. (1991). Real pole approximations to the exponential function. *BIT*, **31**, 144–159.

Ortega, J. M. (1988). *Introduction to parallel and vector solution of linear systems.* Plenum Press, New York.

Owren, B. and Zennaro, M. (1991). Order barriers for continuous explicit Runge-Kutta methods. *Math. Comp.*, **56**, 645–661.

Owren, B. and Zennaro, M. (1992). Derivation of efficient continuous explicit Runge-Kutta methods. *SIAM J. Sci. Stat. Comput.*, **13**, 1488–1501.

Paley, R.E.A.C. and Wiener, N. (1934). *Fourier transforms in the complex domain.* Amer. Math. Soc., Providence, R.I.

Paprzycki, M. and Gladwell, I. (1989). *Solving almost block diagonal systems on parallel computers.* Rept. 89-18, Math. Dept., SMU, Dallas, Texas.

Paprzycki, M. and Gladwell, I. (1990a). *A parallel chopping algorithm for ODE Boundary Value problems.* Rep. 90-9, Math. Dept., SMU, Dallas, Texas.

Paprzycki, M. and Gladwell, I. (1990b). *Solving almost block diagonal systems using level 3 BLAS.* Rep. 90-4, Math. Dept., SMU, Dallas, Texas, U.S.A.

Paul, C.A.H. (1992). Runge-Kutta methods for functional differential equations. Ph.D. Thesis, Dept. of Math., Univ. of Manchester, U.K.

Paul, C.A.H. (1994). *Performance and properties of a class of parallel continuous explicit Runge-Kutta methods for ordinary and delay differential equations.* N.A. Rep. 260, Dept. of Math., Univ. of Manchester, U.K.

Petersen, L. (1992). Dynamic circuit partitioning for concurrent waveform relaxation-based circuit simulation. Ph.D. Thesis, Dept. of Applied Electronics, Lund Inst. of Tech., Lund, Sweden.

Petzold, L.R. (1981a). An efficient numerical method for highly oscillatory ordinary differential equations. *SIAM J. Numer. Anal.*, **18**, 455–479.

Petzold, L.R. (1981b). *Differential-algebraic equations are not ODEs.* Sandia National Lab. Livermore, Rpt. Sand 81–8668.

Petzold, L.R. (1982). A description of DASSL: a differential-algebraic system solver. In *Proc. 10th IMACS World Congress on system simulation and scientific computing.* Montreal, Canada.

Pohl, B. (1992). Ein Algorithmus zur Lösung von Anfangswertproblemen auf Parallelrechnern. Ph.D Thesis, Seminar für Angewandte Mathematik, ETH, Zürich, Switzerland.

Pohl, B. (1993). On the convergence of the discretized multisplitting waveform relaxation algorithm. *App. Num. Math.*, **11**, 251–258.

Prince, P.J. and Dormand, J.R. (1981). High order embedded Runge-Kutta formulae. *J. Comput. Appl. Math.*, **7**, 67–75.

Prokopakis, G.J. and Seider, W.D. (1981). Adaptive semi-implicit Runge-Kutta method for solution of stiff ordinary differential equations. *Ind. Eng. Chem.*, **20**, 255.

Prothero, A. and Robinson, A. (1974). On the stability and accuracy of one-step methods for solving stiff systems of ordinary differential equations. *Math. Comp.*, **28**, 145–162.

Quinn, M.J. (1987). *Designing efficient algorithms for parallel computers.* McGraw Hill, New York.

Rauber, T. and Rünger, G. (1994a). Hypercube implementation and performance analysis for extrapolation methods. *CONPAR 94, Linz, Austria.*

Rauber, T. and Rünger, G. (1994b). Load balancing for extrapolation methods on distributed memory multiprocessors. *PARLE'94, Athens, Greece.*

Rauber, T. and Rünger, G. (1994c). *Parallel iterated Runge-Kutta methods and applications.* Report, Comp. Sci. Dep., Universität des Saarlandes, Saarbrücken, Germany.

Reed, D.A. and Patrick, M.L. (1985). Parallel iterative solution of sparse linear systems, models and architectures. *Par. Comp.*, **2**, 45–67.

Reichelt, M.W., White, J.K. and Allen, J. (1993). *Optimal convolution SOR acceleration of waveform relaxation with application to parallel simulation of semiconductor devices.* Rep., Dept. Elec. Eng. and Comp. Sci., MIT, Cambridge, U.S.A.

Roche, M. (1989). Runge-Kutta methods for differential algebraic equations. *SIAM J. Numer. Anal.*, **26**, 963–975.

Roose, D. and Vandewalle, S. (1991). Efficient parallel computation of periodic solutions of parabolic partial differential equations. *Int. Ser. of Num. Math.*, **97**, Birkhäuser Verlag, Basel.

Rosenbrock, H.H. (1963). Some general implicit processes for the numerical solution of differential equations. *Comput. J.*, **5**, 329–330.

Rosser, J.B. (1967). A Runge–Kutta for all seasons. *SIAM Review*, **9**, 417–452.

Ruehli, A. and Zukowski, C.A. (1992). Convergence of waveform relaxation for RC circuits. In *Semiconductors* (ed. T. Odeh et al.). Springer-Verlag.

Runge, C. (1895). Über die numerische Auflösung totaler Differentialgleichungen. *Nachr. Gesel. Wiss., Göttingen*, 252–257.

Saad, Y. (1985). Practical use of polynomial preconditionings for the conjugate gradient method. *SIAM J. Sci. Statist. Comput.*, **6**, 865–881.

Saad, Y. (1986). Communication complexity of the Gaussian elimination algorithm on multiprocessors. *Lin. Alg. Appl.*, **77**, 315–340.

Saad, Y. (1993). A flexible inner-outer preconditioned GMRES algorithm. *SIAM J. Sci. Stat. Comput.*, **14, 2**, 461–469.

Saad, Y. and Schultz, M.H. (1986). GMRES: A generalized minimal residual algorithm for solving nonsymmetric linear systems. *SIAM J. Sci. Statist. Comput.*, **7**, 856–869.

Santo, M. (1991). One-step continuous methods for the numerical solution of ordinary differential equations. Ph.D Thesis, Univ. of Udine, Italy.

Sarafyan, D. (1966). *Error estimation for Runge-Kutta methods through pseudo-iterative formulas.* Tech. Rep. 14, Louisiana State Univ., New Orleans, U.S.A.

Scherer, R. and Türke, H. (1987). *Algebraic characterization of A-stable Runge-Kutta methods.* Inst. for Practical Mathematics, Univ. of Karlsruhe, Germany.

Schneider, C. (1985). Generalized singly implicit Runge-Kutta methods with arbitrary knots. *BIT*, **25**, 401–414.

Schneider, S. (1993). Numerical experiments with a multistep Radau method. *BIT*, **33**, 332–350.

Shampine, L.F. (1985). Some practical Runge-Kutta formulas. *Math. Comp.*, **46**, 135–150.

Shampine, L.F. and Gordon, M.K. (1975). *Solution of ordinary differential equations – the initial value problem.* W.H. Freeman and Co., San Francisco.

Shampine, L.F. and Watts, H.W. (1969). Block implicit one-step methods. *Math. Comp.*, **23,** 731-740.

Shampine, L.F. and Watts, H.A. (1971). Comparing error estimators for Runge-Kutta methods. *Math. Comp.*, **25**, 445–455.

Shao, J. and Kang, L. (1987). An asynchronous parallel mixed algorithm for linear and nonlinear equations. *Par. Comp.*, **5**, 313–321.

Sharp, P.W. (1991). Numerical comparisons of some explicit Runge-Kutta pairs of orders 4 through 8. *TOMS*, **17**, 387–409.

Sharp, P.W., Fine, J.M. and Burrage, K. (1990). Two stage and three stage diagonally implicit Runge-Kutta-Nyström methods of orders three and four. *IMA J. Numer. Anal.*, **10**, 489–504.

Shroff, G.M. and Keller, H.B. (1993). Stabilization of unstable procedures: the recursive projection method. *SIAM J. Numer. Anal.*, **30**, 4, 1099–1120.

Sidje, R.B. (1994). Parallel algorithms for large sparse exponentials. Ph.D Thesis, INRIA, Rennes, France.

Siemieniuch, J.L. (1976). Properties of certain rational approximations to e^{-z}. *BIT*, **16**, 172–191.

Skålin, R. (1988). *Numerisk løsning av ordinaere differensialligninger ved parallellisering på systemnivå et studium av Runge-Kutta metoder.* Student report, Mathematical Sciences Div., Norwegian Inst. of Tech., Trondheim, Norway.

Skeel, R.D. (1976). Analysis of fixed stepsize methods. *SIAM J. Num. Anal.*, **13**, 664–685.

Skeel, R.D. (1989). Waveform iteration and the shifted Picard splitting. *SIAM J. Sci. Stat. Comp.*, **10**, 756–776.

Skeel, R.D. and Tam, H.W. (1989). *Potential for parallelism in explicit linear methods.* Report. Dep. Comp. Sci., Univ. of Illinois at Urbana-Champaign, U.S.A.

Skelboe, S. (1989). Stability properties of backwards differentiation multirate formulas. *App. Num. Math*, **5**, 151–160.

Skelboe, S. (1991). *Parallel integration of stiff systems of ODES.* Rep., Dept. Comp. Sci., Univ. of Copenhagen, Denmark.

Skelboe, S. and Andersen, P.U. (1989). Stability properties of backward Euler multirate formulas. *SIAM J. Scientific and Statistical Computing*, **10**, 1000–1009.

Simonsen, H.H. (1990). Extrapolation methods for ODEs: continuous approximations, a parallel approach. Ph.D Thesis, Mathematical Sciences Div., Norwegian Inst. of Tech., Trondheim, Norway.

Smith, B.T., Boyle, J.M., Dongarra, J.J., Garbow, B.S., Ikebe, Y., Klema, V.C. and Moler, C.B. (1976). *Matrix Eigensystem Routines - EISPACK Guide, 2nd ed.* Springer-Verlag, New York.

Sommeijer, B.P. (1992). *Parallelism in the numerical integration of initial value problems.* CWI Tract, Amsterdam, The Netherlands.

Sommeijer, B.P., Couzy, W. and van der Houwen, P.J. (1992). A-stable parallel block methods for ordinary and integro-differential equations. *Appl. Numer. Math.*, **9**, 267–281.

Spijker, M.N. (1983). Contractivity in the numerical solution of initial value problems. *Numer. Math.*, **42**, 271–290.

Spilling, R.E. (1993). *Dynamic iteration and the pseudospectral method.* Student report, Mathematical Sciences Div., Norwegian Inst. of Tech., Trondheim, Norway.

Stetter, H.J. (1973). *Analysis of discretization methods for ordinary differential equations.* Springer-Verlag, Berlin.

Stetter, H.J. (1979). Modular analysis of numerical software. In *Numerical Analysis, Dundee 1979, Lecture Notes in Mathematics* **773** (ed. G.A. Watson), pp 133–145. Springer, Berlin.

Steihaug, T. and Wolfbrandt, A. (1979). An attempt to avoid exact Jacobian and nonlinear equations in the numerical solution of stiff differential equations. *Math. Comp.*, **33**, 521–534.

Stone, H.S. (1973). An efficient parallel algorithm for the solution of a tridiagonal linear system of equations. *J. Assoc. Comput. Mach.*, **20**, 27–38.

Stüben, K. and Trottenberg, U. (1982). Multigrid methods: fundamental algorithms, model problem analysis and applications. In *Multigrid Methods*, (ed. W. Hackbusch and U. Trottenberg), Springer Lecture Notes in Mathematics 960. Springer, Berlin.

Szyld, D.B. and Jones, M.T. (1992). Two stage and multisplitting methods for the parallel solution of linear systems. *SIAM J. Matrix and Appl.*, **13, 2,** 671–679.

Ta'asan, S. and Zhang, H. (1994). *On multigrid waveform relaxation method.* Rep. Inst. for Comp. Apps. in Science and Engineering, NASA Langley Research Center, VA23665, U.S.A.

Tam, H.W. (1989). *Parallel methods for the numerical solution of ordinary differential equations.* Rep. No. UIUCDCS-R-89-1516, Comp. Sci. Dept. Univ. of Illinois at Urbana-Champaign, U.S.A.

Tam, H.W. (1992a). One-stage parallel methods for the numerical solution of ordinary differential equations. *SIAM J. Sci. Stat. Comput.*, **13**, 1039–1061.

Tam, H.W. (1992b). Two-stage parallel methods for the numerical solution of ordinary differential equations. *SIAM J. Sci. Stat. Comput.*, **13**, 1062–1084.

Tendler, J.M., Bickart, T.A. and Picel, Z. (1978). A stiffly stable integration process. *ACM Trans. Math. Software*, **4**, 339–368.

Tinney, W.F. and Walker, J.W. (1967). Direct solutions of sparse network equations by optimally ordered triangular factorization. *IEEE Proc.*, **55**, 1801–1809.

Trefethen, L.N. (1992). Pseudospectra of matrices. In *Numerical Analysis* (ed. D.F. Griffiths and G.A. Watson). Longman Scientific and Technical.

Turolte, L.H. (1993). *A survey of software environments for exploiting networked computing resources.* Rep. Engineering Research Centre for Computational Field Simulation, USAE Waterways Experiment Station, Vicksburg, MS 39180-6199, U.S.A.

Valiant, L.G. (1988). Optimally universal parallel computers. *Phil. Trans. Roy. Soc. London,* **A 326**, 373–376.

Valiant, L.G. (1990). Bulk-synchrony: a bridging model for parallel computation. In *Proceedings of DMCCS, Charleston.*

van der Houwen, P.J. (1991). *Implicit initial value problem methods on parallel computers.* CWI Report, Amsterdam, The Netherlands.

van der Houwen, P.J. and Cong, N.H. (1993). Parallel block predictor-corrector methods of Runge-Kutta type. *Appl. Numer. Math.*, **13**, 109–123.

van der Houwen, P.J. and Sommeijer, B.P. (1982). A special class of multistep Runge-Kutta methods with extended real stability interval. *IMA J. of Numer. Anal.*, **2**, 183–209.

van der Houwen, P.J. and Sommeijer, B.P. (1988). *Variable step integration of high order Runge-Kutta methods on parallel computers.* Report NM-R8817, CWI, Amsterdam, The Netherlands.

van der Houwen, P.J. and Sommeijer, B.P. (1991). Iterated Runge-Kutta methods on parallel computers. *SISC*, **12, 5**, 1000–1028.

van der Houwen, P.J. and Sommeijer, B.P. (1992). Block Runge-Kutta methods on parallel computers. *Z. Angew. Math. Mech.*, **72, 1**, 3–18.

van der Houwen, P.J. and Sommeijer, B.P. (1993). Analysis of parallel diagonally implicit iteration of Runge-Kutta methods. *Appl. Numer. Math.*, **11**, 169–188.

van der Houwen, P.J., Sommeijer, B.P. and Cong, N.H. (1991). Stability of collocation-based Runge-Kutta-Nyström methods. *BIT*, **31**, 469–481.

van der Houwen, P.J., Sommeijer, B.P. and Cong, N.H. (1992a). Parallel diagonally implicit Runge-Kutta-Nyström methods. *Appl. Numer. Math.*, **9**, 111–131.

van der Houwen, P.J., Sommeijer, B.P. and Couzy, W. (1992b). Embedded diagonally implicit Runge-Kutta algorithms on parallel computers. *Maths. of Comp.*, **58**, 135–159.

van der Houwen, P.J., Sommeijer, B.P. and van der Ween, W.A. (1993a). *Parallel iteration across the steps of high order Runge-Kutta methods for nonstiff initial value problems.* Rep. NM-R9313, CWI, Amsterdam, The Netherlands.

van der Houwen, P.J., Sommeijer, B.P. and van der Ween, W.A. (1993b). *Parallelism across the steps in iterated Runge-Kutta methods for stiff initial value problems.* Rep. NM-R9322, CWI, Amsterdam, The Netherlands.

van der Houwen, P.J., Sommeijer, B.P. and de Swart, J.J.B. (1994). *Parallel predictor-corrector methods.* Rep. NM-R9408, CWI, Amsterdam, The Netherlands.

van der Vorst, H.A. (1987). Large tridiagonal and block tridiagonal linear systems on vector and parallel computers. *Parallel Computing*, **5**, 45–54.

van der Vorst, H.A. and Dekker, K. (1989). Vectorization of linear recurrence relations. *SIAM J. Sci. Stat. Comput.*, **10**, 27–35.

van der Vorst, H.A. and Vuik, C. (1991). *GMRESR: a family of nested GMRES methods.* Rep 91-80, Faculty of Technical Mathematics and Informatics, Delft, The Netherlands.

van der Vorst, H.A. and Vuik, C. (1993). The superlinear convergence behaviour of GMRES. *Journal of Computational and Applied Mathematics*, **48**, 327–341.

Vandewalle, S. (1993). *Parallel multigrid waveform relaxation for parabolic problems.* B.G. Teubner Verlag, Stuttgart.

Vandewalle, S. and Piessens, R. (1990). *Efficient parallel algorithms for solving initial-boundary value and time-periodic parabolic partial differential equations.* Rep. TW139, Dept. Comp. Sci., Univ. of Leuven, Belgium.

Vandewalle, S. and Piessens, R. (1991). Numerical experiments with nonlinear multigrid waveform relaxation on a parallel processor. *J. App. Num. Math.*, **8**, 149–161.

Vandewalle, S. and Roose, D. (1989). Parallel processing for scientific computing. In *Proceedings of Third SIAM Conf. on Parallel Processing for Scientific Computing* (ed. G. Rodrigue). SIAM, Philadelphia.

Vandewalle, S. and Van de Velde, E.F. (1993). *Space-time concurrent multigrid waveform relaxation.* Report CRPC-93-2, Centre for Research on Parallel Computation, Caltech, U.S.A.

van Dorsselaer, J.L.M. and Spijker, M.N. (1992). *The error committed by stopping Newton iterations in the numerical solution of stiff initial value problems.* Rep. TW–92–02, Dept. Math. and Comp. Sci., Leiden University, The Netherlands.

Vaneslow, R. (1979). *Stabilitäts-und Fehleruntersuchungen bei numerischen Verfahren zur Lösnug steifer nichtlinearer Anfangswertprobleme.* Diplomarbeit, Sektion Mathematik, TU-Diesden, Germany.

Varah, J.M. (1979). On the efficient implementation of implicit Runge-Kutta methods. *Math. Comp.*, **33**, 557–561.

Varga, R.R. (1962). *Matrix iterative analysis.* Prentice Hall, Englewood Cliffs, New Jersey.

Verner, J.H. (1978). Explicit Runge-Kutta methods with estimates of the local truncation error. *SIAM J. Numer. Anal.*, **15**, 772–790.

Verner, J.H. (1992). *A classification scheme for studying explicit Runge-Kutta pairs.* Rep. 1992-04, Dept. Math. and Stat., Queen's Univ., Kingston, Canada.

Verner, J.H. and Sharp, P.W. (1991). *Completely imbedded Runge-Kutta formula pairs.* Rep. 1991-01, Dept. Math. and Stat., Queen's Univ., Kingston, Canada.

Verwer, J.G. (1981). On the practical value of the notion of BN-stability. *BIT*, **21**, 355–361.

Verwer, J.G. (1993). Gauss-Seidel iteration for stiff ODEs from chemical kinetics. To appear, *SIAM J. Sci. Comput.*

Verwer, J.G. and Simpson, D. (1994). *Explicit methods for stiff ODEs from atmospheric chemistry.* Report NM-R9409, CWI, Amsterdam, The Netherlands.

Vinograd, R.E. (1952). On a criterion of instability in the sense of Lyapunov of the solutions of a linear system of ordinary differential equations. *Dokl. Akad. Nauk. SSSR,* **84**, 201–204.

Voss, D.A. and Khaliq, A.Q. (1989). *L-stable parallel methods for parabolic problems.* Report, Dept. of Math., Western Illinois University, U.S.A.

Wang, H.H. (1981). A parallel method for tridiagonal equations. *ACM Trans. Math. Software,* **7**, 170–183.

Wanner, G., Hairer, E. and Nørsett, S.P. (1978). Order stars and stability theorems. *BIT*, **18**, 475–489.

White, J. and Sangiovanni-Vincentelli, A.L. (1987). *Relaxation techniques for the simulation of VLSI circuits.* Kluwer Academic Publishers, Boston.

White, J., Sangiovanni-Vincentelli, A., Odeh, F. and Ruehli, A. (1985). Waveform relaxation: theory and practice. *Trans. of Soc. for Computer Simulation*, **2**, 95–133.

Widlund, O. (1967). A note on unconditionally stable linear multistep methods. *BIT*, **7**, 65–70.

Willé, D.R. and Baker, C.T.H. (1992), The tracking of derivative discontinuities in systems of delay differential equations. *Appl. Numer. Math.* **9**, 209-222.

Williams, A. and Burrage, K. (1994a). *The implementation of a GCV algorithm in a high performance computing environment.* Rep., CIAMP, Univ. of Queensland, Australia.

Williams, A. and Burrage, K. (1994b). *The implementation of a linear solver algorithm with deflation in a parallel environment.* Res. Rep. 94-10, Seminar für Angewandte Mathematik, ETH Zürich, Switzerland.

Wilson, G.V A. (1993). A glossary of parallel computing terminology. In *IEEE Parallel and Distributed Technology*, **February**, IEEE.

Wolfbrandt, A. (1977). A study of Rosenbrock processes with respect to order conditions and stiff stability. Ph.D Thesis, Chalmers Univ. of Technology, Göteborg, Sweden.

Worland, P.B. (1976). Parallel methods for the numerical solution of ordinary differential equations. *IEEE Trans. Comp.*, **C-25**, 1045–1048.

Worley, P.H. (1992). Parallelizing across time when solving time-dependent partial differential equations. Manuscript.

Wright, K. (1970). Some relationships between implicit Runge-Kutta collocation and Lanczos T methods, and their stability properties. *BIT*, **10**, 217–227.

Wright, S.J. (1990). *Stable parallel algorithms for two-point boundary value problems.* Preprint MCS-P178-0990, Argonne National Laboratory, U.S.A.

Xu, J. (1992). Iterative methods by space decomposition and subspace correction. *SIAM Review*, **34, 4**, 581–613.

Young, D.M. and Gregory, R. J. (1972). *A survey of numerical methods.* Addison-Wesley.

Yuba, T., Shimada, T., Hiraki, K. and Kashiwagi, H. (1985). SIGMA-1: a dataflow computer for scientific computations. *Computer Physics Communications*, **37**.

Zennaro, M. (1986). Natural continuous extensions of Runge-Kutta methods. *Math. Comp.*, **46**, 119–133.

Zlatev, Z. (1988). Treatment of some mathematical models describing long-range transport of air pollutants on vector processors. *Parallel Computing*, **6**, 87–98.

Zlatev, Z. (1990). Computations with large and band matrices on vector processors. *Advances in Parallel Computing*, **1**, 7–37.

Zlatev, Z. and Berkowicz, R. (1988). Numerical treatment of large-scale air pollution models. *Comput. Math. Applic.*, **16**, 93–109.

Zukowski, C.A., Gristede, G. and Ruehli, A. (1990). *Measuring error propagation in waveform relaxation algorithms.* Rep., Dept. Elec. Eng., Columbia Univ., N.Y., U.S.A.

Zygmund, A. (1959). *Trigonometric series.* Cambridge University Press.

INDEX

LIBRARY, UNIVERSITY COLLEGE CHESTER